FINITE ELEMENT MODELING
IN ENGINEERING PRACTICE

FINITE ELEMENT MODELING
IN ENGINEERING PRACTICE
Includes Examples with ALGOR®

CONSTANTINE SPYRAKOS

West Virginia University

Distributed by Algor Publishing Division
Pittsburgh, PA

Spyrakos, Constantine C.
Finite Element Modeling in Engineering Practice
Includes Examples with Algor ®

Library of Congress Catalog Card No: 94-60886

ISBN: 0-9652806-0-8

Editorial/Production
Christine Peyton-Jones
John Luchok
Richard Beto

Algor, Inc.
Publishing Division
Pittsburgh, PA

Printed in the United States of America.

To Christina and Athanasia

Table of Contents

Preface

There are numerous excellent textbooks on the finite element theory that provide rigorous theoretical treatments and can be very useful to researchers and finite element program developers. Unfortunately, other than what is in occasional papers and specialty publications, little information on the use of finite element analysis has been collected in the form of a textbook.

Most engineers, however, do not get involved in finite element code development and are basically using finite elements as an analysis tool. Their primary concerns are related to the proper use of the method in assessing the adequacy of systems subjected to a variety of loads. They need to acquire the necessary knowledge to use the method without resorting to references elaborating on specialized topics of finite element theory.

This book is intended to help engineers involved with finite element analysis. It assumes no previous exposure to finite element analysis and attempts to develop the necessary background to perform static and dynamic analysis of systems commonly used in mechanical and structural engineering. It can be a valuable source of information to engineering students that wish to know the basics of finite element theory and the principles of finite element modeling. This text should be of particular interest to both the novice and the advanced users of Algor since the majority of the worked out problems have been solved with the Algor software. The user should be aware that several commands in the examples could be changed with further enhancement of the Algor computer programs.

The emphasis has been placed on presenting the basic concepts and theory that are essential to effectively perform a finite element analysis. We attempted to present the material in a clear manner for both the engineer in the industry and the university student. Some of the frequently asked questions answered in this book include:

a. What is finite element analysis?
b. Should I perform a static or a dynamic analysis?
c. In case of dynamic analysis, which type is the most appropriate for my application?
d. What is the best element for the application?
e. How can I overcome difficulties in combining different types of elements?
f. How can I model cracks, holes, and discontinuities?
g. When and where should I use a finer mesh?
h. How can I identify modeling errors?
i. Are my results accurate and how do I interpret them?

Chapter 1 presents some basic definitions and the necessary steps to apply the finite element method. The procedure is illustrated with a statics and a dynamics problem. Chapter 2 is a review of selected strength of materials and theory of elasticity concepts, such as stresses, yield criteria, isotropy and orthotropy.

While there is little difficulty in reaching an agreement on the theoretical aspects of finite element analysis, it is rather difficult to reach a consensus on many issues of finite element modeling. In fact, many modeling aspects are difficult to address with strict rules applicable to all cases. Many features of modeling are more an art and experience than rigorous engineering science. Recognizing these difficulties, yet willing to assist the analyst in developing an acceptable model, Chapter 3 provides general modeling guidelines as well as a brief description of the most commonly used types of finite elements.

Chapter 4 is an introduction to structural dynamics. Some of the topics include free and forced vibration of single degree-of-freedom systems subjected to harmonic, general, and random loads as well as response spectra. Types and use of damping are also presented. Chapter 5 introduces the most common types of dynamic analysis, their proper use, advantages, and limitations. Emphasis is placed on modal analysis because of its paramount importance in structural dynamics. Each method is illustrated with an example that has been worked out in detail.

Chapters 6 through 9 present examples of static and dynamic analysis, which have been solved with Algor. The procedure and suggestions, however, are valid for most commercial finite element programs. Closed form solutions of the examples are either presented in Chapter 5 or can be found in quoted references. The examples have been carefully selected to address questions that are commonly encountered in practice, e.g., development of proper model, application of boundary conditions, use of symmetry, modeling of "rigid" and "soft parts." Emphasis is placed on the interpretation of the results, use of yield criteria, and assessment of the solution accuracy.

This text has been benefitted from the contributions of many individuals. I wish to thank my colleagues and graduate students at West Virginia University and numerous practicing engineers for their thoughtful comments and detailed reviews; Michael Bussler and Charles Paulsen of Algor who encouraged me in pursuing this effort; The Algor application engineers Mark Decker, Allen Fowkes, and Alfred Snow for their valuable suggestions. Needless to mention my wife Christina, also an engineer, for her patience and technical suggestions. I would also like to acknowledge Jason Cunningham for the diligent drawing of several figures; Christine Peyton-Jones for her excellent work in giving the final format to the book; John Luchok who did a wonderful editing job; Richard Beto for his attention to printing and binding details; and Dean Ruth Jackson, director of West Virginia University Press, for her generous cooperation.

CONSTANTINE SPYRAKOS
Professor, Ph.D., P.E.

1

Basics of Finite Element Analysis

1.1 INTRODUCTION

The finite element method is a computer based procedure that can be used to analyze structures and continua. It is a versatile numerical method that is widely applied to solve problems covering almost the whole spectrum of engineering analysis. Common applications include static, dynamic, and thermal behavior of physical systems, and their components. Advances in computer hardware have made it easier and very efficient to use finite element software for the solution of complex engineering problems on personal computers.

The results obtained with a finite element analysis are rarely "exact." Nevertheless, a very accurate solution can be obtained if a proper finite element model, based on principles of finite element analysis, is used.

If the objective of the engineer is the development of a finite element code, then a thorough understanding of the finite element theory is essential. If the objective, however, is the use of a finite element code, then it is necessary for the analyst to have:

a. rudimentary understanding of the fundamental concepts of the finite element method, and

b. training -including knowledge of the capabilities and limitations- of the computer program that will be used.

In the following we will present some basic concepts of the finite element theory. This discussion will help clarify how finite element codes work and how they are used. Our discussion will be restricted to solid mechanics applications.

1.2 GENERAL DESCRIPTION OF THE FINITE ELEMENT METHOD AND STEPS OF FINITE ELEMENT ANALYSIS

Calculation of deformations, strains, and stresses with classical methods of analysis is achieved through longhand solution of the governing equations and boundary conditions describing the problem. Use of classical methods is probably the best way to analyze simple structures; nevertheless, their use is prohibitive when the physical system is complex. In such cases the best alternative is usually a solution obtained with the finite element method.

The primary differences between classical methods and finite elements are the way they "view" the structure and the ensuing solution procedure. Classical methods consider the structure as a continuum whose behavior is governed by partial or ordinary differential equations. The finite element method considers the structure to be an assembly of small finite-sized particles. The behavior of the particles and the overall structure is obtained by formulating a system of algebraic equations that can be readily solved with a computer. The finite-sized particles are called *finite elements*. The points where the finite elements are interconnected are known as *nodes* or *nodal points*, and the procedure in selecting the nodes is termed *discretization or modeling*, see Fig. 1.2.1.

Typically, a finite element analysis involves seven steps. Steps 1,2,4,5, and 7 require decisions made by the user of the finite element program. The rest of the steps are automatically performed by the computer program.

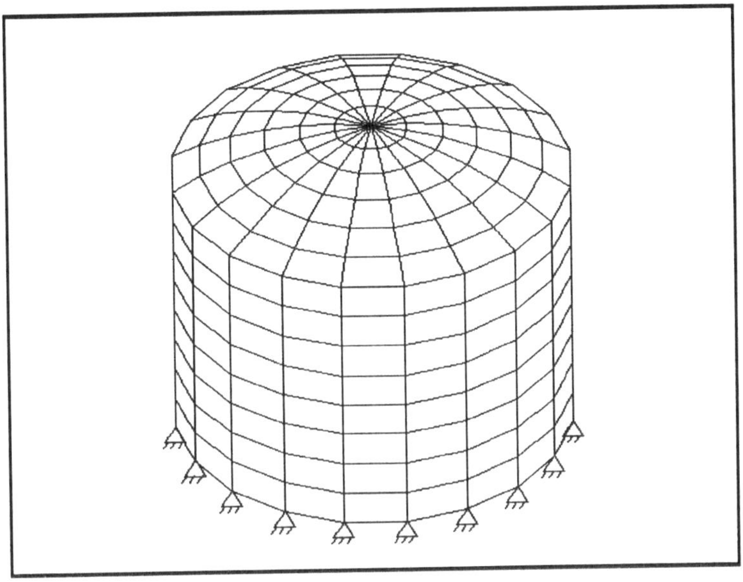

Figure 1.2.1 *Cylindrical tank modeled with plate elements.*

Steps of Finite Element Analysis:

1. **Discretize or model the structure**: The structure is divided into finite elements. Preprocessors, such as SuperDraw of Algor, help the user create the finite element mesh. This step is one of the most crucial in determining the solution accuracy of the problem. A thorough description of the rules for appropriate discretization is given in Chapter 3.

2. **Define the element properties**: At this step, the user must define the element properties and select the types of finite elements that are the most suitable to model the physical system.

3. **Assemble the element stiffness matrices**: The stiffness matrix of an element consists of coefficients which can be derived from equilibrium, a weighted residual, or an energy method. The element stiffness matrix relates the nodal displacements to the applied forces at the nodes. Assembling of the element stiffness matrices implies application of equilibrium for the whole structure.

4. **Apply the loads**: Externally applied concentrated or uniform forces, moments, and ground motions are provided at this step. As explained in Chapter 5, loads are not specified on a structure when we perform modal analysis.

5. **Define boundary conditions**: The support conditions must be provided, i.e., several nodal displacements must be set to known values. Use of *boundary elements* and reaction force processors allow the evaluation of reactions, which otherwise may not be provided as a part of the solution output.

6. **Solve the system of linear algebraic equations**: The sequential application of the above steps leads to a system of simultaneous algebraic equations where the nodal displacements are the unkowns.

7. **Calculate stresses**: At the users discretion, the programs can also calculate stresses, reactions, mode shapes or other pertinent information. Postprocessors, such as SuperView of Algor, help the user display the output in a graphical form.

The following example illustrates all the steps of a typical finite element analysis.

1.3 INTRODUCTORY EXAMPLE

As a simple example that can demonstrate how the finite element method can be used to solve engineering mechanics problems, we will analyze a bar axially loaded with two concentrated forces acting at the free end and at mid-length. The bar and its modeling into two equal length uniform *truss* elements is shown in Fig. 1.3.1. Even though the bar is a very simple structure, the method is applicable to any structure analyzed with finite elements.

Our objective in this example is twofold: the first one is to perform a static finite element analysis to determine the axial displacements, internal forces at node 2, and the reaction force at node 3; and the second one is to perform a dynamic analysis in order to calculate the response of the cantilever bar subjected to dynamic forces.

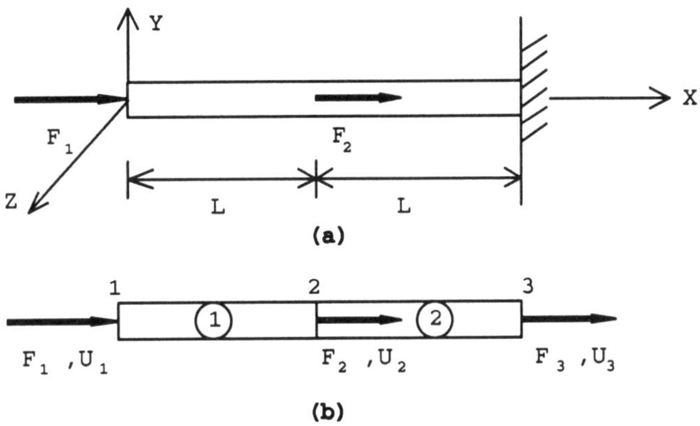

Figure 1.3.1 *(a) Axially loaded bar; (b) Finite element model.*

Static Analysis

Before we proceed to solve the problem, we will examine a single truss element as the one shown in Fig. 1.3.2. A node is placed at each end of the bar and is denoted with numbers 1 and 2. Further, for the axially loaded element, no displacements can develop in the y and z-directions, leaving u_1 and u_2 as the only possible *degrees-of-freedom* of the truss element.

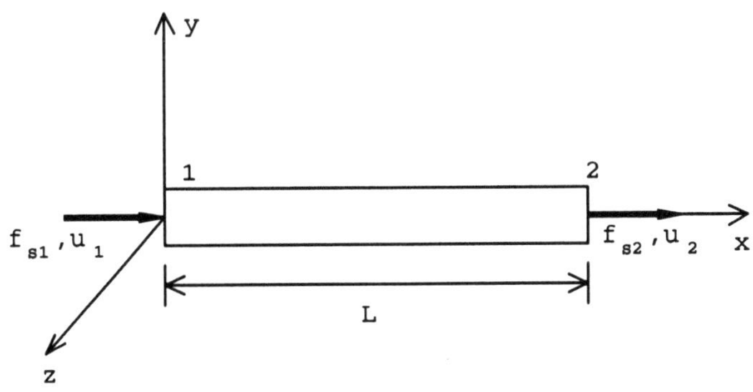

Figure 1.3.2 *Two-degree-of-freedom truss element.*

Hooke's law states that if an axial load in the member is f_s, then the axial deformation of the same truss element restrained as shown in Fig. 1.3.3 is given by

$$u = \frac{f_s L}{AE} \tag{1.3.1}$$

where

L = length of the member
A = cross-sectional area
E = Young's modulus of elasticity.

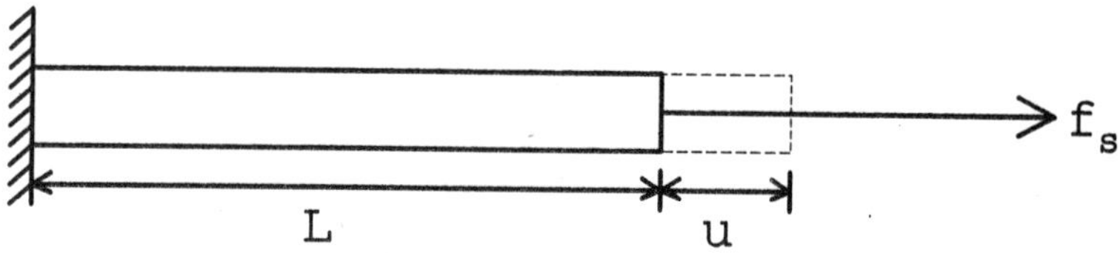

Figure 1.3.3 Axial deformation of truss element.

Equation (1.3.1) can be rewritten as

$$f_s = ku \tag{1.3.2}$$

where k, known as *stiffness* of the bar element, is given by

$$k = \frac{AE}{L} \tag{1.3.3}$$

To introduce the notation used in finite elements, consider again the bar element shown in Fig. 1.3.2. Recognizing the same deformation with the truss of Fig. 1.3.3, equilibrium of the element implies that

$$f_{s1} + f_{s2} = 0 \quad or \quad -f_{s1} = f_{s2} = f_s \tag{1.3.4}$$

The relative displacement of the element nodes u can be expressed in terms of the nodal displacements as

$$u = u_2 - u_1 \qquad (1.3.5)$$

Combining eqns. (1.3.2), (1.3.4) and (1.3.5), the f_{s1} and f_{s2} can be expressed as

$$f_{s1} = \frac{AE}{L}(u_1 - u_2)$$

and (1.3.6)

$$f_{s2} = \frac{AE}{L}(u_2 - u_1)$$

Equations (1.3.6) can be written in matrix form:

$$\begin{Bmatrix} f_{s1} \\ f_{s2} \end{Bmatrix} = \frac{AE}{L}\begin{bmatrix} 1 & -1 \\ -1 & 1 \end{bmatrix}\begin{Bmatrix} u_1 \\ u_2 \end{Bmatrix} \qquad (1.3.7)$$

Equation (1.3.7) is called the truss *element stiffness equation* and refers to the local element system x, y, z shown in Fig. 1.3.2, while the symmetric matrix [k] given by

$$[k] = \frac{AE}{L}\begin{bmatrix} 1 & -1 \\ -1 & 1 \end{bmatrix} \qquad (1.3.8)$$

is known as the *element stiffness matrix* of the truss element.

Having obtained the stiffness equation of one element, we can now proceed to obtain the stiffness equation for the two-element model of the cantilever bar. Our objective can be met if we develop a stiffness equation so that

$$\{F\} = [K]\ \{U\} \tag{1.3.9}$$

or more explicitly

$$\begin{Bmatrix} F_1 \\ F_2 \\ F_3 \end{Bmatrix} = \begin{bmatrix} K_{11} & K_{12} & K_{13} \\ K_{21} & K_{22} & K_{23} \\ K_{31} & K_{32} & K_{33} \end{bmatrix} \begin{Bmatrix} U_1 \\ U_2 \\ U_3 \end{Bmatrix} \tag{1.3.10}$$

where F_i and U_i (i = 1,2,3) denote the *global* external forces and displacements of the whole bar at nodes i. Notice that all quantities in eqn. (1.3.10) refer to the *global system* X,Y,Z shown on Fig. 1.3.1.

At this point we must make a few remarks about the nomenclature that will be used in the following to arrive at eqn. (1.3.10): Truss elements are labeled by numbering in parentheses. Global nodal forces and displacements are written with capital letters with a lower case index indicating the node they refer to, see Fig. 1.3.1. Small letters are used for the element nodal forces and displacements. An upper index in parenthesis and a lower index are used to indicate the element and the node, respectively. The element nodal forces and displacements for the two-element discretization of the cantilever bar are shown in Fig. 1.3.4, and the free-body diagrams of the elements and their common nodes are shown in Fig. 1.3.5.

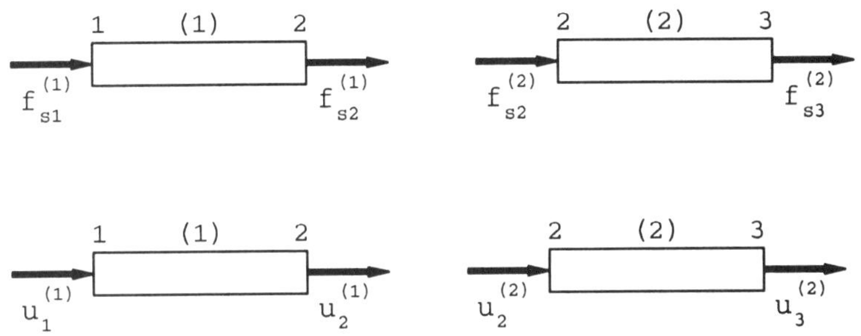

Figure 1.3.4 Element nodal elastic forces and displacements.

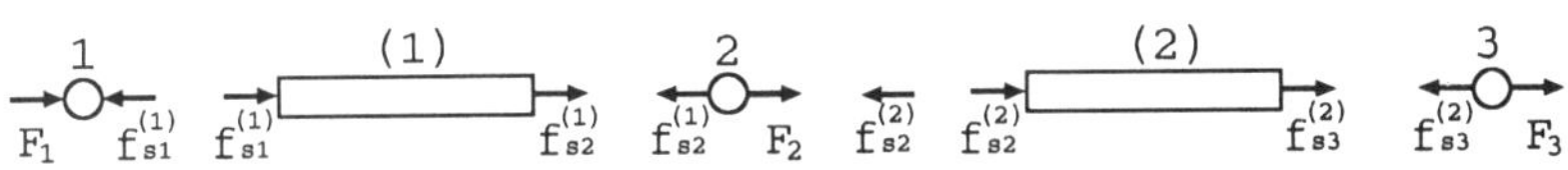

Figure 1.3.5 Free-body diagram of elements and nodes.

Notice that the free-body diagrams for nodes 1,2 and 3 include all forces acting on the nodes, i.e., both the internal (element) and the external (global) forces. In Fig. 1.3.5 also notice that because of the law of action-reaction the element nodal forces are opposite to the forces acting on the nodes. Equilibrium at the three nodes implies

$$
\begin{aligned}
F_1 &= f_{s1}^{(1)} \\
F_2 &= f_{s2}^{(1)} + f_{s2}^{(2)} \\
F_3 &= f_{s3}^{(2)}
\end{aligned}
\tag{1.3.11}
$$

Equations (1.3.11) can be written in the following matrix form:

$$\begin{Bmatrix} F_1 \\ F_2 \\ F_3 \end{Bmatrix} = \begin{Bmatrix} f_{s1}^{(1)} \\ f_{s2}^{(1)} \\ 0 \end{Bmatrix} + \begin{Bmatrix} 0 \\ f_{s2}^{(2)} \\ f_{s3}^{(2)} \end{Bmatrix} \tag{1.3.12}$$

By using eqn. (1.3.7) for each element and by expanding the size of the matrices with the addition of zero columns and rows, the vectors on the right side of eqn. (1.3.12) can be written as

$$\begin{Bmatrix} f_{s1}^{(1)} \\ f_{s2}^{(1)} \\ 0 \end{Bmatrix} = \frac{AE}{L} \begin{bmatrix} 1 & -1 & 0 \\ -1 & 1 & 0 \\ 0 & 0 & 0 \end{bmatrix} \begin{Bmatrix} u_1^{(1)} \\ u_2^{(1)} \\ 0 \end{Bmatrix}$$

$$\begin{Bmatrix} 0 \\ f_{s2}^{(2)} \\ f_{s3}^{(2)} \end{Bmatrix} = \frac{AE}{L} \begin{bmatrix} 0 & 0 & 0 \\ 0 & 1 & -1 \\ 0 & -1 & 1 \end{bmatrix} \begin{Bmatrix} 0 \\ u_2^{(2)} \\ u_3^{(2)} \end{Bmatrix} \tag{1.3.13}$$

Nodal compatibility enforces, see Figs. 1.3.1 and 1.3.4,

$$\begin{aligned} u_1^{(1)} &= U_1 \\ u_2^{(1)} = u_2^{(2)} &= U_2 \\ u_3^{(2)} &= U_3 \end{aligned} \tag{1.3.14}$$

Thus, by substituting eqns. (1.3.14) into eqns. (1.3.13) and the resulting expressions in eqn. (1.3.12) we obtain

$$\begin{Bmatrix} F_1 \\ F_2 \\ F_3 \end{Bmatrix} = \frac{AE}{L} \begin{bmatrix} 1 & -1 & 0 \\ -1 & 2 & -1 \\ 0 & -1 & 1 \end{bmatrix} \begin{Bmatrix} U_1 \\ U_2 \\ U_3 \end{Bmatrix} \tag{1.3.15}$$

Since node 3 is not permitted to move along the X-axis, $U_3 = 0$, and eqn. (1.3.15) can be written as

$$\begin{Bmatrix} F_1 \\ F_2 \end{Bmatrix} = \frac{AE}{L} \begin{bmatrix} 1 & -1 \\ -1 & 2 \end{bmatrix} \begin{Bmatrix} U_1 \\ U_2 \end{Bmatrix}$$

$$F_3 = -\frac{AE}{L} U_2 \tag{1.3.16}$$

Solution of eqns. (1.3.16) yields

$$U_1 = (2F_1 + F_2)\frac{L}{AE}$$

$$U_2 = (F_1 + F_2)\frac{L}{AE} \tag{1.3.17}$$

Finally, in view of eqns. (1.3.16) and (1.3.17) the reaction force F_3 can be expressed as $F_3 = -(F_1 + F_2)$, a result which is expected from equilibrium of the forces shown in Fig. 1.3.1.

Dynamic Analysis

Consider now the cantilever bar vibrating under the effect of the time dependent forces $F_1(t)$ and $F_2(t)$ shown in Fig. 1.3.1. As in the static case, before we proceed to solve the problem we will examine a single truss element. Since the element vibrates under the action of the applied forces, we must include the effect of the inertia forces in the element stiffness equation.

In this introductory example, we ignore the presence of damping. The simplest way to incorporate the effect of the inertia forces is to divide the total mass of the element $m=\varrho AL$, where ϱ is the mass density of the element, into two equal lumped masses m/2 and assign them to each node as shown in Fig. 1.3.6.

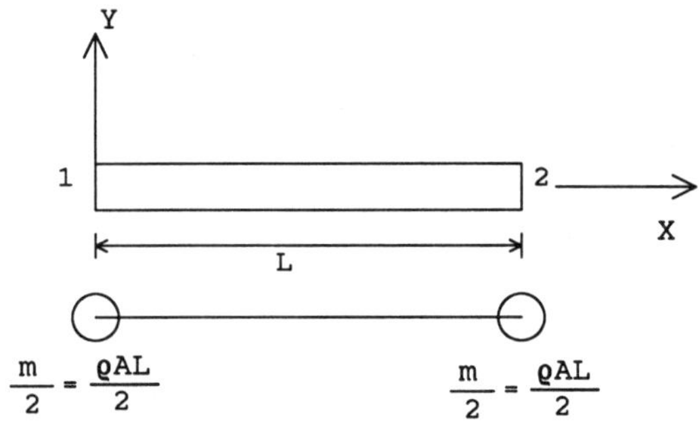

***Figure 1.3.6** Lumped mass discretization of truss element.*

To include the inertia effects we apply D'Alembert's principle, see Section 4.2. This implies that equilibrium is satisfied by adding the inertia forces $f_{I1} = -\frac{m}{2}\ddot{u}_1$ and $f_{I2} = -\frac{m}{2}\ddot{u}_2$ to the stiffness forces f_{s1} and f_{s2} at each node, where $\ddot{u}_1$ and $\ddot{u}_2$ are the accelerations at nodes 1 and 2, respectively, see Fig. 1.3.7. As a result, the element stiffness equation (1.3.7) becomes

$$\begin{Bmatrix} f_{s1} - f_{I1} \\ f_{s2} - f_{I2} \end{Bmatrix} = \frac{AE}{L}\begin{bmatrix} 1 & -1 \\ -1 & 1 \end{bmatrix}\begin{Bmatrix} u_1 \\ u_2 \end{Bmatrix} + \frac{m}{2}\begin{bmatrix} 1 & 0 \\ 0 & 1 \end{bmatrix}\begin{Bmatrix} \ddot{u}_1 \\ \ddot{u}_2 \end{Bmatrix} \tag{1.3.18}$$

The negative sign of the inertia forces indicates that they act in the opposite direction to resist the elastic forces.

We may now proceed to obtain the *dynamic stiffness equation* or equation of motion of the cantilever bar. The element nodal forces and displacements for the two-element model of the bar are shown in Fig. 1.3.8.

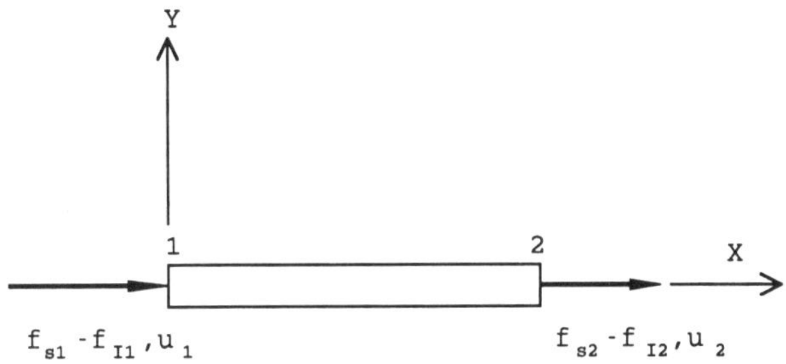

Figure 1.3.7 Elastic and inertia element nodal forces.

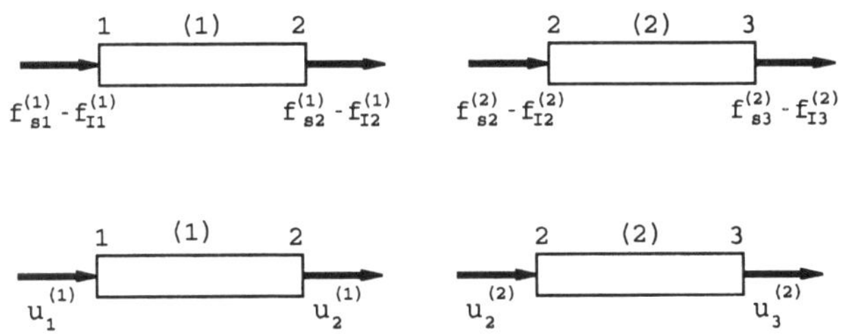

Figure 1.3.8 Element nodal dynamic forces and displacements.

The free-body diagram of the two-element and three-node assemblage is presented in Fig. 1.3.9. Notice that the free-body diagram for nodes 1,2 and 3 includes all forces acting on the nodes, i.e., both the internal (element) and the external (global) forces. In Fig. 1.3.9 also notice that because of the law of action-reaction the element nodal forces are opposite to the ones acting on the nodes.

Equilibrium at the three nodes implies that

$$
\begin{aligned}
F_1 &= f^{(1)}_{s1} - f^{(1)}_{I1} \\
F_2 &= f^{(1)}_{s2} + f^{(2)}_{s2} - f^{(1)}_{I2} - f^{(2)}_{I2} \\
F_3 &= f^{(2)}_{s3} - f^{(2)}_{I3}
\end{aligned}
\tag{1.3.19}
$$

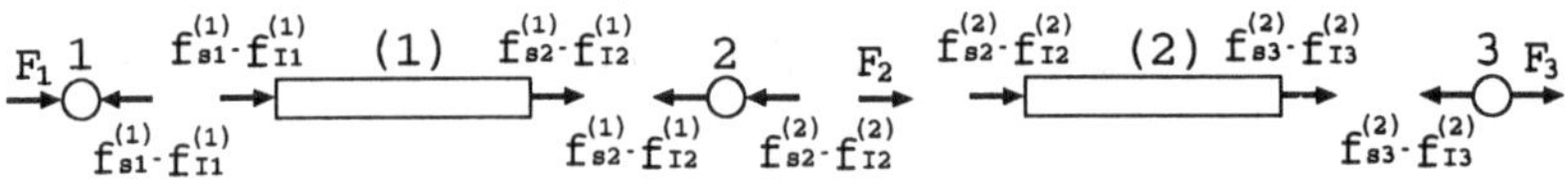

Figure 1.3.9 Free-body diagram of elements and nodes.

Equations (1.3.19) can be written in matrix form as

$$\begin{Bmatrix} F_1 \\ F_2 \\ F_3 \end{Bmatrix} = \begin{Bmatrix} f_{s1}^{(1)} - f_{I1}^{(1)} \\ f_{s2}^{(1)} - f_{I2}^{(1)} \\ 0 \end{Bmatrix} + \begin{Bmatrix} 0 \\ f_{s2}^{(2)} - f_{I2}^{(2)} \\ f_{s3}^{(2)} - f_{I3}^{(2)} \end{Bmatrix} \tag{1.3.20}$$

By using eqn. (1.3.18) for each element and by expanding the size of the matrices with the addition of zero columns and rows, the vectors on the right side of eqn. (1.3.20) can be written as

$$\begin{Bmatrix} f_{s1}^{(1)} - f_{I1}^{(1)} \\ f_{s2}^{(1)} - f_{I2}^{(1)} \\ 0 \end{Bmatrix} = \frac{AE}{L}\begin{bmatrix} 1 & -1 & 0 \\ -1 & 1 & 0 \\ 0 & 0 & 0 \end{bmatrix}\begin{Bmatrix} u_1^{(1)} \\ u_2^{(1)} \\ 0 \end{Bmatrix} + \frac{m}{2}\begin{bmatrix} 1 & 0 & 0 \\ 0 & 1 & 0 \\ 0 & 0 & 0 \end{bmatrix}\begin{Bmatrix} \ddot{u}_1^{(1)} \\ \ddot{u}_2^{(1)} \\ 0 \end{Bmatrix} \tag{1.3.21}$$

$$\begin{Bmatrix} 0 \\ f_{s2}^{(2)} - f_{I2}^{(2)} \\ f_{s3}^{(2)} - f_{I3}^{(2)} \end{Bmatrix} = \frac{AE}{L}\begin{bmatrix} 0 & 0 & 0 \\ 0 & 1 & -1 \\ 0 & -1 & 1 \end{bmatrix}\begin{Bmatrix} 0 \\ u_2^{(2)} \\ u_3^{(2)} \end{Bmatrix} + \frac{m}{2}\begin{bmatrix} 0 & 0 & 0 \\ 0 & 1 & 0 \\ 0 & 0 & 1 \end{bmatrix}\begin{Bmatrix} 0 \\ \ddot{u}_2^{(2)} \\ \ddot{u}_3^{(2)} \end{Bmatrix} \tag{1.3.22}$$

As in the static case, nodal compatibility enforces

$$\begin{aligned} u_1^{(1)} &= U_1 \\ u_2^{(1)} = u_2^{(2)} &= U_2 \\ u_3^{(2)} &= U_3 \end{aligned} \qquad (1.3.14)$$

The displacement compatibility expressed by eqns. (1.3.14) also implies compatibility of the nodal velocities and accelerations.

By substituting eqns. (1.3.14) into eqns. (1.3.21) and (1.3.22), and the resulting expressions into eqn. (1.3.20), we obtain

$$\begin{Bmatrix} F_1 \\ F_2 \\ F_3 \end{Bmatrix} = \frac{AE}{L} \begin{bmatrix} 1 & -1 & 0 \\ -1 & 2 & -1 \\ 0 & -1 & 1 \end{bmatrix} \begin{Bmatrix} U_1 \\ U_2 \\ U_3 \end{Bmatrix} + \frac{\varrho AL}{2} \begin{bmatrix} 1 & 0 & 0 \\ 0 & 2 & 0 \\ 0 & 0 & 1 \end{bmatrix} \begin{Bmatrix} \ddot{U}_1 \\ \ddot{U}_2 \\ \ddot{U}_3 \end{Bmatrix} \qquad (1.3.23)$$

It can be readily shown that if we ignore the inertia effects, that is, the vector of the accelerations is set equal to zero, eqn. (1.3.23) reduces to eqn. (1.3.15).

Since node 3 is not allowed to displace along the X-axis, $U_3 = \ddot{U}_3 = 0$, and eqn. (1.3.23) yields

$$\begin{Bmatrix} F_1 \\ F_2 \end{Bmatrix} = \frac{AE}{L} \begin{bmatrix} 1 & -1 \\ -1 & 2 \end{bmatrix} \begin{Bmatrix} U_1 \\ U_2 \end{Bmatrix} + \frac{\varrho AL}{2} \begin{bmatrix} 1 & 0 \\ 0 & 2 \end{bmatrix} \begin{Bmatrix} \ddot{U}_1 \\ \ddot{U}_2 \end{Bmatrix}$$

$$F_3 = \frac{AE}{L}(-U_2 + U_3) + \frac{\varrho AL}{2}\ddot{U}_3 \qquad (1.3.24)$$

Notice that since $U_3 = \ddot{U}_3 = 0$, F_3 reduces to

$$F_3 = -\frac{AE}{L}U_2$$

Equations (1.3.24) are the equations of motion of the two-element cantilever bar. Contrary to eqns. (1.3.16), which have been solved for the nodal displacements, eqns. (1.3.24) represent a system of three coupled ordinary differential equations that can be solved with one of the methods presented in Chapter 5.

As we shall see in Example 5.4.1, the two-element model of the cantilever bar is not sufficient to provide accurate results, and has been selected here only for computational simplicity. Example 5.10.1 also uses the two-element cantilever bar to determine its response to a suddenly applied constant load with time history modal superposition.

2

Stresses and Failure Criteria

2.1 CONCEPT OF STRESS

The main objective of finite element analysis is to calculate the displacements and stresses for given loads. Another important objective is to determine if the structure or a portion of it has failed under the effect of the applied forces. Both objectives require understanding of the notion and representation of stresses as they develop in the various types of finite elements. It is expected that the reader has a basic background on strength of materials, therefore, the discussion here will summarize the salient concepts rather than explain them.

In order to completely define the state of stress at a point in a structure, we must specify the components of stress at the given point acting on three mutually orthogonal planes passing through it. The three orthogonal planes as well as the stresses acting on them are shown in Fig. 2.1.1. Normal stresses are indicated with σ_{ij} (i,j = x,y,z), while τ_{ij} is used to denote in-plane or shear stresses. The first subscript specifies the direction of the normal to the plane on which the stress acts and the second indicates the direction of the stress itself. All stresses shown in Fig. 2.1.1 are positive according to standard engineering mechanics convention.

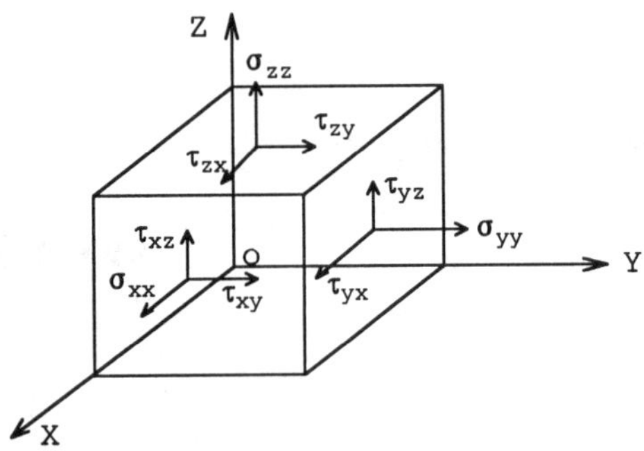

Figure 2.1.1 *Stress state at a point.*

2.2 PRINCIPAL STRESSES

In practice, the state of stress at a point with reference to a Cartesian system is not significant, since maximum stresses and failure of a structure may occur due to the state of stress on a plane inclined to the Cartesian system. In engineering applications, the interest lies in determining the maximum normal and shear stresses as well as their orientation at every location in the structure in order to determine safety against failure. At every point in a loaded structure, the maximum normal stresses develop on three mutually perpendicular planes (*principal planes*). Briefly stated, the *principal normal stresses* represent the maximum, minimum, and intermediate normal stresses at a given point. The shear stresses that develop on the principal planes are zero. Shear stresses also have limiting values. They can be determined in a manner similar to that of principal stresses. In a three-dimensional case, the maximum and minimum shear stresses are generally referred to as *principal shear stresses*.

Evaluation of the principal stresses in truly three-dimensional problems is computationally rather involved. Many practical problems, however, can be analyzed as two-dimensional, for example the case of plane stress where $\sigma_3=0$. In this case, the principal stresses σ_1 and σ_2 can be calculated in terms of the normal and shear stresses from

$$\sigma_{1,2} = \frac{\sigma_{xx}+\sigma_{yy}}{2} \pm \sqrt{\frac{(\sigma_{xx}-\sigma_{yy})^2}{4} + \tau_{xy}^2} \qquad (2.2.1)$$

For the plane stress case with $\sigma_2<0$ and $\sigma_1>0$, the maximum shear stress τ_{max} is given by

$$\tau_{max} = \frac{\sigma_1 - \sigma_2}{2} \tag{2.2.2}$$

2.3 STRESSES IN TRUSSES AND BEAMS

In trusses and beams *axial* or *uniform stresses* develop when their members are subjected to either tensile or compressive loads, and are defined by the well known expression:

$$\sigma = \frac{P}{A} \tag{2.3.1}$$

where

P = the applied axial load,
A = the cross-sectional area.

Contrary to uniform stresses, *bending stresses* vary linearly through the cross section of beams and are calculated by the equation

$$\sigma_{xx} = \pm\frac{M_z y}{I_z} \tag{2.3.2}$$

where

M_z = the bending moment,
y = the distance of a fiber from the neutral axis,
I_z = the moment of inertia about the neutral axis normal to the plane of the applied loads.

It should be noted that eqn. (2.3.2) is valid for beams with a symmetric cross section subjected to transverse external loads acting along the plane of symmetry, see Fig. 2.3.1. If the beam length L, width b, and depth h are such that both b and h are at least ten times less than L, then the deformation developed in the beam is primarily attributed to bending stresses and can be determined from eqn. (2.3.2).

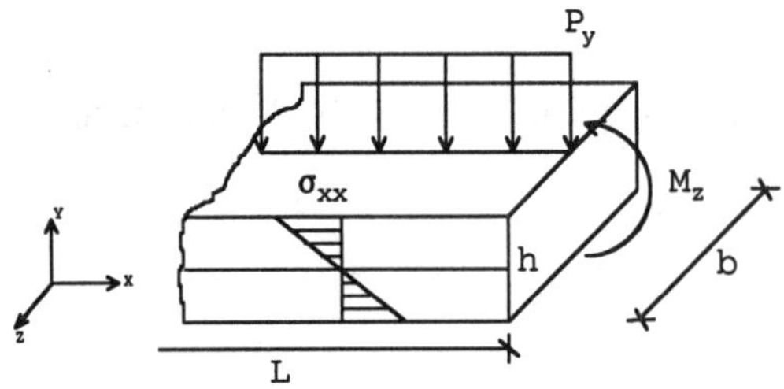

Figure 2.3.1 Moment and bending stress distribution.

Besides bending stresses, *shear stresses* develop in beams. These stresses undertake the shear forces V at a given cross section. Contrary to bending stresses, which obtain their maximum values at the outer fibers of the section, shear stresses vary from zero at the outer fibers to their maximum at the neutral axis of the beam. For a beam with a rectangular cross section, the distribution of shear stresses along the Y-axis is shown in Fig. 2.3.2.

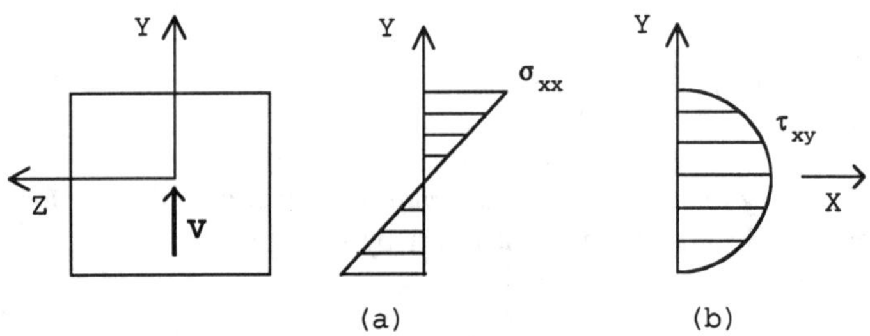

Figure 2.3.2 Bending and shear stress distribution.

The shear stress at a cross section of the beam is given by

$$\tau_{xy} = \frac{V_y Q_z}{I_z b} \tag{2.3.3}$$

where

I_z = moment of inertia about the neutral Z-axis,
Q_z = first moment of the area about the Z-axis.

Figures 2.3.3(a) and (b) show the variation of bending and shear stresses in a wide-flange beam. The distribution of the shear stresses clearly indicates that in wide-flange beams the flanges undertake the bending, while shear forces are carried by the web.

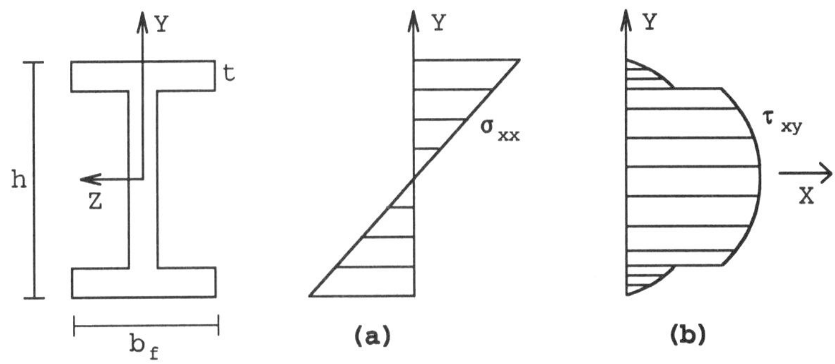

Figure 2.3.3 *Stresses in I-section: (a) Bending; (b) Shear.*

It should be noted that the variation of shear stresses in a cross section cannot be captured with beam elements. Finite element analysis of beams can provide the shear forces at any desired section, from which the variation of shear stresses can be obtained from eqn. (2.3.3). The deflection of a beam due to shear can be considered if appropriate values for the beam shear areas are given non-zero values. In Section 3.2 additional information on how to consider the effects of shear is provided. The significance and contribution of shear to the dynamic response of structures is examined in Chapter 7.

2.4 PLANE STRESS

Three-dimensional problems can be reduced to two-dimensional if they fulfill conditions of plane stress, plane strain, or axisymmetry. In such case modeling of the structure is greatly simplified and the computational effort is substantially reduced.

A typical plane stress problem is a thin plate loaded with a uniform load along any of its edges, as shown in Fig. 2.4.1. Due to the small thickness of the plate no stresses develop along the X-axis.

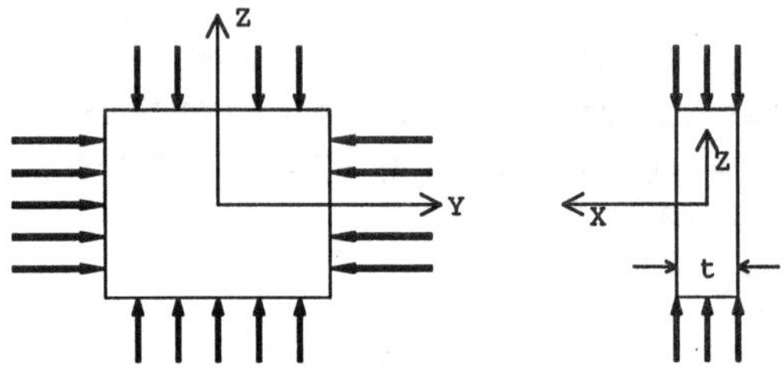

Figure 2.4.1 Thin plate under plane stress conditions.

In general, a structure is under conditions of plane stress if the stresses that develop along one of the three orthogonal axes are so small compared to the rest of the stresses that they can be approximated as being zero:

$$\sigma_{xx} = \tau_{yx} = \tau_{zx} = 0 \tag{2.4.1}$$

The non-zero stresses that develop at a point are shown in Fig. 2.4.2. It should be noted, however, that the shear strains along the X-axis are zero, while the normal strain along the X-axis is non-zero due to Poisson's effect:

$$\begin{aligned} \gamma_{yx} &= \gamma_{zx} = 0 \\ \epsilon_{xx} &= -\frac{\nu}{E}(\sigma_{yy} + \sigma_{zz}) \end{aligned} \tag{2.4.2}$$

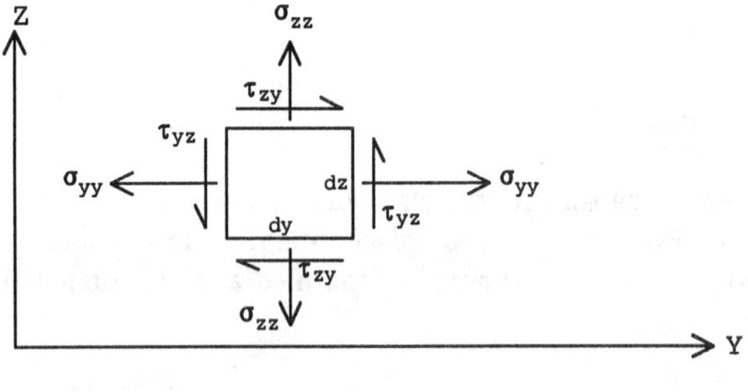

Figure 2.4.2 Non-zero stresses at a point.

2.5 PLANE STRAIN

The long cylinder shown in Fig. 2.5.1 is a representative example of a plane strain problem. The cylinder has uniform thickness t, its two ends are confined between two smooth planes, and is subjected to an interior pressure of constant intensity p. A strip-foundation such as the one shown in Fig. 2.5.2 is another typical example of a plane strain problem. Apparently, all cross sections of the cylinder and the strip-foundation that are away from the boundaries along the X-axis are under the same loading conditions and experience the same deformation.

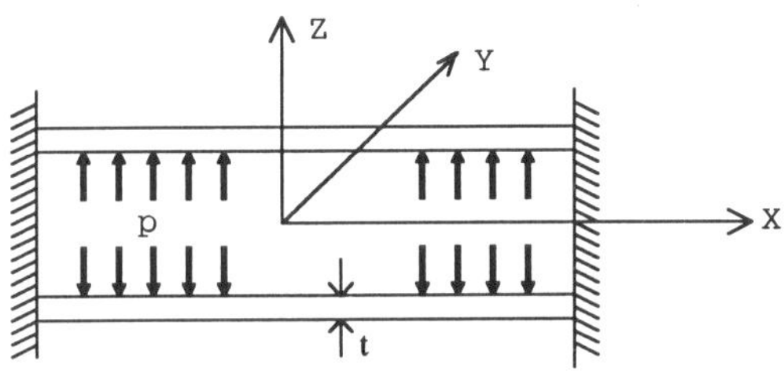

Figure 2.5.1 *Cylinder under internal uniform pressure.*

Since geometric, material, and loading conditions are the same at all cross sections, it is sufficient to analyze only one slice of the structure a unit distance apart between two cross sections. The deformation and stresses obtained for this slice are representative of all sections along the X-axis except those sections that are close to the boundaries. In general, a structure is under conditions of plane strain, if the strains that develop along one of the three orthogonal axes, say the X-axis, are zero

$$\epsilon_{xx} = \gamma_{zx} = \gamma_{yx} = 0 \qquad \textit{(2.5.1)}$$

The longitudinal normal stress along the X-axis is not zero and can be evaluated in terms of σ_{zz} and σ_{yy}, while the shear stresses along the X-axis are zero:

$$\begin{aligned} \sigma_{xx} &= \nu(\sigma_{yy} + \sigma_{zz}) \\ \tau_{yx} &= \tau_{zx} = 0 \end{aligned} \qquad \textit{(2.5.2)}$$

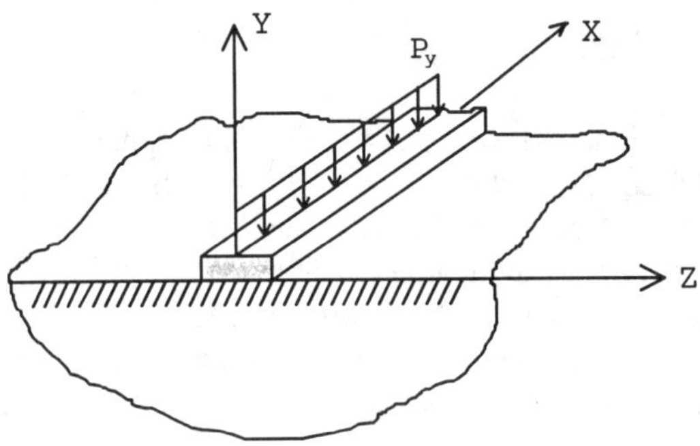

Figure 2.5.2 Strip-foundation subjected to a line load.

2.6 AXISYMMETRIC SOLIDS

Typical axisymmetric structures are pressure vessels and solid or hollow spherical structures, see Fig. 2.6.1. In axisymmetric structures both the geometry and applied loading are symmetric about an axis. Usually the geometry of these structures is specified in a polar coordinate system r-θ on the YZ plane. The X-axis is normal to the YZ plane. Figure 2.6.1 shows the cross section of a hollow cylinder subjected to a normal internal pressure. Every point of the cylinder can be specified in terms of the r and θ coordinates.

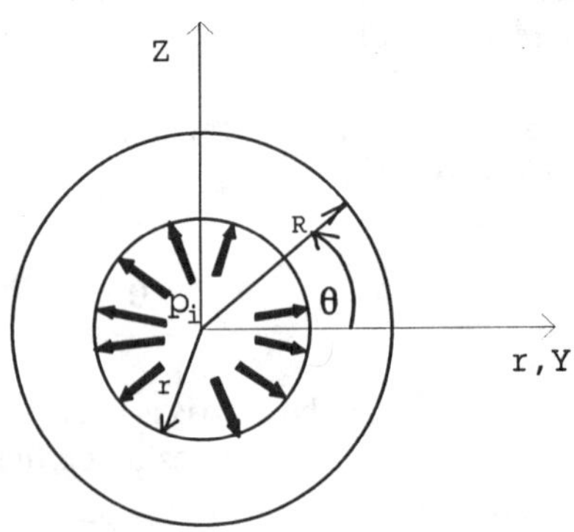

Figure 2.6.1 Thick-wall cylinder subjected to internal pressure.

The non-zero stresses in an axisymmetric structure, such as the cylinder of Fig. 2.6.1, are shown in Fig. 2.6.2. Specifically, Fig. 2.6.2 shows two of the four non-zero stress components that develop on an infinitesimal element of the cylinder. The other non-zero stresses are the σ_x and τ_{xr}.

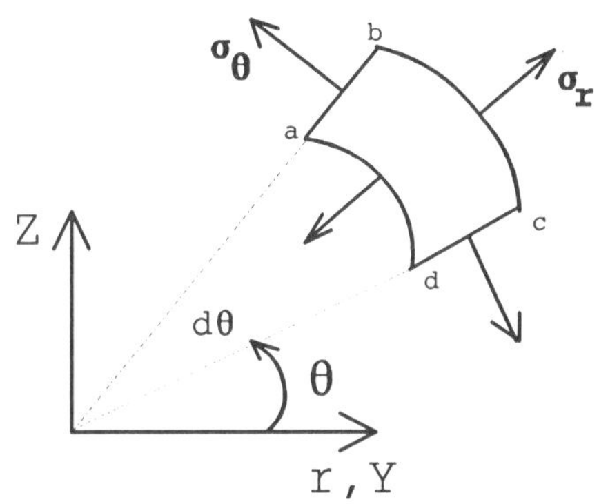

Figure 2.6.2 *Stresses on axisymmetric element.*

Several simplifications are possible in reducing the number of the unknown stresses and strains in axisymmetric problems; such is the case of a *thin-wall cylinder* subjected to an internal pressure p. In this case the tangential stress σ_θ can be considered as constant within the thickness and is given by

$$\sigma_\theta = \frac{pr}{t} \qquad \textit{(2.6.1)}$$

where t is the thickness and r is the mean radius.

A cylinder is considered thin-walled if the wall thickness t is at least less than 10% of the inner radius r. Cylinders that do not satisfy the above criterion are classified as *thick-wall cylinders* and the variation of σ_θ within the thickness should be included in the analysis. As presented in more detail in Section 3.2, thin-shell structures should be modeled with either axisymmetric or plate/shell elements, while thick-shell structures should be analyzed with brick/solid, tetrahedral, and hexahedral elements. It is understood that thin cylinders can be modeled with axisymmetric elements if both their geometry and loading are axisymmetric.

2.7 PLATES AND SHELLS

Thin plates support transverse loads by bending action and in-plane loads by membrane action. A plate is classified as thin if it fulfills the geometric relationship:

$$\frac{h}{L} \leq \frac{1}{10} \tag{2.7.1}$$

where

h = plate thickness,
L = is the minimum of the other two plate dimensions.

The primary difference between plates and shells is that shells have curvature. As a consequence a shell is capable of resisting transverse loads not only with bending but also with membrane action. In fact, efficient use of shells is achieved when most of the transverse loads are carried through membrane action. Figure 2.7.1 shows the cross section of a thin shell circular roof subjected to a pressure p and normal forces N. A shell structure is classified as thin if the ratio of its thickness to the radius of curvature is equal to or less than 1/10.

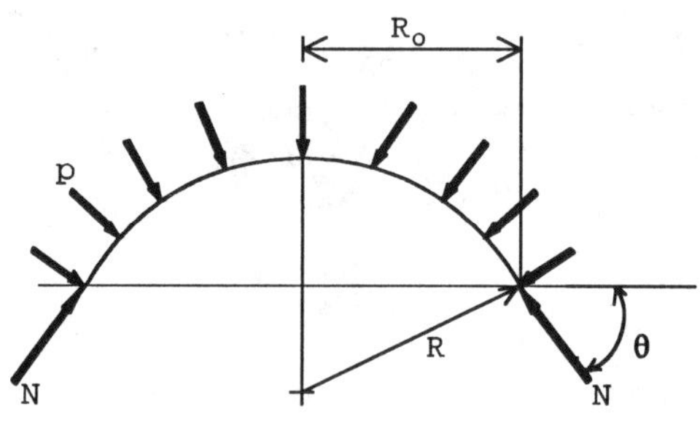

Figure 2.7.1 Thin shell circular roof.

In linear small deformation analysis of plates, membrane and bending actions are decoupled, that is, the mechanisms they manifest themselves are not related to each other, and as such they can be studied independently. Since membrane action can be viewed as plane stress, which has been presented in Section 2.4, the following discussion will be limited to the bending behavior of thin plates. The *flexural rigidity* of a thin plate is given by

$$D = \frac{Eh^3}{12(1-\nu^2)} \tag{2.7.2}$$

The variation of stresses that develop on a differential element of a homogeneous, isotropic, thin plate is shown in Fig. 2.7.2. The moments and shear forces per unit length corresponding to the stresses are also shown in the same figure. The relationship between maximum stresses and moments/shears per unit length are given by

$$\sigma_{xx,\max} = \frac{6M_x}{h^2}, \quad \sigma_{yy,\max} = \frac{6M_y}{h^2}, \quad \tau_{xy} = \frac{6M_{xy}}{h^2}$$
$$\tau_{xz,\max} = 1.5\frac{Q_x}{h}, \quad \tau_{yz,\max} = 1.5\frac{Q_y}{h} \tag{2.7.3}$$

where M_x, M_y and M_{xy} are moments per unit length along the sides of the plate, and Q_x and Q_y are the shears per unit length normal to the plane of the plate.

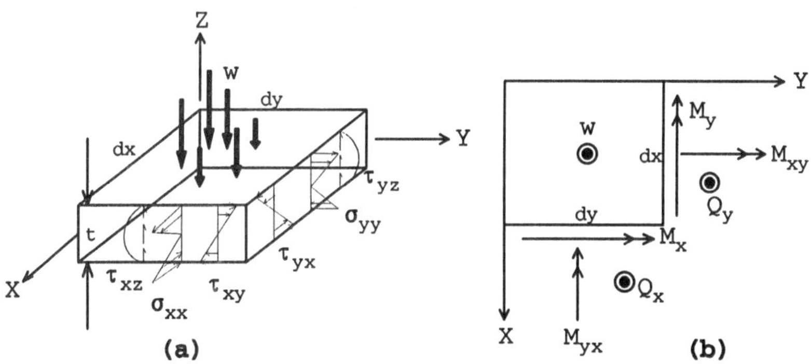

Figure 2.7.2 *(a) Stresses on differential plate element; (b) Moments and shear forces per unit length.*

2.8 ISOTROPY-ORTHOTROPY-ANISOTROPY

In linear elasticity, strains $\{\epsilon\}$ and stresses $\{\sigma\}$ are related through Hooke's law:

$$\{\sigma\} = [D]\{\varepsilon\} \tag{2.8.1}$$

where [D] is the *elasticity matrix* with entries D_{ij} (i,j = 1,2,...,6) that are independent of time and any applied loading.

For a uniaxial case, [D] is simply the modulus of elasticity E, and eqn. (2.8.1) takes the simple form

$$\sigma = E\,\varepsilon \tag{2.8.2}$$

For the general case though, the elasticity matrix [D] relates six stress to their corresponding strain components and is given by

$$[D] = \begin{bmatrix} D_{11} & D_{12} & D_{13} & D_{14} & D_{15} & D_{16} \\ D_{21} & D_{22} & D_{23} & D_{24} & D_{25} & D_{26} \\ D_{31} & D_{32} & D_{33} & D_{34} & D_{35} & D_{36} \\ D_{41} & D_{42} & D_{43} & D_{44} & D_{45} & D_{46} \\ D_{51} & D_{52} & D_{53} & D_{54} & D_{55} & D_{56} \\ D_{61} & D_{62} & D_{63} & D_{64} & D_{65} & D_{66} \end{bmatrix} \tag{2.8.3}$$

Depending on the relationship between the 36 elastic constants of the elasticity matrix, we can distinguish three different classes for the material of the system:

a. *anisotropic*: if the 36 elastic constants are independent from each other. Note that for anisotropic materials D_{ij} are functions of the coordinate system.

b. *orthotropic*: if it exhibits elastic symmetry about the three planes XY, YZ and XZ. In this case, the number of the independent constants reduces to 12, that is, D_{11}, D_{12}, D_{13}, D_{21}, D_{22}, D_{23}, D_{31}, D_{32}, D_{33}, D_{44}, D_{55}, D_{66}.

c. *isotropic*: if the material properties are the same at every point of the structure, i.e., they are independent of the coordinate system. In this case the independent D_{ij} reduce to only three: D_{11}, D_{12}, D_{44}, which are given by

$$D_{11} = \lambda + 2\mu \qquad D_{12} = \lambda \qquad D_{44} = \mu \tag{2.8.4}$$

The Lame constants λ and μ can be expressed in terms of the modulus of elasticity E and Poisson's ratio ν as given by

$$\lambda = \frac{\nu E}{(1+\nu)(1-2\nu)} \qquad \mu = \frac{E}{2(1+\nu)} \tag{2.8.5}$$

Thus, for an isotropic material, [D] is a function of only the two independent constants E and ν.

2.9 FAILURE CRITERIA

For all types of analysis that are presented in this text, the structure is assumed to behave elastically, that is, the deformation it experiences is

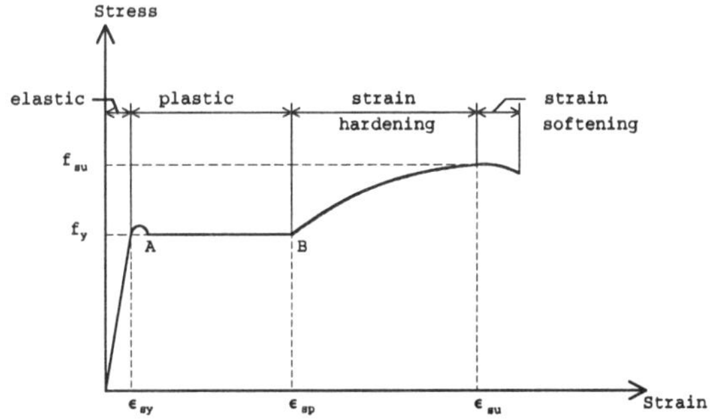

Figure 2.9.1 Steel under tensile static load.

recoverable. In addition, it is also assumed that under the effect of the applied forces, the system experiences small deformations and small strains. The word "small" implies infinitesimal changes in the geometry of the system. In the elastic state the stresses and strains that develop in a structure are uniquely defined regardless of the spatial and time variation of the applied forces. Once the elastic state is exceeded, even in a small region of the structure, deformation is no longer recoverable and its further development depends on whether the structural material is either *ductile* or *brittle*. Common ductile materials are mild steel, aluminum, copper, and in general most metals. Representative brittle materials are unreinforced concrete, glass, and cast iron. The stress-strain relationship for a ductile material, e.g., mild steel, in simple tension is given in Fig. 2.9.1. Figure 2.9.2 depicts the stress-strain relationship of unreinforced concrete.

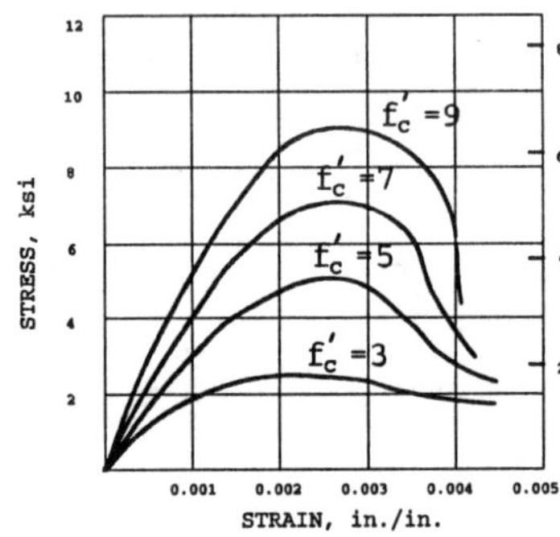

Figure 2.9.2 *Typical stress-strain relationship of concrete for varying compressive strengths in ksi.*

For ductile materials, when the magnitude of the external forces is substantial, the structure experiences a permanent deformation known as *plastic deformation*. Plastic deformations are permanent changes in atomic positions of the structural material. They are irrecoverable and depend on both the spatial and the time variation of the applied forces. Contrary to the elastic state, the plastic state cannot be defined uniquely. Treatment of nonlinear statics and dynamics problems is beyond the scope of this text and can be found in various references.

Plastic deformations develop when the combination of stresses at a region in the structure exceeds a critical value. The critical combination of the stress components, that initiates plastic deformations for ductile materials and cracks for brittle materials, is mathematically expressed as a failure or yield criterion. When regions of a structure are stressed beyond the elastic limit, the results obtained through a linear analysis could be invalid. In linear static and dynamic analysis we do not evaluate deformations in the plastic state; nevertheless, we are concerned that the system experiences only linear deformations. The task of identifying regions that have yielded is accomplished with the aid of *failure* or *yield criteria*.

It should be pointed out that the word "failure" as used for the failure criteria is rather misleading. The development of either plastic stresses or cracks at a point or section in a structure does not necessarily imply that the structure has failed and cannot carry any further load. It simply means that further analysis using linear elastic theory is no longer valid, at least for the vicinity of the region of the structure that "failed."

Several competing theories propose different failure criteria. In general, all these theories provide fairly similar results. The most widely used failure criteria are the *Tresca* and the *von Mises* for ductile materials, and the *maximum normal stress criterion* for brittle materials. For composite materials appropriate criteria must be used, e.g., Tsai-Wu.

In applying the Tresca and von Mises yield criteria one should bear in mind:

a. both the Tresca and von Mises stresses are scalar quantities, see eqns. (2.9.1) and (2.9.2), and when compared with the yield stress of the structural material under uniaxial tension σ_y they indicate whether a part of the structure is stressed beyond the elastic limit.

b. As discussed in Chapter 9, yield criteria should be applied with caution in dynamic analysis.

Finally, it should be emphasized that besides securing that the structure remains in the elastic range for the applied loading, there are other considerations that must be examined to confirm correctness of the results. Besides yield stresses, large displacements and buckling of flexural members such as beams, thin plates and thin shells under compression must be checked because they can also invalidate the analysis results, see Section 7.7.

Tresca Yield Criterion

The Tresca criterion states that yielding initiates when one-half the largest difference between the principal stresses reaches one-half the yield stress of the structural material under pure tension. Assuming that $\sigma_3 < \sigma_2 < \sigma_1$ then Tresca's yield criterion can be expressed as

$$\frac{1}{2}(\sigma_1 - \sigma_3) = \frac{1}{2}\sigma_y \qquad (2.9.1)$$

Notice that this criterion is not affected by the intermediate principal stress σ_2. The difference 0.5 (σ_1 - σ_3) is the Tresca stress. If the Tresca stresses in the system are less than $0.5\sigma_y$, linear analysis is valid. Use of the Tresca yield criterion is elucidated in examples presented in Chapter 6 through 9.

Von Mises Yield Criterion

According to von Mises, plastic deformation at a point initiates when the principal stresses at that point satisfy the relationship

$$\sqrt{0.5[(\sigma_1 - \sigma_2)^2 + (\sigma_2 - \sigma_3)^2 + (\sigma_3 - \sigma_1)^2\,]} = \sigma_y \qquad (2.9.2)$$

As for the Tresca yield criterion, σ_y is the yield stress of the structural material under uniaxial tension.

The quantity on the left-hand side of eqn. (2.9.2) is the *von Mises stress*. Thus, direct comparison of the von Mises stress with σ_y allows identification of the areas that have yielded. Notice that contrary to Tresca, the von Mises criterion involves all principal stresses.

The Tresca criterion predicts yielding at a generally lower stress level than does the von Mises. However, the von Mises criterion is more widely accepted than the Tresca for ductile materials. A detailed discussion on the use of von Mises stresses is presented in several examples in Chapters 6 through 9.

Maximum Normal Stress Criterion

For brittle materials, e.g., unreinforced concrete, cast iron, porcelain, the *maximum principal stress theory* is used to predict brittle fracture. This theory proposes that fracture initiates when the maximum principal stress at a point, if tensile, is equal to the ultimate stress from a uniaxial tension test and, if compressive, equals the ultimate stress from a uniaxial compression test.

For completeness we should also mention that there are other failure criteria beyond the three criteria that we discussed so far. One such criterion is the Mohr-Coulomb failure envelope that is widely used in soil mechanics.

3

Finite Element Modeling Fundamentals

3.1 MODELING CONSIDERATIONS

The objective of finite element analysis is to accurately determine the response of a system modeled with finite elements and subjected to given loads. In creating a finite element model, always keep in mind that we are developing a model which is an idealization of the real physical system. With very few exceptions, such as the static analysis of simple truss, frame, and membrane systems, finite element analysis does not provide "exact" answers. However, with proper modeling, it yields accurate solutions. When an analytical formulation of a problem is difficult to develop, FEM has proven to be one of the most reliable methods to tackle the problem.

In creating a FEM model one should strive for accuracy and computational efficiency. In most cases, use of a complex and very refined model is not justifiable since it most likely provides computational accuracy at the expense of unnecessarily increased processing time. The type and complexity of the model is dependent upon the type of results required. As a general rule, finite element modeling should start with a simple model. The results from the simple model combined with an understanding of the behavior of the system will help us decide whether and at which part of the model further refinement is needed.

3.2 TYPES OF FINITE ELEMENTS

This section describes several salient features of the most commonly used types of finite elements; namely, *truss, beam, plane stress, plane strain, axisymmetric, membrane, plate, shell, solid or brick, tetrahedral, hexahedral, boundary, and gap* elements. The reader may be aware that several commercial finite element programs have numerous types of finite elements in their library. Nevertheless, the majority of structural and mechanical engineering applications can be solved with the aforementioned *basic element* types. Besides the basic elements, other special types, such as *pipe* and *rigid* elements are available in finite element program libraries and can be used for specialized applications.

Depending on their dimensions, basic elements can be divided into three categories: *line, area, and volume elements.* Truss, beam, and boundary elements are line elements. Plane stress, plane strain, axisymmetric, membrane, plate and shell are area elements. Solid or brick, tetrahedral, and hexahedral are volume elements. Criteria for selecting the appropriate elements for the application are provided in Section 3.3. At this point it should be mentioned that there are considerable differences between proper modeling for static and dynamic analysis. These differences as well as their similarities are discussed in Sections 3.4 through 3.10 and in Section 7.1.

Truss Elements

Truss structures such as mechanical linkages, roof frames, and bridges can be modeled with truss elements. In general, a structural member can be modeled with truss elements if it fulfills the following three requirements:

a. its length is much greater than its width and depth (the loosely defined term *much greater* can be quantified from 8 to 10 times greater for most applications);

b. it is connected with the rest of the structure with hinges that do not allow transfer of moments; and

c. it is loaded with external loads applied only at the joints.

Truss elements can only undertake tension or compression. Thus, the only cross-sectional property that needs to be specified is their axial area. The geometrical profiles of truss elements are identical to beam elements. Figure 3.2.1 shows the geometry and nodal forces of a three-dimensional truss element. As depicted in this figure, a three-dimensional truss element has three degrees-of-freedom per node, that is, three displacements along the global X, Y and Z-axes.

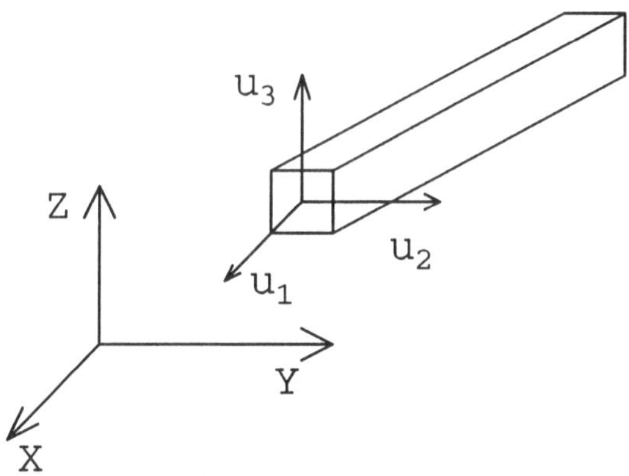

Figure 3.2.1 Three-dimensional truss element.

Beam Elements

Beam elements are probably the most commonly used elements. Besides their obvious application in frame structures, many other systems such as mechanical linkages, piping systems, conduits, and bridge girders can be modeled with beam elements.

For a structural member to be modeled with beam elements, one of its dimensions must be much greater, as a general rule at least 10 times greater, than the other two. Contrary to truss elements, beam elements can undertake shear and moment in addition to tension and compression. The geometry and the displacements/rotations at a node are shown in Fig. 3.2.2. Notice that a three-dimensional beam element has six degrees-of-freedom per node, that is, three displacements along and three rotations about the global X, Y and Z-axes.

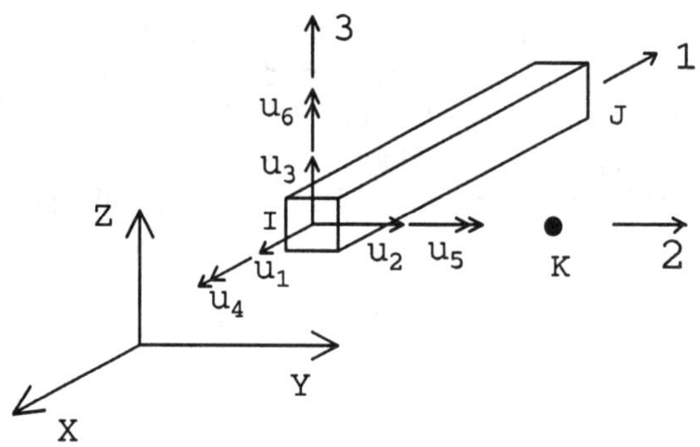

***Figure 3.2.2** Three-dimensional beam element.*

Common beam profiles include I-sections, T-sections, box, circular, and channel sections. The cross-sectional properties that must be specified for a beam element include the axial area, the shear areas, the torsional resistance, the flexural moments of inertia, and the section moduli. Table 3.2.1 presents the shear areas of several common beam profiles.

Unless the beam cross-sectional area is axisymmetric, e.g., a circular or a square cross section, the orientation of the beam element with respect to the global system XYZ is critical. Most finite element programs orient a beam element with the aid of an additional node "K," which together with the two end-nodes "I" and "J," specify the local coordinate system 123, see Fig. 3.2.2. Note that the 123 and the XYZ systems in Fig. 3.2.2 coincide. In the general case, however, the two systems are not identical.

It is recommended that the appropriate non-zero value for the shear areas is used in the analysis, since accounting for shear deformations improves the solution accuracy. Usually, non-zero values for shear areas must be assigned if $L < 10\,D$, where L and D are the longest and the shortest dimensions of the beam, respectively. A more detailed discussion on the effects of shear on the system response and numerical examples are given in Section 7.2.

The torsional constant is equal to the polar moment of inertia only for circular sections. Use of the wrong torsional constant is a common source of error in frame analysis. The torsional constant for non-circular uniform sections is available in various sources in the literature.

Table 3.2.1

Shear Area of Common Cross Sections

Cross Section	Type	Shear Area
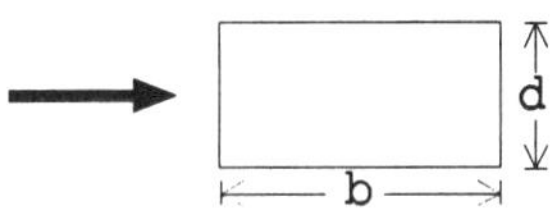	Rectangular Section b : Width d : Thickness	$\frac{5}{6}$ bd
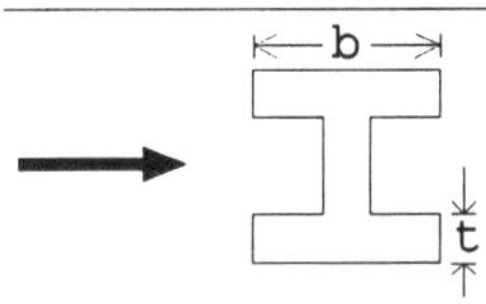	Wide Flange Section b : Width t : Thickness of flange	$\frac{5}{3}$ bt
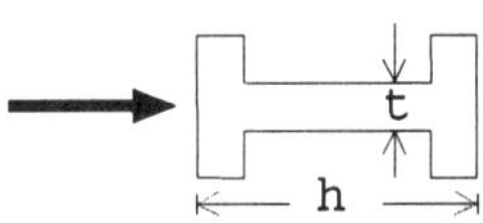	Wide Flange Section h : Height t : Thickness of web	≈ th
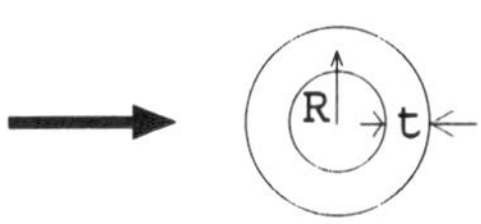	Thin-Wall Circular Tube Section R : Mean radius t : Thickness	$\frac{\pi Rt}{2}$
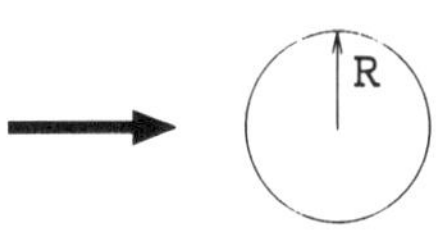	Solid Circular Section R : Radius	$0.9\pi R^2$
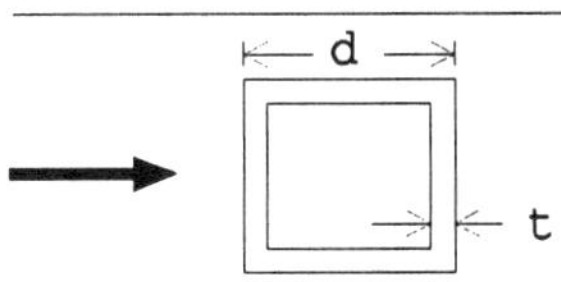	Thin-Wall Rectangular Tube Section d : Width t : Thickness	2td

* Arrows indicate direction of shear forces

Two-Dimensional Solid Elasticity Elements

There are three types of two-dimensional elements:

1. Plane Stress Elements
2. Plane Strain Elements
3. Axisymmetric Elements.

For consistency with the notation in Chapter 2, all two-dimensional elements are defined in the global YZ plane. Their use and a description of their configuration and properties is provided in the following.

Plane Stress Elements

A brief presentation of plane stress as treated by the theory of elasticity is given in Section 2.4. When a structural system is under conditions of plane stress all stress components normal to the YZ plane vanish:

$$\sigma_{xx} = \tau_{yx} = \tau_{zx} = 0 \tag{2.4.1}$$

A representative example of a plane stress problem is a thin plate loaded along its edges with a uniform load in the YZ plane. Due to the small thickness of the plate, the stresses that develop along its thickness are very small

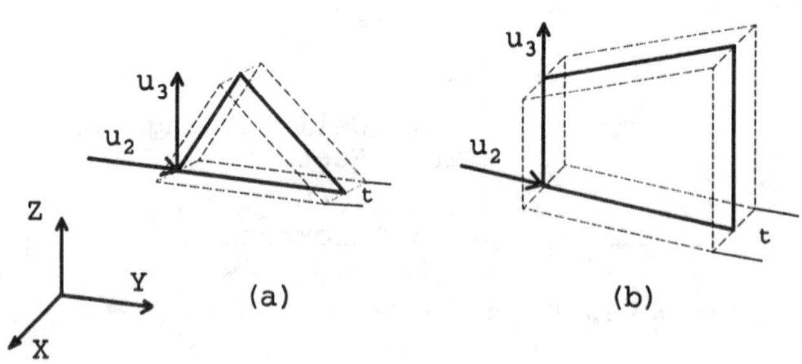

Figure 3.2.3 Two-dimensional plane stress elements: (a) Triangular; (b) Quadrilateral.

compared with the stresses in the YZ plane and they can be ignored in the analysis, see Fig. 2.4.1. Most commonly used plane stress elements are either quadrilateral or triangular. In either case, every node has two translational degrees-of-freedom, as shown in Fig. 3.2.3.

As a rule, prefer quadrilateral over triangular elements for reasons of geometric isotropy. However, it is suggested that triangular elements be used to better model irregular boundaries as well as areas around crack tips, see Fig. 3.2.4.

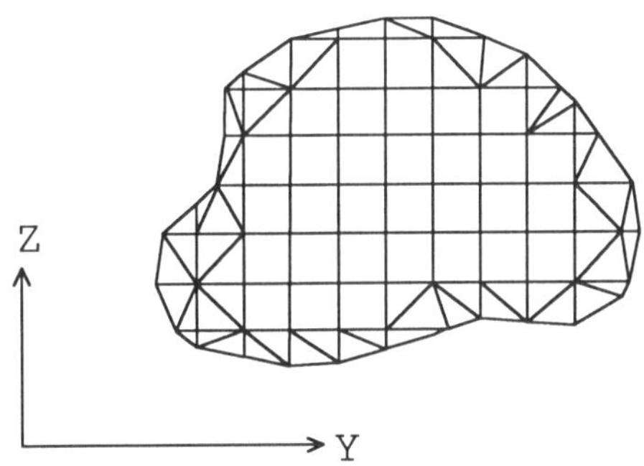

Figure 3.2.4 *Quadrilateral and triangular plane stress elements.*

Depending on the approximation of stresses within the element, two different plane stress elements are commonly used: elements with a constant stress variation within the element (CST) also known as *compatible*, and elements with a linear stress variation (LST) known as *incompatible.* Incompatible elements are usually effective in modeling regions with significant changes in stress gradient. A more refined mesh is required if compatible elements are used in the analysis. As the best modeling approach that combines the computational efficiency of compatible elements with the computational accuracy of incompatible elements, it has been suggested to use incompatible elements only in those regions of the structure where good solution accuracy is desired, for example, areas of high variation in stress gradient, e.g., around holes and cracks. In practice, however, incompatible elements are used for the whole model. An effort should also be made to keep the *aspect ratio*, i.e., the ratio between the element's longest to shortest dimensions, close to one. For a discussion on the aspect ratio refer to Section 3.4.

In plane stress analysis one should:

a. keep the aspect ratio as close to unity as possible, i.e., strive for either equilateral triangular elements or square elements.

b. keep internal angles close to 90^0 in quadrilateral elements.

Additional recommendations on proper modeling with plane stress elements are given in Section 3.4.

Plane Strain Elements

As discussed in Section 2.5, a system is considered to be under conditions of plane strain if one of its dimensions, say along the X-axis, is much greater than the other two and is subjected to a load that does not vary along the X-axis, see Fig. 2.5.2. Typical examples are dams, tunnels, retaining walls, but also small scale systems such as rollers and bars compressed by forces normal to their cross section. If the structure fulfills conditions of plane strain, and in addition it is restrained to deform along its longest dimension, say along the X-axis, then the strain ϵ_{xx} is zero, see Fig. 2.5.1. Note, however, that σ_{xx} is not zero. If the structure is not restrained along the X-axis, then the value of ϵ_{xx} is a non-zero constant. The latter case is known as *generalized plane strain*. In either case, however, we can determine the stresses by simply modeling and analyzing a *typical cross section* in the YZ plane that carries the load applied on a unit length along the X-axis.

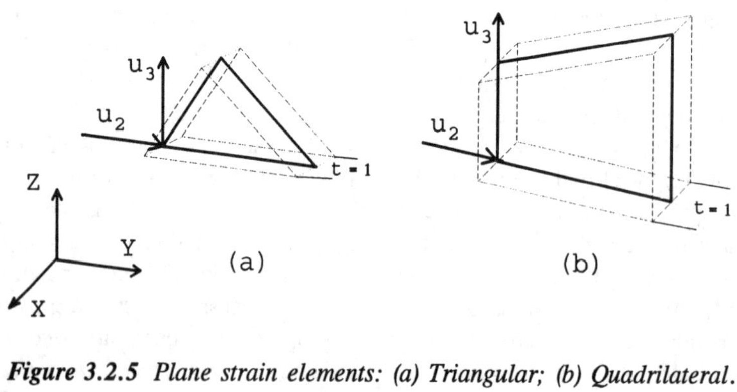

Figure 3.2.5 *Plane strain elements: (a) Triangular; (b) Quadrilateral.*

The most commonly used plane strain elements are either triangular or quadrilateral with two translational degrees-of-freedom per node, as shown in

Fig. 3.2.5. They have the same geometry and degrees-of-freedom with plane stress elements, their difference being that plane strain elements have unit thickness, while the thickness of plane stress elements is t. Finally, the remarks made on compatible and incompatible as well as the use of a proper aspect ratio for plane stress elements are also valid for plane strain elements.

Axisymmetric Elements

Steel and concrete storage tanks, rotors, shells, nozzles, and nuclear containment vessels are some representative axisymmetric structures. Similar to three-dimensional structures that are under conditions of plane stress or plane strain, axisymmetric structures loaded with axisymmetric loads can be analyzed as two-dimensional systems. In order to analyze an axisymmetric structure, such as a cylinder with thickness t subjected to a uniform internal pressure p, what we model is the intersection of the cylinder with the plane YZ in the positive quadrant of the XYZ Cartesian system drawn with continuous lines in Fig. 3.2.6(b). Note that Z is the axis of load and geometric symmetry and the radial coordinates are specified along the positive Y-axis. Several software do not allow negative Y coordinates, see the Algor Processor Reference Manual. The loading p that is applied to the finite element model shown in Fig. 3.2.6(b) corresponds to a slice of the structure defined by rotating a cross-section of the structure about the Z-axis by a unit angle $\theta = 1$ rad. Axisymmetric quadrilateral and triangular elements have two translational degrees-of-freedom per node, see Figs. 3.2.6(c) and (d).

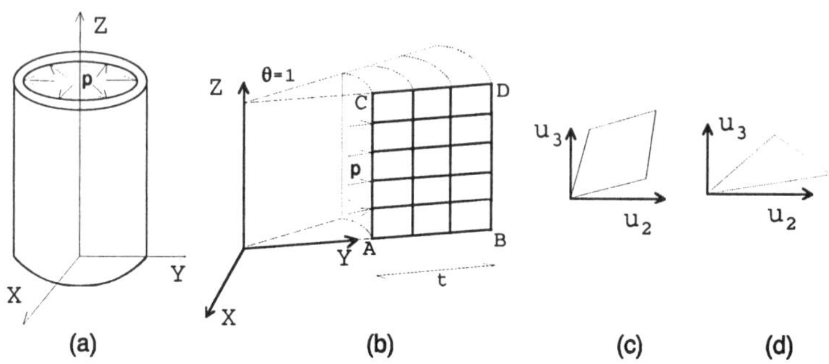

Figure 3.2.6 *(a) Cylinder under internal pressure; (b) Axisymmetric model on YZ plane; (c) Quadrilateral element; (d) Triangular element.*

Both triangular and quadrilateral elements have proved to be very highly accurate in comparison with classical analytical solutions. The comments made in the discussion on plane stress compatible and incompatible elements are also valid for axisymmetric elements.

Three-Dimensional Plane Stress or Membrane Elements

Examples of structures that can be analyzed with membrane elements include roofs made of fabric-like materials used to cover sports stadiums. Because of their small thickness such structures can develop only in-plane stresses.

Membrane elements can be "viewed" as three-dimensional plane stress elements. The difference is only that membrane elements have three instead of two translational degrees-of-freedom per node, as shown in Fig. 3.2.7. It should be noted that a membrane element does not necessarily lie in the YZ plane which is the case for the two-dimensional plane stress elements shown in Fig. 3.2.3.

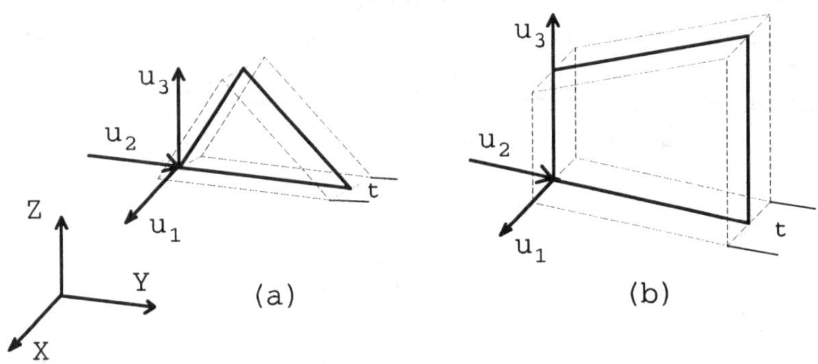

Figure 3.2.7 Membrane elements: (a) Triangular; (b) Quadrilateral.

All comments made regarding compatible and incompatible triangular and quadrilateral plane stress elements are also applicable to membrane elements.

Plate and Shell Elements

Use of plate and shell structures is extensive. It includes architectural structures, containers, airplanes, ships, missiles, and machine parts. Plates are plane, flat surface structures with thickness that is very small compared to their other dimensions, see Section 2.7.

The load-carrying behavior of plates resembles that of beams. In fact, plate behavior can be approximated by a gridwork of beams running in two perpendicular directions on the same plane. The two-dimensional load-carrying action of plates results in light and economical structures. In finite element modeling of a plate, what we model is the *mid-surface* which is a plane parallel to the two plate surfaces that divides the thickness into equal halves.

Modeling of plate structures is usually done with either triangular or quadrilateral plate elements. Each nodal point of either element type possesses five degrees-of-freedom in the element or local system xyz: three displacements along the x, y, and z-axes and two rotations about the x and y-axes. No rotational degree-of-freedom is specified about the direction z normal to the plate element, see Fig. 3.2.8. Nevertheless, a very small value for the rotational stiffness is assigned about the local z-axis to eliminate numerical instability. It

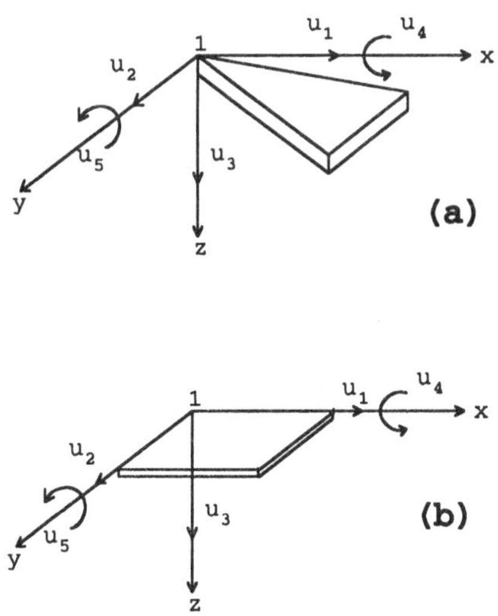

***Figure 3.2.8** Plate elements and degrees-of-freedom at node 1; (a) Triangular; (b) Quadrilateral.*

should be noted that plate and shell elements used in finite element programs combine bending and membrane actions.

Quadrilateral elements exhibit higher accuracy than triangular elements, and thus they should be preferred. Use of triangular elements should be limited to the part of the structure that is close to irregular boundaries.

Thin shells can be "viewed" as plates with a curved surface. For the definition of thin shell elements refer to Chapter 2. In this text shell elements are understood as flat plate elements used to model curved surfaces. In fact, most commercial finite element programs rely on plate elements to model shells and curved surface structures. As for thin plates, the mid-surfaces of thin shell structures are modeled with plate elements.

Two aspects should be kept in mind when modeling thin shells with plate elements:

a. for good solution accuracy, it is recommended that the plate elements should be approximating the shell surface at angles between 4^0 and 10^0 degrees, see Fig. 3.2.9.

b. lack of a rotational degree-of-freedom about the local z-axis may cause numerical instability. Such case could appear when plate elements intersect at an angle less than 4^0, see Fig. 3.2.10. The numerical instability might not allow processing of the model and could be reported by the processor as an "error message." A remedy to this problem is to use rotational elastic elements with very small stiffness about the axes parallel to the common sides of the elements.

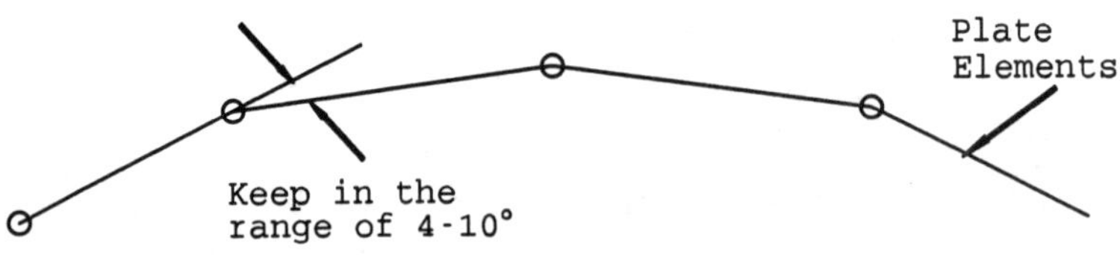

Figure 3.2.9 Modeling of thin shell with plate elements.

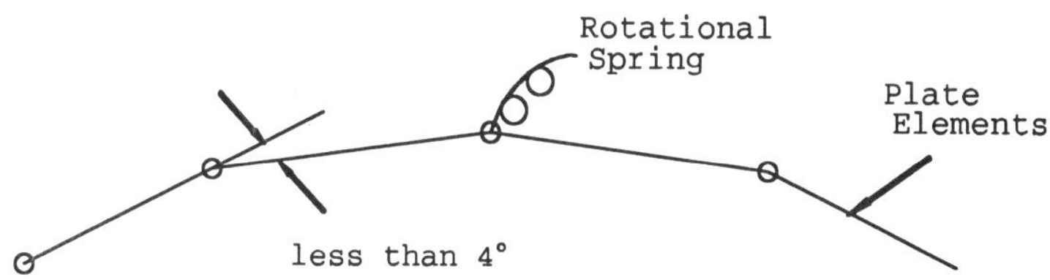

Figure 3.2.10 *Small stiffness rotational spring to avoid numerical instability.*

Three-Dimensional Solid Elasticity or Brick Elements

Thick plates, thick cylindrical or spherical components, thick joints, and gear housings are a few examples where solid elements can be used to perform a finite element analysis. In general, structures or structural components with a thickness comparable to the other two dimensions can be modeled with brick elements.

Solid elements are three-dimensional elements with three translational degrees-of-freedom per node. They can be viewed as membrane elements that can also account for variation of stresses along the thickness, see Fig. 3.2.11. Nodes are usually introduced at either the intersections of three planes or the mid-sides of the intersection of two planes. An eight-node brick element with the associated degrees-of-freedom at one node is shown in Fig. 3.2.11.

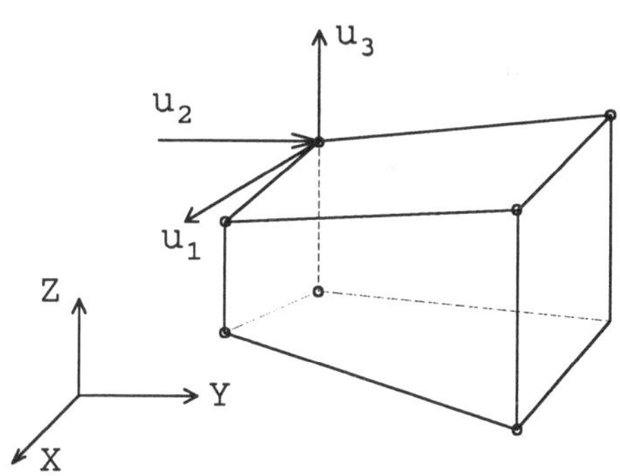

Figure 3.2.11 *Brick element and degrees-of-freedom at a node.*

The advantage of using solid or brick elements instead of plate or beam elements to model components with large thickness is that solid elements can provide information about the three-dimensional variation of stresses and deformations within the component. Plate and beam elements cannot provide such information and should be used with caution only for a preliminary analysis of systems having comparable sizes in all three dimensions.

When solid elements are interconnected with beam or plate and shell elements, attention should be paid to preserve the rotational degrees-of-freedom at the common nodes. This can be easily achieved by artificially extending the beam and plate elements into the solid elements, see Fig. 3.3.4. The issue of combining different element types is further discussed in Section 3.3.

Tetrahedral and Hexahedral Elements

Instead of brick elements, tetrahedral and hexahedral elements can be used to model three-dimensional structures. The tetrahedron can be viewed as an extension of the triangle in the third dimension, see Fig. 3.2.12, while the hexahedron can be considered as the counterpart of the planar quadrilateral extended in the third dimension. It should be noted that the hexahedral element has the same geometric shape as the eight-node brick element of Fig. 3.2.11. They differ, however, in their theoretical formulation and computational accuracy. Commonly used tetrahedral and hexahedral elements have only three translational degrees-of-freedom per node, see Fig. 3.2.12(a). The accuracy of tetrahedral and hexahedral elements is increased by introducing additional nodes at mid-sides, see Fig. 3.2.12(b).

Automatic mesh generation techniques favor tetrahedral and hexahedral elements over brick elements to create detailed models of three-dimensional systems with complex geometries, e.g., Supergen, Supersurf, Hypergen, and Hexagen of Algor. In practical applications, use of tetrahedral elements requires some caution to develop models without leaving any "holes" in them. In general, hexahedral elements are more accurate than tetrahedral elements and should be preferred. Nevertheless, tetrahedral elements are more versatile since they allow modeling of intricate geometries and facilitate transition from coarsely meshed regions to finely meshed regions in a model. In most cases, combination of tetrahedral and hexahedral provides the optimum mesh. Further elaborations on these elements, including the methods of formulation and comparative studies on their efficiency are given in several references, e.g., Gallagher, Yang, Zienkiewicz and Taylor.

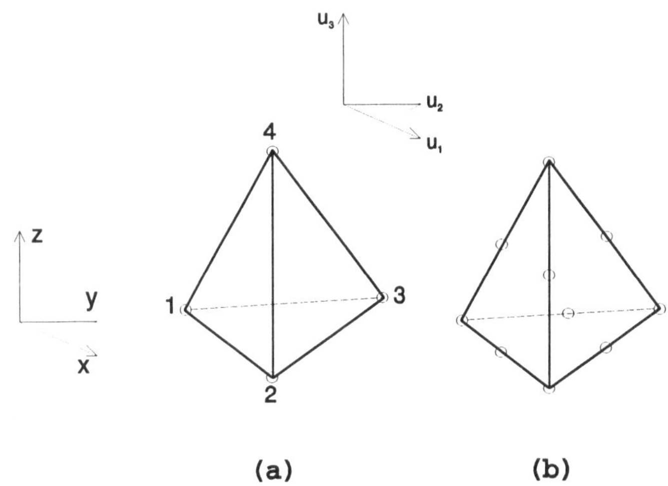

Figure 3.2.12 (a) Four-node tetrahedral; (b) Ten-node tetrahedral.

Boundary Elements

Boundary elements are used to model boundary conditions and links between structural components. As such, they can be useful in either evaluating reactions at rigid, flexible supports, and elastic links or specifying non-zero displacements and rotations to nodes. Boundary elements are two-node elements. The line defined by the two nodes indicates the direction along or about which the reaction (force or moment) is evaluated or the displacement (translation or rotation) is specified, see Fig. 3.2.13.

Boundary elements that are used to obtain reaction forces *(rigid boundary elements)* or specify translational displacements *(displacement boundary elements)* can be considered as truss elements with only one non-zero translational stiffness. When they are employed to either evaluate reaction moments or specify rotations, they resemble beam elements with only one non-zero stiffness, that is, the rotational stiffness about the user specified axis.

Elastic boundary elements are used to model flexible supports and to calculate reactions at skewed boundaries, see Fig. 3.8.1. The stiffness of elastic boundary elements is defined by the user. Caution is needed when one wants to model skewed supports. In this case, the user should employ either translational or rotational elastic boundary elements and assign moderately high stiffness values to the elastic boundary elements. The process of specifying appropriate stiffness to boundary elements is presented in Section 3.8.

As mentioned in the Section on "Plate and Shell Elements," elastic boundary elements with a small stiffness may also be used in models of thin shell structures, see Fig. 3.2.10.

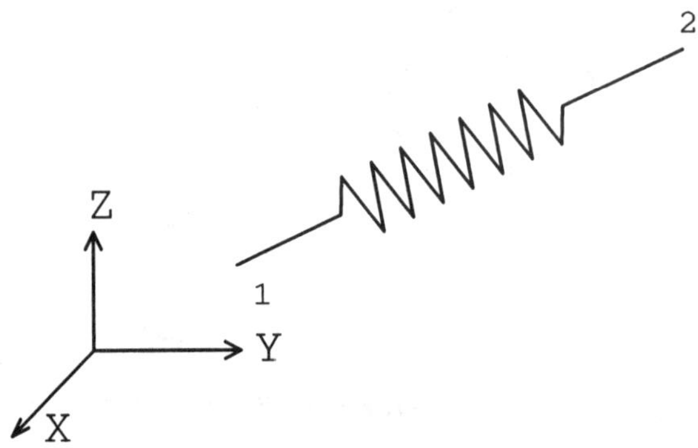

Figure 3.2.13 *Boundary element.*

Gap Elements

These elements are used to model cables, gaps between structural components, and gaps between a structure and its supports. Gap elements can be either compressive or tensile. Compressive gap elements become active when compressive forces develop on the boundary element after the gap is closed. Similarly, tensile gap elements resist tensile forces after the gap is closed. Figure 3.2.14 shows a frame structure with diagonal cable bracings that can be modeled with gap elements. Since in the braced frame there is no gap between the cables and the other members, the gap elements become active immediately after the load is applied.

It should be mentioned that most structural problems involving the use of gap elements are *nonlinear*. The nonlinearity is attributed to the change of the system's stiffness which occurs when the gap elements are either activated or de-activated. Thus, all the dynamic analysis methods presented for linear problems in Chapters 5, 7, 8, and 9 cannot be applied when modeling requires use of gap elements.

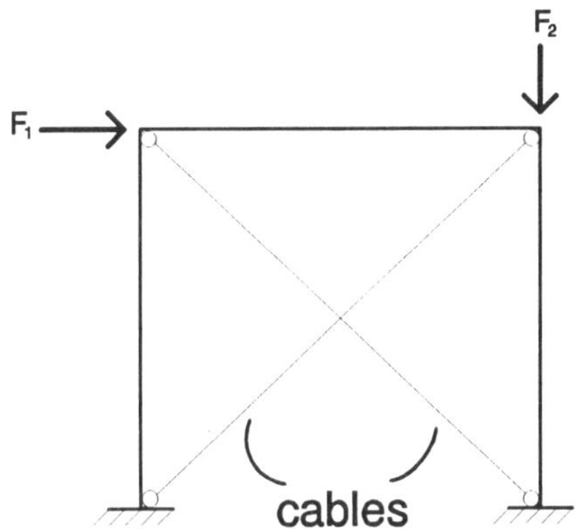

Figure 3.2.14 Braced frame with diagonal cables.

3.3 SELECTION OF ELEMENT TYPES

Before selecting the types of elements to model a structure, we must first draw an outline of the physical system indicating the overall geometry, boundary conditions, loads, and regions of material and geometric discontinuities. The sketch should also include the global coordinate system and the dimensions of the structure.

Next, we should examine whether the size of the model can be reduced. Plane strain, plane stress, and axisymmetry allow reduction of three-dimensional problems to two-dimensional. In addition, the presence of planes of symmetry allows modeling of only a part of the structure. Appropriate use of symmetry in static and dynamic analysis is discussed in Section 3.9.

The outline of the physical system will guide us in selecting the appropriate types of elements. For example, to model transversely or axially loaded components in mechanical, electrical, and civil structures, we could use beam and truss elements. Plane stress elements are appropriate to model in-plane action of plates and short beams. Plane strain elements are usually employed to model retaining walls and long dams. Axisymmetric elements are used to model structures that are rotationally symmetric about an axis and symmetrically or antisymmetrically loaded about the same axis, such as cylinders subjected to internal pressure.

The choice of the element type(s) also depends on the kind of results requested. For example, the cylinder shown in Fig. 3.3.1 with both ends clamped and subjected to a point load can be modeled in several different ways:

a. A beam model can be used if we are interested in an approximate calculation of the deflections.

b. A model using shell/plate elements can be used if an accurate calculation of the stresses is our objective.

Choice between plate or solid elements is not always straightforward. As a general rule, use plate elements to model non-axisymmetric thin shell and plate structures, and brick, tetrahedral, and hexahedral elements to model thick plate and shell structures. Support movements and elastic supports can be modeled with displacement and elastic boundary elements, respectively. Gap elements are appropriate to model cables and gaps between structures and structural components.

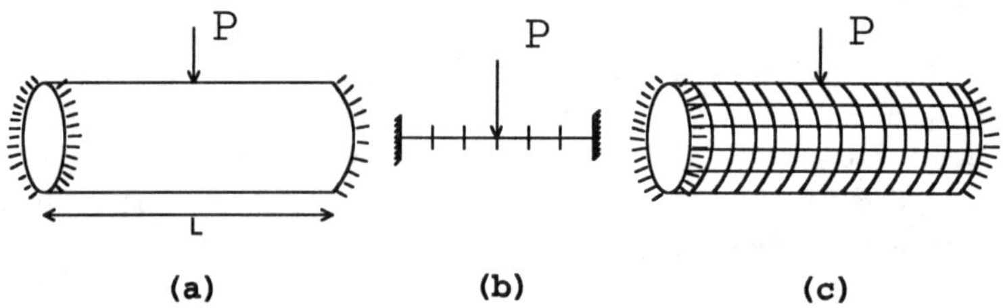

Figure 3.3.1 *(a) Cylinder with built-in ends subjected to point load; (b) Beam model; (c) Plate/shell element model.*

Certain problems can be dealt with more than one type of analysis. For example, the cylinder under internal pressure shown in Fig. 2.5.1 can be analyzed in at least three different ways. It can be solved as:

a. a plane strain problem. In this case the geometry of the model consists of the circular section obtained by intersecting the cylinder with the YZ plane. By using plane strain elements, we obtain results that are valid for any section of the cylinder that is away from the clamped ends.

b. an axisymmetric problem. The model using axisymmetric elements is shown in Fig. 3.2.6(b) with continuous lines. In this model all nodes on sides AB and CD are constrained, and the pressure p is applied on the side AC. The deformation and stresses obtained from the analysis of the axisymmetric model are valid for the whole structure. Note that we can analyze the cylinder with this model because the loading is also axisymmetric.

c. a truly three-dimensional structure. The cylinder is modeled using plate elements as shown in Fig. 3.3.1(c). This is the most elaborate and less efficient model; it would be our choice if the load is not axisymmetric.

Combination of Different Types of Elements

In most cases, more than one type of elements must be selected to model the system, e.g., for a bridge structure, plate elements can be used to model the deck, beam elements the piers and girders, solid elements the abutments, and spring elements the foundations. In order to model intersections of pipes and bottom and side tank walls one may use plate elements with a refined mesh at the intersections, see Fig. 3.3.2. However, one may obtain better accuracy for critical areas around the intersections by using plate elements for the pipes or the walls and solid elements for regions at the intersections, see Fig. 3.3.3. For every component modeled with the appropriate element type, the recommendations corresponding to that element type must be followed.

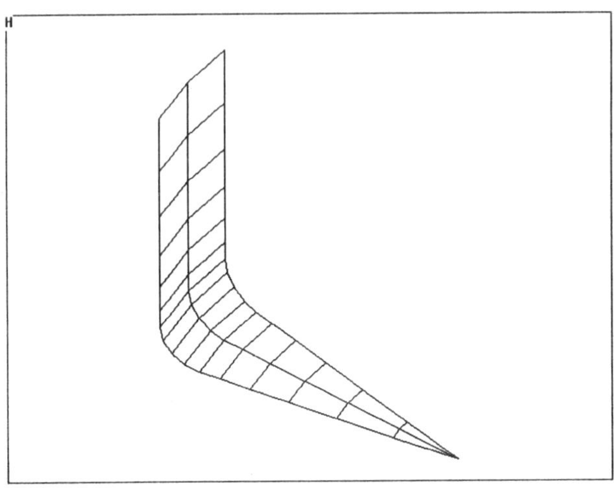

Figure 3.3.2 Plate element model for section of bottom and side tank walls.

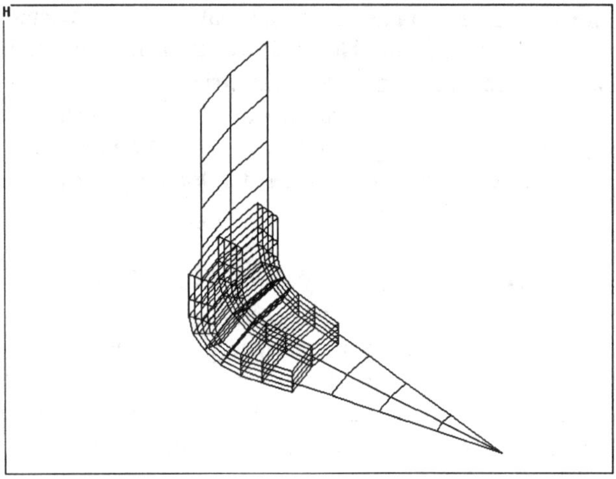

Figure 3.3.3 Combined plate and solid element model for section of bottom and side tank walls.

When combining different types of elements that do not have the same degrees-of-freedom, for example, plate and solid elements or beam and tetrahedral elements, attention should also be paid to the degrees-of-freedom at

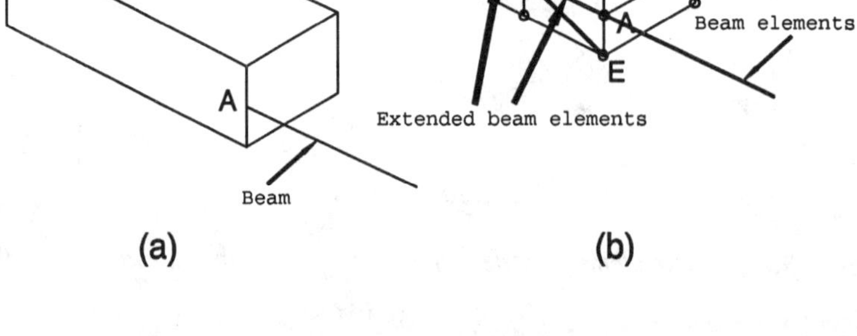

Figure 3.3.4 Modeling of solid-beam intersection.

the common nodes of the different element types. This point is illustrated in Fig. 3.3.4(a). In order to simulate the continuity of stresses and strains at joint A, the beam elements AB, BC, BD, and BE have been added to the model, see Fig. 3.3.4(b).

3.4 MODELING GUIDELINES

Creating the proper model is the most crucial step in finite element analysis. The goal is to develop the most suitable nodal pattern that provides enough elements to obtain accurate results without wasting data interpretation and processing time. Modeling should always be based on conceptual understanding of the physical system and judgment on the anticipated behavior of the structure. Besides understanding the system's behavior, the analyst should make the effort to comprehend the fundamental concepts underlying the pertinent finite element theory. Failure to do so will most likely lead to meaningless results.

The following list of guidelines, accumulated from modeling experience and selected from various sources, can be used as a basis to develop reliable models for static and to a great extent for dynamic analysis. It must be noted that the recommendations may not apply to all cases. No guidelines can substitute for expert use of engineering knowledge and judgement.

General Recommendations

1. Define nodes at or near load points, points of geometric discontinuity, at supports, and at those regions where you require information about stresses or displacements.

2. A finite element mesh should be uniform whenever practical. However, non-uniformity is often required to obtain accurate results in regions of rapid changes in geometry and loading. Only at those parts of the structure where the geometry, stresses, or loading change dramatically should the mesh be refined, see Fig. 3.4.1.

3. Prefer to use quadrilateral, six-sided solid, and hexahedral elements, except where triangular, four-sided solid, and tetrahedral elements are necessary to accommodate irregularities in geometry and loading.

4. When using plate elements to model curved surfaces, maintain a subtended angle of less than 10 degrees, see Fig. 3.2.9.

5. A more refined mesh is required to accurately obtain the stresses than the displacements. If necessary perform a convergence study, i.e., start with a simple model with a relatively small number of elements and progressively move to more refined models. Figure 3.4.2 illustrates the procedure with two plane stress models of a short cantilever beam modeled in the YZ plane. The procedure is also demonstrated through a numerical example in Section 6.6.

Aspect Ratio

The element aspect ratio is defined as the ratio between the element's longest to shortest dimensions. Figure 3.4.3 shows configurations of three two-dimensional element types classified as "good," "poor," and "illegal." Good elements are characterized by an aspect ratio close to unity and angles which are about 90^0. Poor elements should be avoided, since they may lead to inaccurate results. Illegal elements are unacceptable, and should not be used in finite element modeling. When illegal element shapes are inadvertently drawn, they create voids in finite element models. Several postprocessors of commercial finite element programs, such as Algor, allow identification of the voids.

6. If the stress field has similar gradients in all directions, try to maintain an aspect ratio of 1. Elements with high aspect ratio should be avoided, see Fig. 3.4.4(a). However, elements characterized as "poor" in Fig. 3.4.3 can be used in the analysis provided that they model regions of the structure where the stress gradient varies very gradually along the longest dimension of the elements. "Illegal" elements should always be avoided. Because of their geometry and location of the nodes, illegal elements cannot be assigned finite element properties.

7. For regions with small variation of stresses, the aspect ratio could be as high as 40 to 1 and still yield good results. As a general rule, however, keep the element aspect ratios under 10 for *deformation analysis* and under 3 for *stress analysis,* see Fig. 3.4.4(b). It should be noted that in deformation analysis the emphasis is placed on the accurate calculation of the nodal displacements, while in stress analysis we are interested in an accurate calculation of both the displacements and stresses.

8. Rapid changes in element size should be avoided. When the use of different element sizes is necessary, model the transitional parts of the model by changing the dimensions of adjacent elements by less than a factor of two.

Skewed Elements

Skewing is defined as the variation of the element vertex angles from 60^0 for triangles, and from 90^0 for quadrilaterals. Notice that two of the "poor" elements in Fig. 3.4.3(b) can be characterized as skewed.

9. For triangular elements avoid acute angles less than 30^0, and for quadrilateral elements avoid obtuse angles greater than 120^0, see Fig. 3.4.3(b). As a general rule, use of skewed elements is acceptable when we are primarily interested in an accurate calculation of the displacements. If stresses, however, must be calculated accurately, then less skewing should be incorporated to the model. For a discussion on skewed elements see the Algor Software Companion with Verification Examples.

Geometric Discontinuities

10. Holes, cracks, and localized changes in geometry can be characterized as geometric discontinuities. Modeling of geometric discontinuities depends on the interest in accurately calculating deformations and stresses in the vicinity of the discontinuity. If we are not concerned with deformations and stresses near the discontinuity, we can use a *global model* of the structure that has a coarse mesh around the discontinuity. If our goal is to obtain the response near the discontinuity, a *local model* should be used. Contrary to the global model, the local model uses a refined mesh in the vicinity of the discontinuity.

There are two basic approaches to model holes, cracks and notches:

a. The first one involves use of a local model with a refined mesh around the discontinuity. This approach usually requires a detailed mesh that can model the region around the discontinuity in very fine detail.

b. If the stress concentration factor K is known from handbooks or experimental data, a model that calculates the primary stress around the discontinuity can be used. Such a model is less refined than the one required in approach (a).

When there are abrupt geometric changes, one should use a more refined mesh in the region where the changes are located, see Figs. 3.3.2 and 3.4.1. A model similar to the one shown in Fig. 3.3.3 that uses a fine mesh of solid elements to connect plate elements in the region of interest is another modeling alternative.

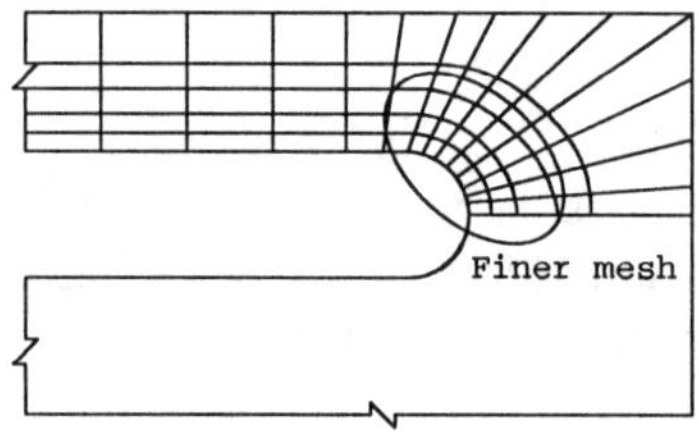

Figure 3.4.1 *Plate with stress concentration at slot.*

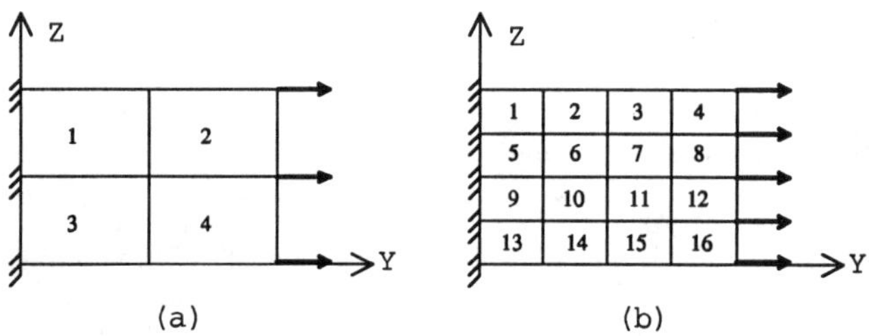

Figure 3.4.2 *Four- and sixteen-element models of a cantilever short beam.*

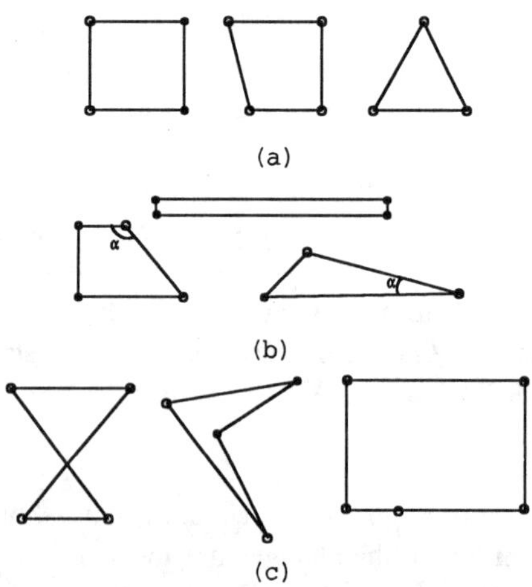

Figure 3.4.3 *(a) Good; (b) Poor; (c) Illegal element shapes.*

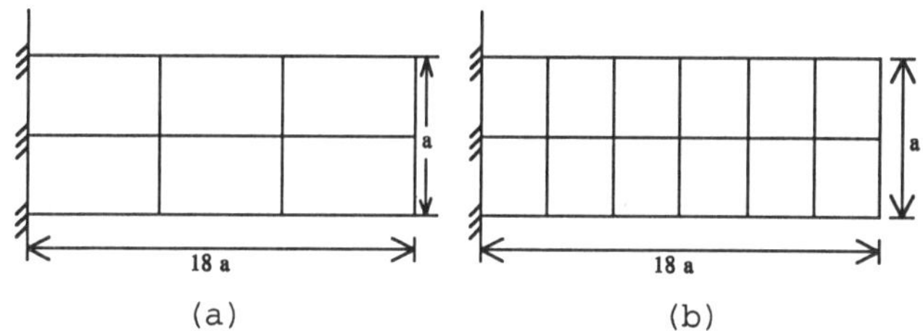

Figure 3.4.4 *(a) Very large aspect ratio; (b) Aspect ratio within acceptable range for deformation analysis.*

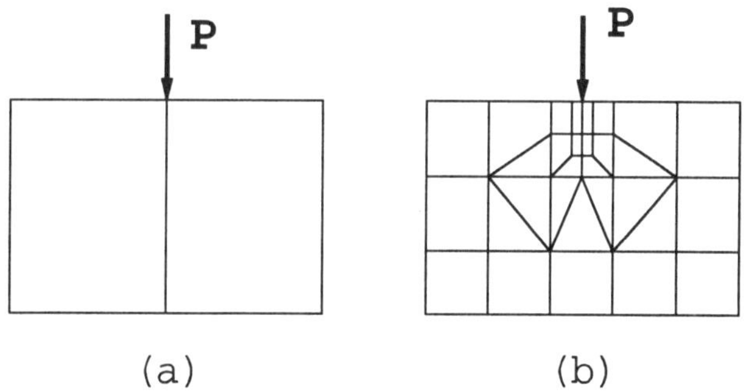

Figure 3.4.5 *Mesh in the vicinity of point loads: (a) Coarse; (b) Refined.*

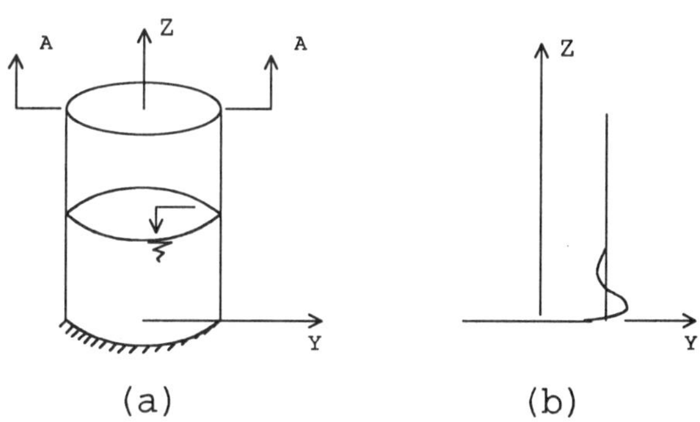

Figure 3.4.6 *(a) Liquid filled cylinder; (b) Moment variation and attenuation length in section A-A.*

Material Discontinuities

11. Abrupt changes in material properties can be handled with the approaches suggested for "geometric discontinuities," that is, with the aid of global and local models.

12. For isotropic materials, Poisson's ratio must not be very close to 0.5. Also, for a Poisson's ratio close to zero, the structure may lose almost all its stiffness, and as a consequence the results could be erroneous.

13. For anisotropic materials, the theoretical limits of Poisson's ratio and Young's modulus along the specified directions must be verified in order to avoid solution inaccuracies.

Abrupt Changes in Stiffness

14. Avoid models with elements that have stiffness ratios differing more than 10^4. The maximum/minimum stiffness ratios are usually provided in the numerical output of commercial software, e.g., see the "stiffness matrix parameters" in Table 6.5.1.1. Large stiffness variations in a model may lead to singularities and *ill-conditioned* matrices that cannot be numerically handled by the processor. Large stiffness variations could occur in models that contain very small and very large size elements, in models simulating either soft or stiff structural parts, and in models that include stiff or soft elastic boundary and gap elements. A discussion on how to address this issue as well as the proper modeling of stiff and rigid parts of structures is presented in Section 3.7.

Abrupt Changes in Loads

15. A refined mesh should be used in the vicinity of abrupt load changes in order to capture the stress variation near the loads. Such a refined mesh is shown in Fig. 3.4.5. The extent of the region that must be modeled with a fine mesh depends on the attenuation lengths from the points of application of the loads, see Fig. 3.4.6. Various design codes provide attenuation lengths for a variety of engineering applications, e.g., ASME Boiler Code.

3.5 APPLICATION OF BOUNDARY CONDITIONS

Typical boundary conditions include roller, hinge, fixed, and free supports. Rollers are simulated by setting all displacements equal to zero except the ones along the direction that the rollers can move. Typical boundary conditions for a two-dimensional beam are shown in Fig. 3.5.1. Boundary conditions are specified in the global coordinate system.

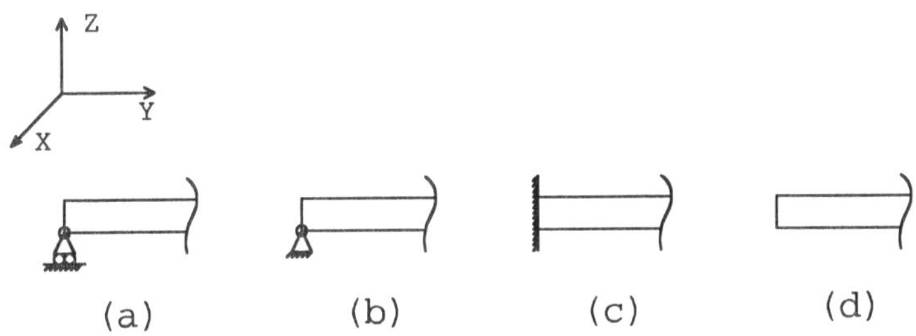

Figure 3.5.1 *(a) Roller; (b) Hinge; (c) Fixed; (d) Free boundary conditions.*

In problems involving beam elements it may be necessary to include hinged connections. Such connections (*releases*) do not allow the transfer of either moments or shear forces between contiguous beam elements.

Consider the frame shown in Fig. 3.5.2, in which the horizontal nodal displacements at nodes 2, 4 and 6 are equal. This is a common case in frame analysis where the axial stiffness of beams is much greater than the flexural stiffness of the columns. Finite element programs usually allow the user to impose equal displacements at nodes along selected directions through the *master-slave* option, e.g., Algor Software Companion with Verification Examples. Use of the master-slave option could lead to substantial reduction of the degrees-of-freedom, and should be taken advantage of, whenever possible.

Displacements of specified magnitudes and directions as well as skewed boundary conditions can be defined along the boundaries of the structure with the aid of boundary elements. A discussion on how to model skewed boundary conditions is provided in Section 3.8.

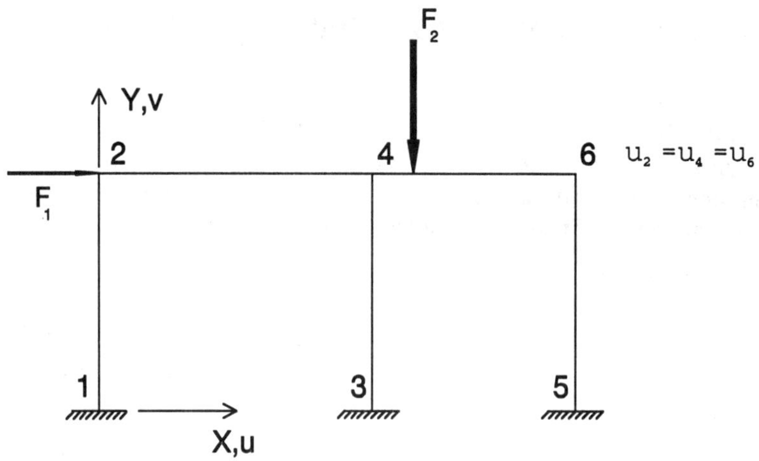

Figure 3.5.2 Master-slave option.

In case of symmetry, boundary conditions must also be specified at the planes of symmetry. Section 3.9 presents pertinent modeling guidelines for static and dynamic analysis.

In any type of analysis except in kinematics and modal analysis, the imposed boundary conditions should not allow rigid body motion. The presence of rigid body motion does not guarantee a unique solution of the problem. Non-uniqueness of solution is reported by the processors of commercial finite element codes.

3.6 APPLICATION OF LOADS

In many statics problems, the contribution of dead loads to the structural deformation could be negligible compared to the effect of live loads. For these cases, setting the mass or the weight density equal to zero not only does not affect the solution but could also save substantial processing time. However, in dynamic analysis either the mass or the weight density of the members must be inputed in order to account for the effects of the inertia forces.

Concentrated loads and reactions cause high stresses in the vicinity of their points of action. Thus, modeling should be very refined close to concentrated loads, if "sufficiently accurate" stresses and deformations are required in the vicinity of the concentrated loads, see Fig. 3.4.5.

In static and dynamic analysis of trusses, only nodal loads can be applied to the truss elements. The nodal loads can be only concentrated forces since no moments can be undertaken by the nodes. In static analysis of beams and frames, commercial finite element programs use *work equivalent* loads to account for the effect of any uniform loads that act on the structure. This implies that the nodal displacements evaluated for frames subjected to either concentrated or uniform loads will be "exact," that is, practically identical to strength of materials solutions, provided that nodes are introduced at the points of application of the concentrated loads and at the ends of the uniform loads, see Fig. 3.6.1. For non-uniform loads, a solution close to the exact can be obtained if additional nodes are introduced between the ends of the non-uniform loads, see Fig. 3.6.2. For dynamic analysis, intermediate nodes are needed for any type of loads acting on a beam, truss, and frame, see Fig. 3.6.3. It should be pointed out that for any other type of elements intermediate nodes are required for both static and dynamic analysis.

For area or volume elements, caution is required in defining the direction of distributed loads, see Fig. 3.6.4. Moments cannot be defined on the nodes of elements that have only translational degrees-of-freedom, such as two- and three-dimensional solid elasticity, tetrahedral, and hexahedral elements. In such cases, if there is a need to define a point moment, M, the moment can be simulated with a couple of opposite forces F acting at a distance d so that M=F*d.

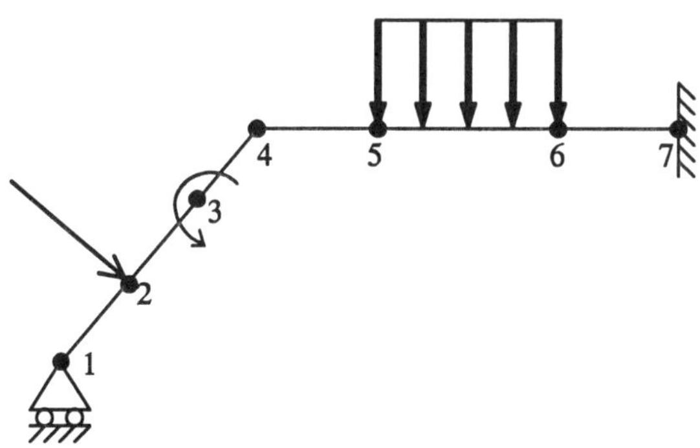

Figure 3.6.1 *Model for uniform static load.*

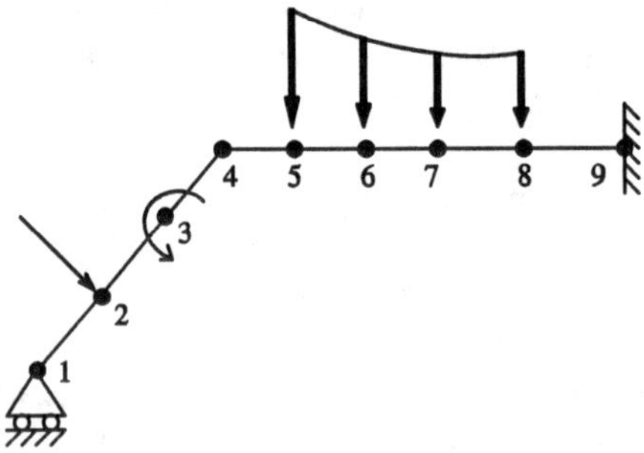

Figure 3.6.2 Model for non-uniform static load.

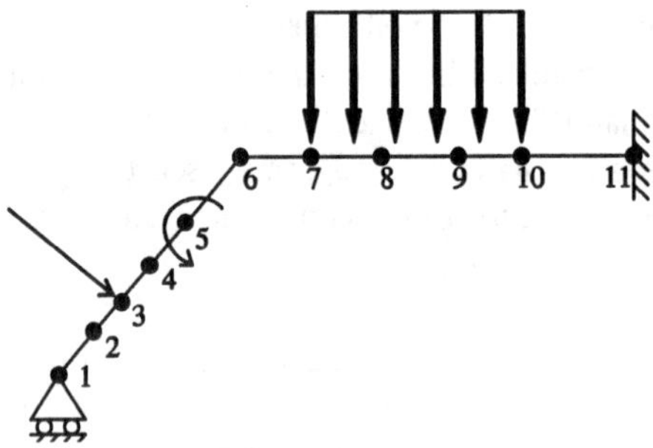

Figure 3.6.3. Model for dynamic analysis of frame.

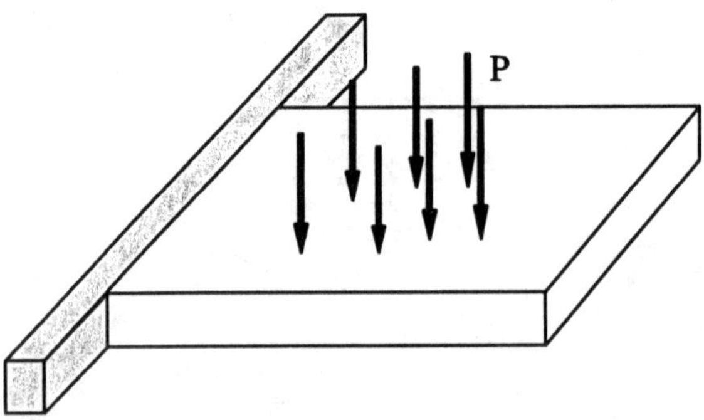

Figure 3.6.4 Plate under uniform load.

3.7 STIFF AND RIGID STRUCTURAL PARTS

Very stiff or almost rigid parts of structures are commonly encountered in practice. For example, such occasions arise when either a portion of the structure or one of its links is much stiffer than the rest of the structure, see Figs. 3.7.1(a) and (b), or when a rather stiff and relatively large object is supported by a flexible structure, see Fig. 3.7.1(c). Detailed modeling of beam-column joints may also require simulation of the joints as rigid, that is assigning an I_R that is much greater than I_1 and I_2, see Fig. 3.7.1(d).

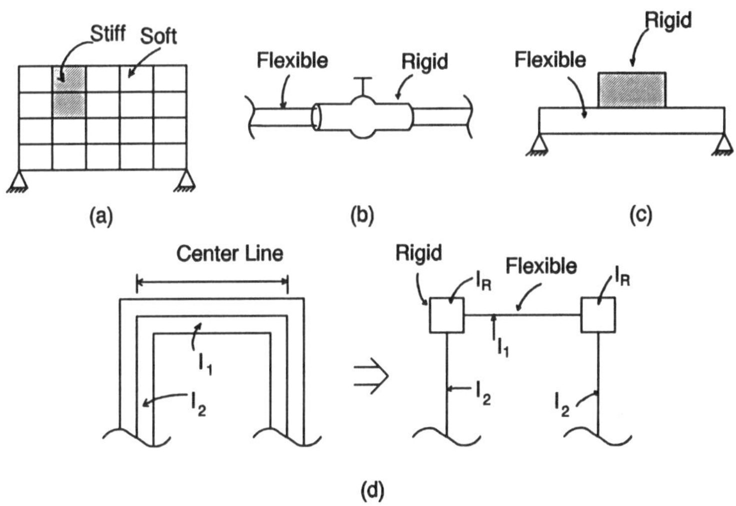

Figure 3.7.1 Modeling of stiff structural parts.

Modeling of very stiff and rigid parts by assigning arbitrarily chosen very high material or cross-sectional properties may lead to either erroneous results or ill-conditioned matrices. A two-step procedure, depicted in Fig. 3.7.2, is recommended in order to avoid such errors. The procedure refers to a structure with elastic properties E_1 that includes stiff and rigid parts with elastic properties E_2. It should be noted that the procedure requires the structure to be analyzed twice.

a. Model the whole structure using a relatively coarse mesh. The whole structure, including the rigid and stiff parts, should be assigned the elastic properties E_1. Then, perform a static analysis of the structure to obtain the maximum value of the structure's stiffness matrix, k_{max}, see the "maximum diagonal element" in Table 6.5.1.1.

b. Analyze the structure "as is" but now use a refined mesh and assign to the flexible part its modulus E_1 and to the rigid part a modulus E_2 such that the maximum stiffness coefficient corresponding to an element of the rigid part is in the range of $4k_{max}$ to $5k_{max}$. In plate and frame structures, instead of using E_2, one may chose to assign a large moment of inertia I to the rigid beam elements or a large D to the rigid plate elements, where D is the flexural rigidity of a plate, see eqn. (2.7.2).

The methodology can be easily extended to analyze structures with "soft" parts.

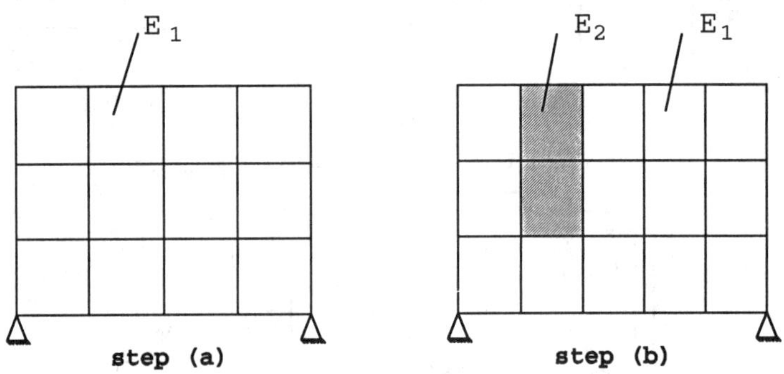

Figure 3.7.2 Two-step approach to model rigid parts.

3.8 SKEWED BOUNDARIES

As mentioned in the Section on "Boundary Elements," modeling and calculation of the reactions at skewed supports can be performed with elastic springs. If the supports are unyielding, a two-step approach can be followed:

a. Model the structure using fixed boundary conditions at the skewed supports. Perform a static analysis to obtain the maximum value of the structure's stiffness matrix, say k_{max}, see "maximum diagonal element" in Table 6.5.1.1.

b. Analyze the structure using translational elastic springs to model the skewed supports. Each spring should be oriented along the direction of the restrained displacement. Also, the stiffness assigned to each spring should be in the range of $4k_{max}$ to $5k_{max}$, in order to warrant sufficient stiffnesses in the springs to restrain the displacements.

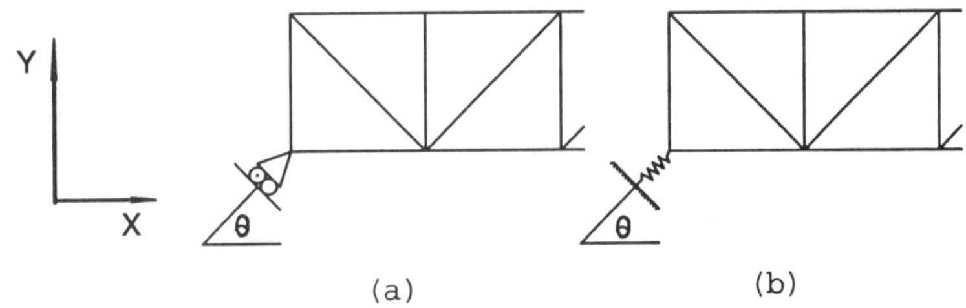

Figure 3.8.1 *(a) Skewed boundary condition; (b) Model with elastic boundary element.*

3.9 SYMMETRY IN STATIC AND DYNAMIC ANALYSIS

This section presents modeling guidelines for the static and dynamic analysis of symmetric structures. There are several differences between static and dynamic analysis in using symmetry. An objective of this section is to identify these differences and demonstrate the proper use of symmetry through examples. First, we introduce the use of symmetry in static analysis followed by a discussion on dynamic analysis.

Static Analysis

If a structure is symmetric, advantage should be taken of symmetry to reduce the model size and the computational effort. The development of the reduced model requires specification of the appropriate boundary conditions at the planes of symmetry as well as application of the appropriate loads.

In order to characterize a structure as symmetric, one should examine the following three parameters:

a. geometry,
b. material properties, and
c. boundary conditions.

An additional parameter that affects modeling of symmetric structures is the loading.

A structure is symmetric if it has a plane such that the reflection of every point in the structure along this plane results in a similar configuration. The number of planes of symmetry specifies whether the structure has a one-, two-, three-, four-way, or cyclic symmetry.

Figures 3.9.1(a) through (c) show representative examples of one-, two- and four-way symmetric structures subjected to symmetric loads as well as the corresponding reduced models. For the cantilever plate with one axis of symmetry shown in Fig. 3.9.1(a), the boundary conditions for the half model on side AB are T_xR_{yz}. The clamped plate in Fig. 3.9.1(b) is subjected to a uniform load normal to the surface of the plate. For the corresponding quarter model, the boundary conditions on sides BC and DC are T_yR_{xz} and T_xR_{yz}, respectively. In Fig. 3.9.1(c), the boundary conditions on the sides OA and OB are T_yR_{xz} and T_nR_{tz}, respectively. The unit vectors **n** and **t** define the boundary conditions on the side OB. The **t** is parallel to OB and the **n** is normal to the plane defined by **t** and the Z-axis.

When a symmetric structure is subjected to a symmetric load, symmetry restrains displacements normal to the plane of symmetry and rotations in the plane of symmetry. The restrained displacement and rotations are indicated with single and double arrows, respectively, in Fig. 3.9.2(a).

It should be noted that the properties of any members lying on the planes of symmetry should be altered to reflect that only half of these members are a part of the reduced model. Specifically, the members in the reduced model should be assigned one-half their rigidities. For example, the columns on the axis of symmetry of the frame shown in Fig. 3.9.3(a) are assigned one-half their rigidities in the half-model shown in Fig. 3.9.3(b). The magnitude of any load acting at the axis of symmetry should also be reduced, see Figs. 3.9.1 and 3.9.3.

A circular plate with holes subjected to a uniform load normal to its plane and simply supported along its outer boundary is a typical structure with cyclic symmetry, see Fig. 3.9.4. The system can be analyzed by modeling only the section shown in Fig. 3.9.4(b). This section extends between the dashed lines of Fig. 3.9.4(a). The boundary conditions on the planes of symmetry are defined according to the nodal restrains shown in Fig. 3.9.2(a), that is, along the OA and BC the translation along the Y-axis and the rotations about the X and Z-axis are restrained (T_yR_{xz}), while the boundary conditions on OD are T_nR_{tz}. The boundary conditions on OD can be applied with elastic boundary elements as discussed in Section 3.8.

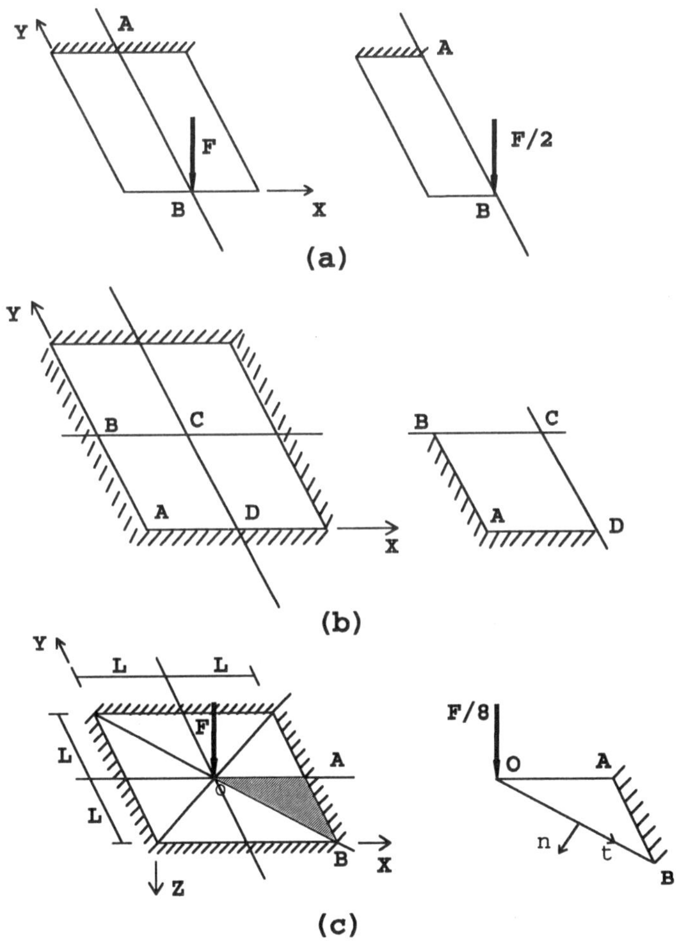

Figure 3.9.1 (a) One-way; (b) Two-way; (c) Four-way symmetry.

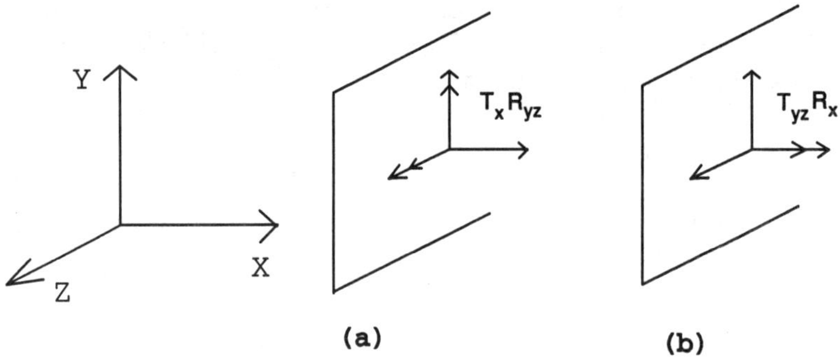

Figure 3.9.2 (a) Restrained degrees-of-freedom for symmetric loads; (b) Restrained degrees-of-freedom for antisymetric loads.

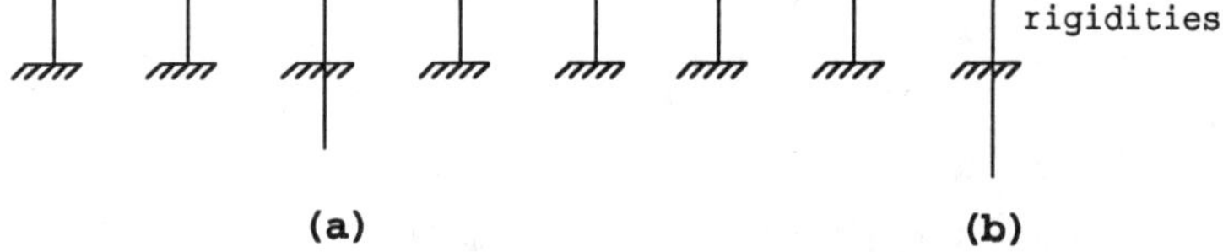

Figure 3.9.3 (a) Symmetric frame; (b) Half structure with reduced rigidities and loads on the axis of symmetry.

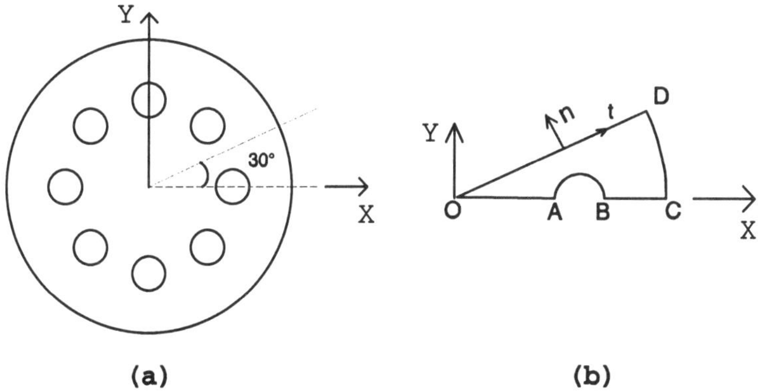

Figure 3.9.4 *(a) Plate with cyclic symmetry; (b) Section to model.*

When a symmetric structure is subjected to an *antisymmetric* load, the rotation about an axis normal to the plane of symmetry and the displacements in the plane of symmetry should be prevented. The restrained displacements and rotation are indicated with single and double arrows, respectively, in Fig. 3.9.2(b). As shown in Figs. 3.9.5(a) and (b), advantage of symmetry can also be taken when a symmetric structure is subjected to antisymmetric loads. For the frame and the cantilever plate, only half models with appropriate boundary conditions on the axes of symmetry can be analyzed.

If a symmetric structure is subjected to an arbitrary loading, the loads can be decomposed into symmetric and antisymmetric components. In this case, the responses to the symmetric and antisymmetric loading components must be added in order to obtain the response of the part of the structure that was analyzed. By the term *response* we mean the displacements, reactions, internal forces, and stresses. As shown in Figs. 3.9.6, more than two load combinations and analyses may be needed to determine the response of the structure. The response of the frame is obtained by combining the results of the two half models, see Fig. 3.9.6(f). For the plate shown in Fig. 3.9.6(a) with fixed supports on sides AB and DC and with simple supports on the other sides, the results are obtained as a superposition of the four quarter models of Figs. 3.9.6(b) through (e) with the corresponding boundary conditions shown in Figs. 3.9.7 (b) through (e).

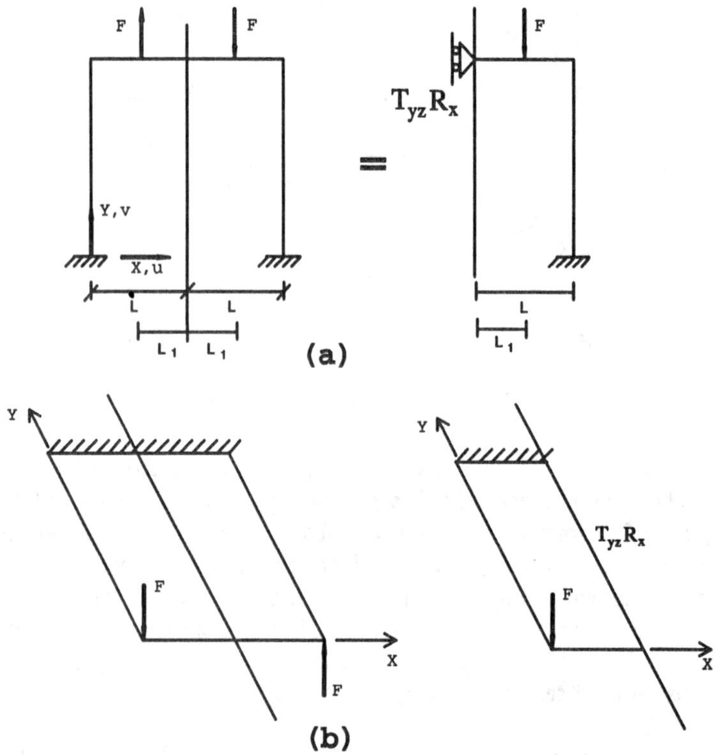

Figure 3.9.5 Symmetric structures subjected to antisymmetric loads: (a) Frame and half model; (b) Plate and half model.

Dynamic Analysis

If a structure fulfills the three requirements of symmetry defined above, then one can also take advantage of symmetry in dynamic analysis. In this section we will illustrate the use of symmetry in modal analysis. Examples on the proper use of symmetry in modal analysis are presented in Chapter 7. Use of symmetry in systems subjected to dynamic loads and analyzed with time history modal superposition is also demonstrated in Chapter 8. Use of symmetry for any other type of dynamic analysis can be deduced from the developments on modal analysis and time history modal superposition.

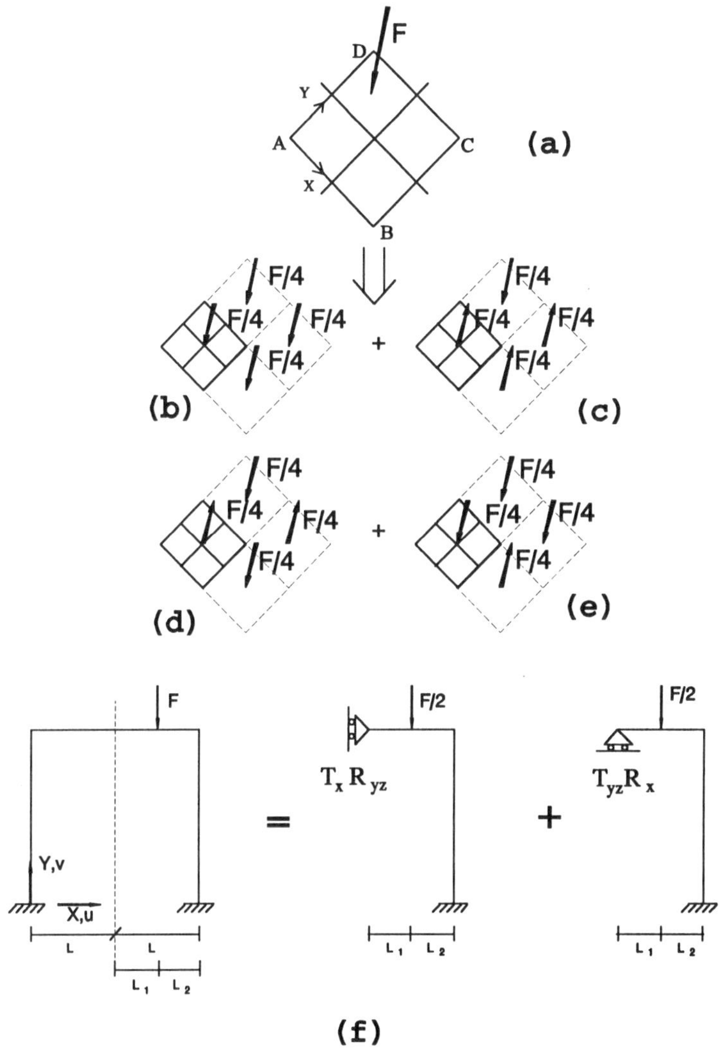

Figure 3.9.6 Symmetric structures subjected to arbitrary loads.

When a system has one or more planes of symmetry, the mode shapes are either symmetric or antisymmetric with respect to the planes of symmetry. Thus, in order to obtain all the mode shapes and natural frequencies we must perform several modal analyses of a part of the structure with proper boundary conditions on the axes of symmetry. The procedure is demonstrated with the modal analysis of a plate that is fully restrained on two opposite sides and simply supported on the others, see Fig. 3.9.7(a). The plate has two planes

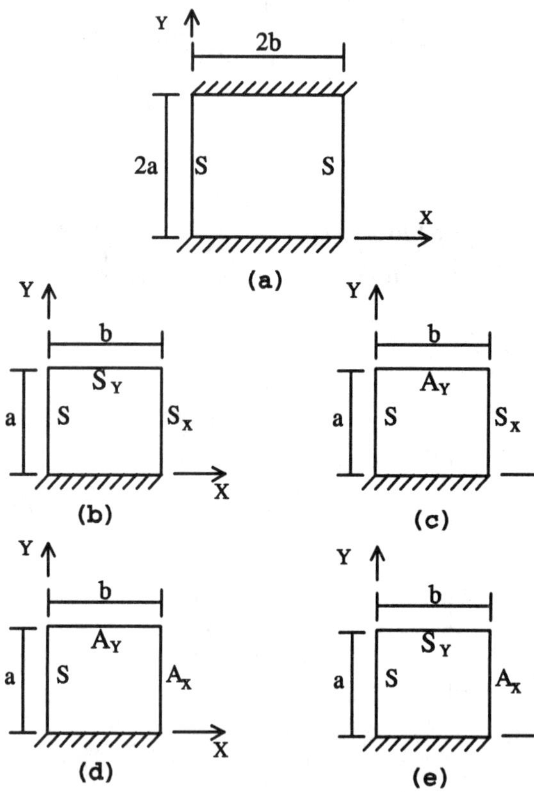

***Figure 3.9.7** (a) Symmetric plate; (b)-(e) Quarter models and boundary conditions. Note: The S, S_x, S_y, A_x, A_y indicate the boundary conditions: $T_{xyz}R_{xz}$, T_xR_{yz}, T_yR_{xz}, $T_{yz}R_x$, and $T_{xz}R_y$, respectively.*

of symmetry, thus we can obtain the natural frequencies and mode shapes of the plate by performing modal analysis on four quarter models. The quarter models and the associated boundary conditions are shown in Figs. 3.9.7(b) through (e). Notice that the only differences between the quarter models are the boundary conditions along the axes of symmetry which are either conditions of symmetry or antisymmetry as specified in Fig. 3.9.2(a) and (b). The corner nodes at the intersection of the planes of symmetry should fulfill the boundary conditions for both intersecting edges, i.e., the conditions at the upper right corner of the plate in Fig. 3.9.7(b) should be $T_{xy}R_{xyz}$. For the modal analysis of this plate with specific geometric and material properties refer to Section 7.3.3.

3.10 RECOMMENDATIONS FOR EVALUATING STRESSES

The evaluation of stresses using finite elements requires a more refined model than the one needed to calculate displacements. This is attributed to the fact that stresses are calculated using derivatives of the displacements. If the finite element mesh is not adequately refined, stresses calculated at adjacent elements may differ substantially. Several programs remove such discrepancies by averaging the stresses in all elements incident to the same node. They also include options to assess the adequacy of the mesh in evaluating stresses, and identifying the parts of the structure that must be modeled with a more refined mesh in order to increase the accuracy of the results.

In evaluating the stress output, the analyst should focus on examining the stress components of interest at *critical locations*. Such critical locations include the boundaries, points of application of concentrated loads, and areas where

a. von Mises, Tresca, or the principal stresses are large,
b. the geometry of the structure changes considerably,
c. the stiffness and/or the mass of the system presents abrupt changes.

Corner nodes are difficult locations to compute stresses. Unfortunately in many cases peak stresses occur at corners, and their magnitudes may govern the design.

As a general rule, the more degrees-of-freedom a node has, the more refined the mesh must be in order to accurately calculate the stresses. For example, the same structure when analyzed under conditions of plane stress, such as a plate loaded parallel to its plane, requires a coarser mesh to calculate the stresses than when analyzed as a plate using plate/shell elements, i.e., loaded normal to its plane. This is attributed to the fact that in the former case plane stress elements with two degrees-of-freedom are used, while in the latter case use is made of plate elements that have five degrees-of-freedom per node. Bending moments in shells and plates can have very high gradients and as a result demand a fine mesh at certain portions of the model, e.g., close to the support of the liquid filled tank shown in Fig. 3.4.6.

4

Concepts of Dynamic Analysis

4.1 INTRODUCTION

Vibration of structural and mechanical systems, can be either undesirable or deliberate. Undesirable vibrations that can adversely affect the operation of structural and mechanical systems include both nature and man/machinery induced dynamic loads, such as seismic, wind, and moving machine part excitations.

Oscillatory systems can be distinguished as either *linear or nonlinear*. This text considers the dynamics of linear systems, that is, systems for which the *principle of superposition* holds. It also considers the *deterministic* and *probabilistic (random)* response of structural and mechanical systems. In the former case, the loads are known as specified functions of time. In the latter case, the loads and the response they induce to the systems are described through statistical parameters, such as the *mean value* and the *power spectral density*.

In general, the response of a system to external loads depends on the *natural frequencies* and *damping* of the system as well as on the frequency content and amplitude of the exciting forces. If one frequency of a harmonic excitation coincides with one of the natural frequencies of the system, the amplitude of vibration may become very large. Such a condition is referred to as *resonance*.

It is always desirable to avoid or at least design for resonance. For this reason it is very important that the analyst knows the natural frequencies of the structure, the frequency content, the amplitude of the exciting forces as well as the damping that the system will exhibit when oscillating. Damping represents all mechanisms that dissipate energy during excitation and is the primary factor in limiting the amplitude of oscillation at resonance.

Two general classes of oscillations are studied: *free* and *forced vibrations*. Free vibration occurs when a system is displaced from its static position and then left free to oscillate. Under free vibration the system oscillates at its *natural frequencies*. The natural frequencies are dynamic characteristics of the system specified by its stiffness and inertia properties. Natural frequencies are calculated with *modal analysis*. Forced vibrations are classified into *periodic and non-periodic*. In a periodic vibration, the response repeats itself at a regular time interval, called *period* T, see Fig. 4.1.1. The sub-class of periodic vibrations studied first in the following Sections is the *harmonic*, because harmonic vibrations play a significant role in vibration theory and allow an easy introduction to structural dynamics concepts.

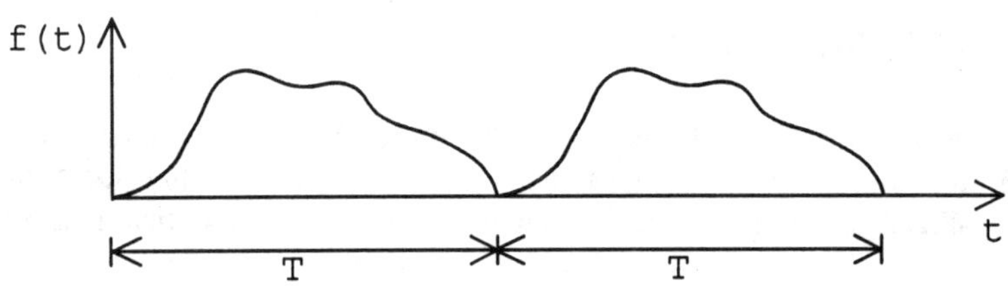

Figure 4.1.1 *Periodic vibration.*

4.2 FREE VIBRATION OF SINGLE DEGREE-OF-FREEDOM SYSTEM (SDOF)

One of the basic concepts in dynamics is the number of degrees-of-freedom of a system that vibrates when subjected to time varying (dynamic) loads. In general, structures have an infinite number of degrees-of-freedom, i.e., we need to know an infinite number of independent coordinates to specify the position of the system at any instant of time. There are several systems, however, whose configurations can be determined with sufficient accuracy in terms of only one coordinate. These systems are called one-degree or *single-degree-of-freedom systems (SDOF)*.

We will start the introduction of structural dynamics with a simple system: a cantilever beam with a uniform cross section and a concentrated mass attached at its tip, as shown in Fig. 4.2.1(a). If we neglect the mass of the cantilever, then the system can be represented as a SDOF system, the degree-of-freedom being the lateral displacement at the tip. The SDOF system shown in Fig. 4.2.1 (b), consists of a spring with a stiffness k expressing the lateral stiffness of the beam at its tip, and a mass m representing the concentrated mass. The horizontal displacement of the SDOF system corresponds to the lateral displacement at the tip of the cantilever beam.

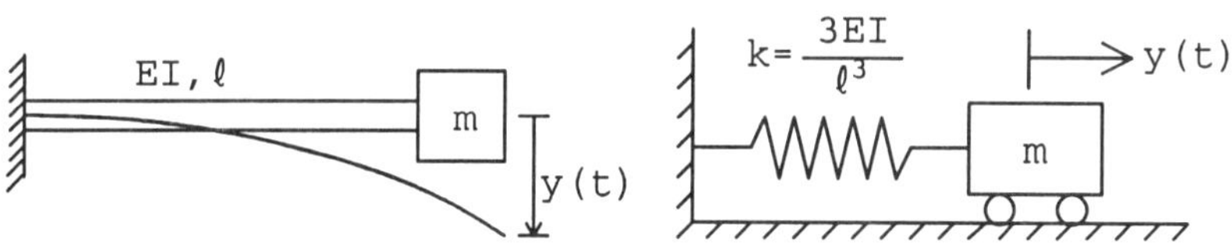

(a) (b)

Figure 4.2.1 *(a) Cantilever beam; (b) Equivalent SDOF system.*

If we displace the system by a distance y_0 and then release it, the mass will start oscillating with a constant response amplitude as shown in Fig. 4.2.2.

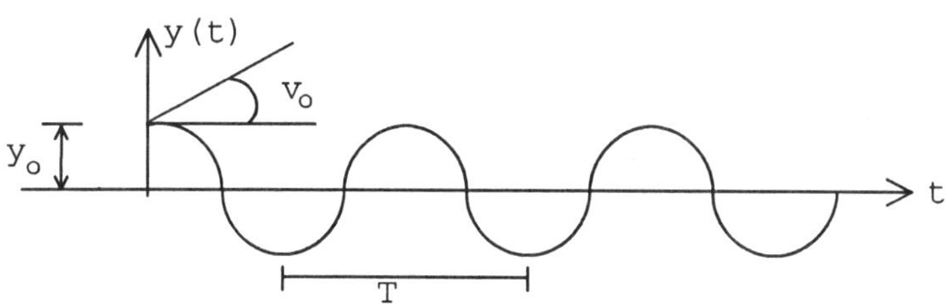

Figure 4.2.2 *Undamped free vibration response of SDOF system.*

Since no external forces are acting on the mass, it oscillates under the effect of the *initial conditions*, i.e., the displacement y_0 and the velocity v_0 at t=0 when the system was let free to vibrate. According to Newton's second law of motion, the force $f_s(t)$ exerted from the spring to the mass is equal to the mass times the acceleration along the direction of the excitation, and is expressed by

$$ky(t) = -m\ddot{y}(t) \tag{4.2.1}$$

As shown in Fig. 4.2.3, $f_I(t)$ should maintain equilibrium with the elastic resisting force f_s developed by the spring at every instant of time. The equilibrium between the inertia and all the other forces acting on the system is known as *D'Alembert's Principle*. It is applicable to any linear system and can be expressed as

$$m\ddot{y}(t) + ky(t) = 0 \tag{4.2.2}$$

Thus, for the undamped SDOF system D'Alembert's principle can be viewed as Newton's second law expressed in the form of static equilibrium between the inertia and the elastic forces.

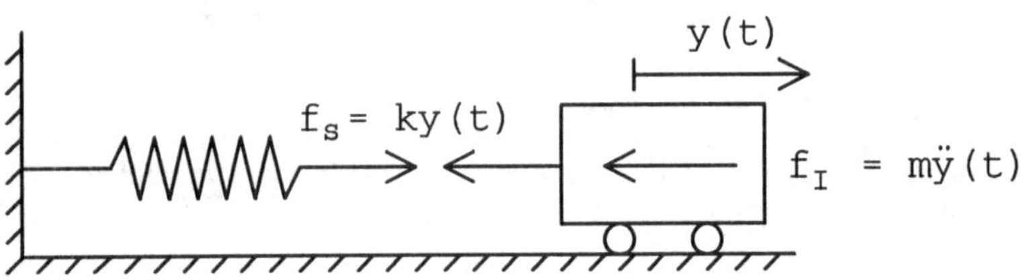

***Figure 4.2.3** Free-body diagram of undamped SDOF system.*

4.3 NATURAL FREQUENCY AND PERIOD

The solution of eqn. (4.2.2) introduces the most significant parameters that characterize vibrating systems, that is, the *natural circular frequency, the natural frequency*, and the *natural period*.

The theory of differential equations states that a solution of eqn. (4.2.2) is given by

$$y(t) = A \cos \omega t \tag{4.3.1}$$

where A is a constant that can be determined from the initial conditions of the system. Substitution of eqn. (4.3.1) into eqn. (4.2.2) yields

$$(-m\omega^2 + k)\, A \cos\omega t = 0 \tag{4.3.2}$$

Equation (4.3.2) is satisfied for any value of time t only if the term in the parenthesis is equal to zero. This implies that

$$\omega = \sqrt{\frac{k}{m}} \tag{4.3.3}$$

where ω is called the *natural circular frequency* of the SDOF system and is usually expressed in units of rad/sec.

Similarly, the theory of differential equations states that a second solution of eqn. (4.2.2) is given by

$$y(t) = B \sin\omega t \tag{4.3.4}$$

and the general solution of eqn. (4.2.2) is the combination of eqns. (4.3.1) and (4.3.4), that is

$$y(t) = A \cos \omega t + B \sin \omega t \tag{4.3.5}$$

It is easy to show that if the initial displacement and velocity at t=0 are y_0 and v_0, respectively, the constants A and B can be expressed in terms of y_0 and v_0 to arrive at the following form of the general solution of eqn. (4.3.5):

$$y(t) = y_o \cos\omega t + \frac{v_o}{\omega} \sin\omega t \tag{4.3.6}$$

Using simple trigonometric transformations eqn. (4.3.6) can be written in the form

$$y(t) = Y_A \cos(\omega t - \theta) \tag{4.3.7}$$

where

$$Y_A = \left[y_0^2 + (v_0/\omega)^2\right]^{1/2} \tag{4.3.8}$$

and the phase angle θ is given by

$$\tan\theta = \frac{v_0}{\omega y_0} \tag{4.3.9}$$

Figure 4.2.2, which is a graphical representation of eqn. (4.3.7), shows that the amplitude of the free vibration response of a SDOF system is determined by the natural frequency and the initial conditions, and remains constant with time.

The *natural period* T of the system is defined by

$$T = 2\pi/\omega \tag{4.3.10}$$

The natural period is the minimum time interval for the system to complete one oscillation, and is expressed in units of time, usually (sec).

The *natural frequency* f is the inverse of the natural period

$$f = \frac{1}{T} = \frac{\omega}{2\pi} \tag{4.3.11}$$

The natural frequency denotes the number of amplitude reversals per unit of time and is usually expressed in cycles per second (cps) or Hertz (Hz).

Equations (4.3.3), (4.3.10) and (4.3.11) indicate that by either increasing the stiffness or decreasing the mass, the natural frequency of the system increases and the natural period decreases. The dependence of frequency on mass and stiffness is the predominant means to affect the response of the system.

4.4 FREE VIBRATION OF DAMPED SDOF SYSTEM

Undamped free vibration never occurs in structural systems. If we displace the mass at the tip of the cantilever beam laterally and then let the beam free to vibrate, the amplitude of the response will gradually decrease with time and eventually vanish. This decrease of the response amplitude is attributed to *damping*. In general, the phenomenon of losing energy stored in a vibrating system is known as damping. What usually happens is that when a system vibrates damping forces develop and convert the mechanical and kinetic energy of the system to thermal energy.

Consideration of damping and the associated damping forces is a rather complex procedure. Due to the paramount significance of damping in vibrating structural systems, a separate section that discusses types and treatment of damping is included later in this chapter. A cursory exposure to the contents of the section on damping is suggested even to the reader who does not intend to incorporate damping in his/her applications.

In most practical applications, damping forces f_D are assumed to be *viscous*. This implies that the amplitude of a damping force is proportional to the velocity of the mass and its direction is opposite to the direction of motion. Such restraining forces develop on a solid that moves in a viscous fluid. A graphical representation of viscous damping is a dashpot characterized by a constant c, as shown in Fig. 4.4.1.

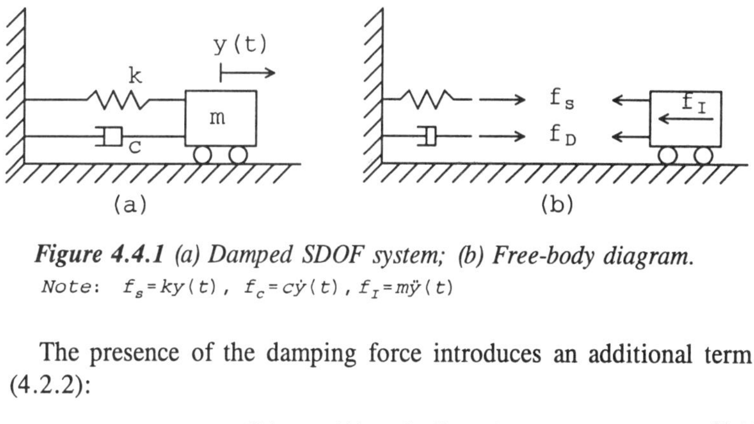

Figure 4.4.1 *(a) Damped SDOF system; (b) Free-body diagram.*
Note: $f_s = ky(t)$, $f_c = c\dot{y}(t)$, $f_I = m\ddot{y}(t)$

The presence of the damping force introduces an additional term in eqn. (4.2.2):

$$m\ddot{y}(t) + c\dot{y}(t) + ky(t) = 0 \tag{4.4.1}$$

Equation (4.4.1) has a solution of the form $y(t)=e^{pt}$. If we substitute y, $\dot{y}$ and $\ddot{y}$ in eqn. (4.4.1) and cancel e^{pt} from both sides of the resulting expression, we obtain the characteristic equation

$$mp^2+cp+k=0 \tag{4.4.2}$$

The roots of the characteristic equation are given by

$$p_{1,2} = -\frac{c}{2m} \pm \sqrt{\left(\frac{c}{2m}\right)^2 - \frac{k}{m}} \tag{4.4.3}$$

We can identify three different cases:

i. the quantity under the radical is equal to zero. In this case the damping is called *critical damping* c_{cr} and is given by

$$c_{cr} = 2\sqrt{km} \tag{4.4.4}$$

ii. the quantity under the radical is positive. In this case the damping coefficient is greater than the critical damping and the system is called *overdamped:*

$$c > c_{cr} \tag{4.4.5}$$

If a system is either critically damped or overdamped no oscillation occurs as shown in Fig. 4.4.2.

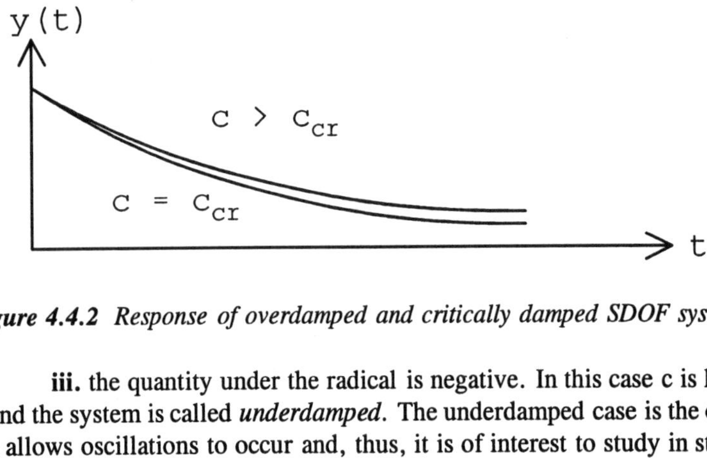

***Figure 4.4.2** Response of overdamped and critically damped SDOF system.*

iii. the quantity under the radical is negative. In this case c is less than c_{cr} and the system is called *underdamped.* The underdamped case is the only one that allows oscillations to occur and, thus, it is of interest to study in structural dynamics.

The theory of ordinary differential equations states that the solution of eqn. (4.4.1) is given by

$$y(t) = e^{-\xi\omega t}(A\cos\omega_D t + B\sin\omega_D t) \tag{4.4.6}$$

Substitution of eqn. (4.4.6) in eqn. (4.4.1) and application of the initial conditions y_0 and v_0 yield

$$y(t) = e^{-\xi \omega t} \left[y_0 \cos \omega_D t + \frac{y_0 \omega \xi + v_0}{\omega_D} \sin \omega_D t \right] \tag{4.4.7}$$

Equation (4.4.7) is the counterpart of eqn. (4.3.6). Notice that due to damping two additional variables have been introduced: ξ and ω_D. The parameter ξ is called *damping ratio* and is defined as

$$\xi = \frac{c}{c_{cr}} = \frac{c}{2\sqrt{mk}} = \frac{c}{2m\omega} \tag{4.4.8}$$

while ω_D is the *damped circular frequency* of the system and is given by

$$\omega_D = \omega \sqrt{(1 - \xi^2)} \tag{4.4.9}$$

A graphical representation of eqn. (4.4.7) is shown in Fig. 4.4.3. As applied to equation (4.3.6), trigonometric manipulations can also be used to express eqn. (4.4.7) in the form

$$y(t) = Y_A e^{-\xi \omega t} \cos \left(\omega_D t - \theta \right) \tag{4.4.10}$$

where

$$Y_A = \left[y_0^2 + (v_0 + y_0 \xi \omega)^2 / \omega_D^2 \right]^{1/2} \tag{4.4.11}$$

and

$$\tan\theta = \left[v_0 + \xi \omega y_0 \right] / (y_0 \omega_D) \tag{4.4.12}$$

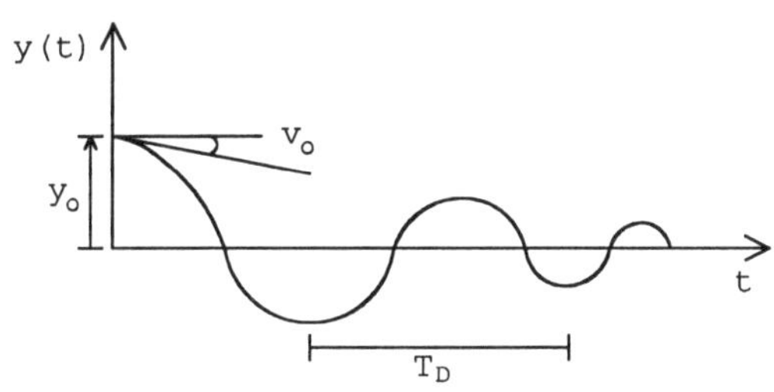

Figure 4.4.3 *Free vibration response of underdamped SDOF system.*

Equations (4.4.10) through (4.4.12) are the counterparts of equations (4.3.7) through (4.3.9). By comparing the corresponding counterparts we can make the following observations on the free vibration of the undamped and the underdamped cases:

a. Both motions are harmonic with circular frequencies ω and ω_D, respectively.

b. The amplitude of the undamped case remains constant with time, see Fig. 4.2.2. In the damped case as shown in Fig. 4.4.3, the term $e^{-\xi\omega t}$ reduces the response amplitude with time.

At a first glance, eqn. (4.4.9) indicates that ω_D is less than ω, which implies that the damped SDOF system vibrates with a frequency that is less than the natural frequency of the undamped SDOF system. The difference between ω and ω_D depends on the value of ξ which for the majority of real structures ranges between 0.01 and 0.1. For the extreme value of ξ=0.1, eqn. (4.4.9) yields

$$\omega_D = 0.99\omega \tag{4.4.13}$$

Equation (4.4.13) indicates that the frequency of the damped SDOF system can be taken as equal to the natural frequency of the corresponding undamped SDOF. This observation can be extended to multi-degree-of-freedom systems and is of paramount importance in the evaluation of the natural frequencies and mode shapes of real structures.

Thus, in practice, damping is ignored in evaluating the natural frequencies and mode shapes of a system. In the following, the underdamped case will be called damped for reasons of simplicity.

4.5 UNDAMPED EXCITATION TO HARMONIC EXTERNAL FORCES

In this section we will study the behavior of a SDOF system subjected to a harmonic excitation, that is, to a force with a magnitude which is either a sine or a cosine function of time. Studying the behavior of a system to a harmonic excitation is essential to comprehend how the system also responds to a general type of excitation.

Consider an undamped SDOF system that is subjected to a harmonic force P(t) with amplitude P_0 and circular frequency $\overline{\omega}$, see Fig. 4.5.1. The equation of motion is given by

$$m\ddot{y} + ky = P_0 \sin\bar{\omega}t \tag{4.5.1}$$

The solution of eqn. (4.5.1) is

$$y(t) = A\cos\omega t + B\sin\omega t + \frac{P_0}{k}\frac{1}{1-r^2}\sin\bar{\omega}t \tag{4.5.2}$$

where r is defined as the ratio of the circular frequency of the externally applied load to the natural circular frequency of the system, that is

$$r = \frac{\bar{\omega}}{\omega} = \frac{\bar{f}}{f} = \frac{T}{\bar{T}} \tag{4.5.3}$$

Notice that r can also be expressed in terms of the fundamental frequency f and period T of the SDOF system, the frequency $\bar{f}$, and the period $\bar{T}$ of the exciting force through eqns. (4.3.10) and (4.3.11).

The solution given by equation (4.5.2) is the superposition of the free vibration problem given by eqn. (4.3.5), and the effect of the exciting force expressed by the last term of eqn. (4.5.2). The coefficients A and B can be evaluated from the initial displacement y_0 and velocity v_0. For zero initial conditions, eqn. (4.5.2) becomes

$$y(t) = \frac{P_0}{k}\frac{1}{1-r^2}(-r\sin\omega t + \sin\bar{\omega}t)) \tag{4.5.4}$$

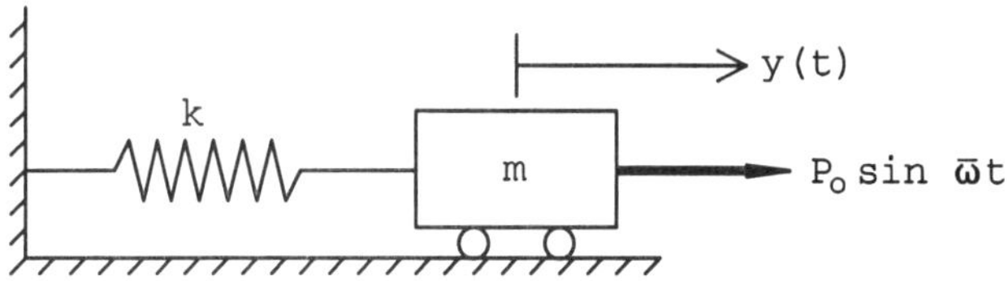

Figure 4.5.1 *Undamped SDOF system subjected to a harmonic load.*

For reasons that will become apparent in the following section on the damped excitation of a SDOF system, we call the first two terms of the response given by eqn. (4.5.2) *transient response*. The last term, which involves only the frequency of the harmonic load, is called the *steady-state response*.

Equations (4.5.2) and (4.5.4) indicate that:

a. contrary to the undamped free vibration, a harmonically excited SDOF system does not vibrate with its natural frequency ω but its response is a combination of two harmonic motions with frequencies $\bar{\omega}$ and ω; and

b. when r=1, i.e., the frequency of the external force is equal to the natural frequency, the response becomes infinitely large.

The special case of r=1 is known as *resonance*. The fact that the response of a SDOF system at resonance becomes infinitely large is true only in the realm of idealized structures. As we will see in subsequent sections, damping does not allow real structures to attain an infinite amplitude at resonance. In fact, even the presence of very little damping can dramatically decrease the response amplitude.

4.6 DAMPED EXCITATION TO HARMONIC FORCES

4.6.1 Transient and Steady-State Response

Consider now a damped SDOF system subjected to a harmonic force, as shown in Fig. 4.6.1. The equation of motion that governs the response of the system is given by

$$m\ddot{y}(t) + c\dot{y}(t) + ky(t) = P_o \sin\bar{\omega}t \tag{4.6.1}$$

As for the undamped harmonic excitation, the solution of eqn. (4.6.1) is obtained as a superposition of the damped free vibration response, eqn. (4.4.6), and the effect of the exciting force, that is

$$\begin{aligned} y(t) &= e^{-\xi\omega t}\left(A\cos\omega_D t + B\sin\omega_D t\right) + \\ &+ Y_0 \frac{\sin(\bar{\omega}t - \theta)}{\sqrt{(1-r^2)^2 + (2r\xi)^2}} \end{aligned} \tag{4.6.2}$$

where $Y_0 = P_0/k$ and the parameters ξ and r have been defined in eqns. (4.4.8) and (4.5.3). The constants A and B can be determined from the initial displacement and velocity, and the *phase angle* θ is given by

$$\tan\theta = \frac{2\xi r}{1 - r^2} \tag{4.6.3}$$

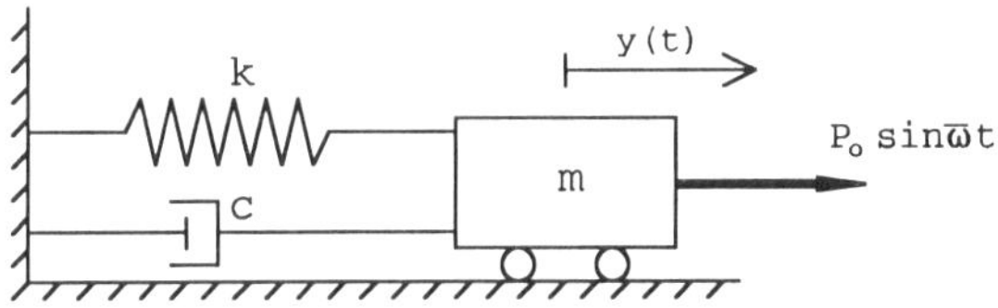

Figure 4.6.1 *Damped SDOF system subjected to harmonic load.*

Notice that eqn. (4.6.2) is the counterpart of eqn. (4.5.2). Similarly, the first two terms of eqn. (4.6.2) are called transient response $y_t(t)$, and the last term is the steady-state response $y_s(t)$. Even though the two equations are very similar, the presence of damping has introduced certain significant differences between the behavior of the undamped and the damped harmonically excited systems:

a. Contrary to the transient response whose amplitude remains constant with time in the undamped case, the amplitude of the transient response in the damped case decreases exponentially with time as the term $e^{-\xi\omega t}$ indicates. This implies that after a few reversals the transient response will diminish. The rate of amplitude decay of the transient response depends on the value of ξ. In fact, for a relatively large value of ξ, say $\xi = 0.1$, a few reversals will be sufficient to justify elimination of the transient response from eqn. (4.6.2). Thus, for most practical applications, the total response y(t) can be considered as equal to the steady-state response after a few reversals of the applied load.

b. Under resonance conditions, (r=1), the steady-state response does not attain an infinitely large value, but rather the limiting value $Y_0/2\xi$.

4.6.2 Dynamic Magnification Factor

The significant role that damping plays on the dynamic response can be visualized through the graphical representation of the *dynamic magnification factor* D_m. As dynamic magnification factor we define the ratio of the steady-state response amplitude Y to the static response amplitude Y_0, that is,

$$D_m = \frac{Y}{Y_0} \tag{4.6.4}$$

By dividing the steady-state amplitude Y with Y_0 in eqn. (4.6.2), the dynamic magnification factor is expressed as

$$D_m = \frac{1}{\left[(1-r^2)^2 + (2\xi r)^2\right]^{1/2}} \tag{4.6.5}$$

The dynamic magnification factor depends on two parameters: i) the damping ratio ξ, and ii) the frequency ratio r. Figure 4.6.2 shows the variation of D_m versus r for several representative values of ξ.

Observing eqn. (4.6.5) and Fig. 4.6.2, we can make the following remarks:

a. for r very close to zero, Y is almost equal to Y_0; that is, when the frequency of the externally applied load is very small compared to the natural frequency, the dynamic nature of the loading can be ignored and the problem can be solved as static;

b. for r close to 1, the response of the system depends on the damping ratio with a negligible effect of both the stiffness and the mass;

c. for r=1, the $D_m=1/2\xi$; that is, at resonance the dynamic magnification factor is only a function of damping;

d. for r greater than 1.45 and for practically any value of damping, the dynamic magnification factor is less than one. Thus, when the frequency of the external load is about 45% greater than the natural frequency, the amplitude of the steady-state response is less than the static displacement. Intuitively, we anticipate such behavior, since the load reversal is so "fast" that the system "experiences" a maximum displacement that is less than the static.

Under working loads most systems exhibit damping in the range of 1% - 10%. Consider a steel structure with 1% critical damping (ξ=0.01), at resonance (r=1). Evaluation of the amplification factor from eqn. (4.6.5) yields D_m=50. That is, at resonance the amplitude of the dynamic response is 50 times greater than the static displacement. For r=1.1 and the same damping ratio, eqn. (4.6.5) yields D_m= 4.76. Thus, if we make a 10% error in evaluating a natural frequency there could be a disproportionally large error in response prediction. This simple calculation clearly indicates the necessity to accurately calculate the natural frequencies in dynamically excited systems, especially when they are lightly damped.

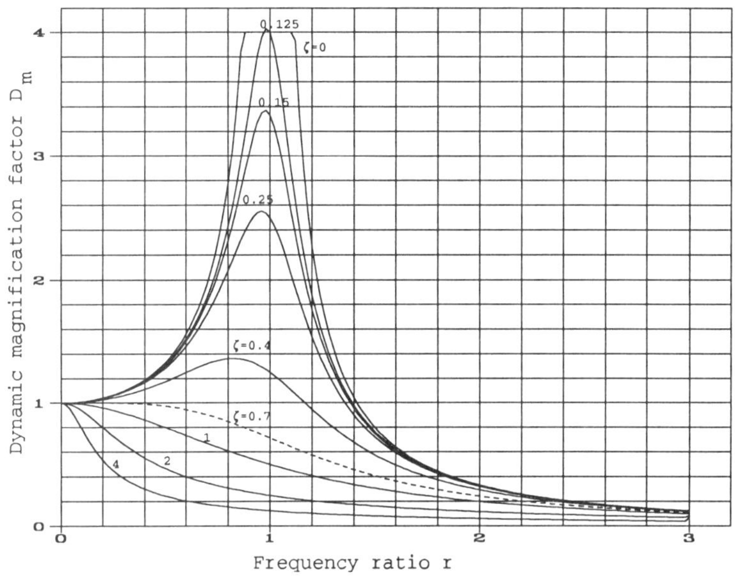

Figure 4.6.2 Dynamic magnification factor versus frequency ratio.

As we will see in Chapter 5, the above four observations can serve as useful guidelines to anticipate and control the dynamic response of multi-degree-of-freedom systems.

4.6.3 Effect of Loading Duration on Response Amplitude

After the lapse of certain time t_d, eqn. (4.6.2) indicates that the transient response diminishes, and the total response is practically equal to the steady-state response. This section determines the duration t_d for a harmonic load acting at resonance. As discussed in Chapter 5, calculation of t_d is valuable in performing time history modal superposition and direct integration.

Assuming that the initial velocity and displacement are zero, eqn. (4.6.2) becomes

$$y(t) = \frac{Y_0}{(1-r^2)^2 + (2\xi r)^2}[e^{-\xi\omega t}(2\xi r\cos\omega_D t + \frac{2\xi^2\omega r - \bar{\omega}(1-r^2)}{\omega_D}\sin\omega_D t) + + (1-r^2)\sin\bar{\omega}t - 2\xi r\cos\bar{\omega}t)] \qquad (4.6.6)$$

For a lightly damped system, the contribution of the term $\omega\xi^2/\omega_D$ is insignificant since ξ is small and according to eqn. (4.4.9) $\omega_D \simeq \omega$. In addition, if the system vibrates at resonance, i.e., r = 1 or $\overline{\omega} = \omega$, eqn. (4.6.6) reduces to

$$y(t) = \frac{Y_0}{2\xi}(e^{-\xi\omega t} - 1)\cos\omega t \tag{4.6.7}$$

Consequently, the amplitude of the response at resonance Y_R is given by

$$\frac{Y_R}{Y_0} = \frac{1}{2\xi}|e^{-\xi\omega t} - 1| \tag{4.6.8}$$

Figure 4.6.3 shows the variation of Y_R/Y_0 versus the number of cycles of the harmonic load. Either eqn. (4.6.8) or Fig. 4.6.3 can be used to estimate the duration that a harmonic load should act on a SDOF system to attain steady-state response. Making use of Fourier series or Fourier transform one could extent the use of Fig. 4.6.3 to systems subjected to transient loads. An introduction to Fourier series and Fourier transform is presented in Section 4.11.

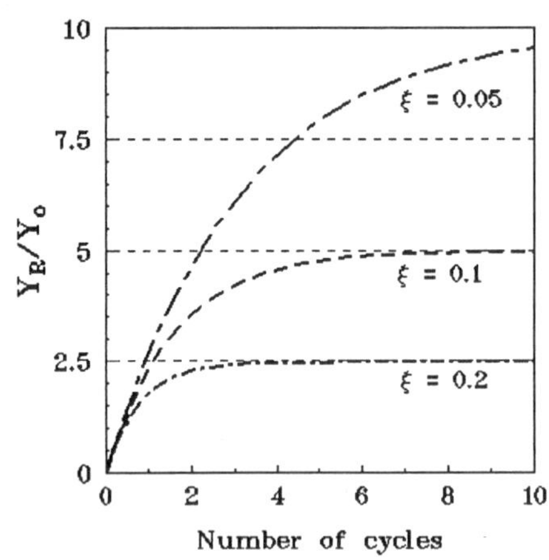

Figure 4.6.3 *Effect of loading duration on response amplitude at resonance.*

4.7 EXCITATION TO HARMONIC GROUND MOTION

A structure can also be excited when its support or foundation is subjected to a time varying motion. Typical ground motions are earthquakes, underground and surface explosions, and vibrations generated by heavy traffic or machinery. In many instances ground excitations can induce the most severe loading to a structure and as a consequence they could govern the design.

Following the procedure presented in Section 4.6, we will now examine the response of a SDOF system subjected to a harmonic ground excitation. The idealized SDOF system and the corresponding free-body diagram are shown in Fig. 4.7.1. Note that $y_g(t)$ represents the ground displacement and y(t) is the total displacement of the mass. If the displacement of the mass relative to the ground is denoted as u(t), then the total displacement can be expressed as

$$y(t) = y_g(t) + u(t) \tag{4.7.1}$$

where

$$y_g(t) = Y_g \sin\bar{\omega}t \tag{4.7.2}$$

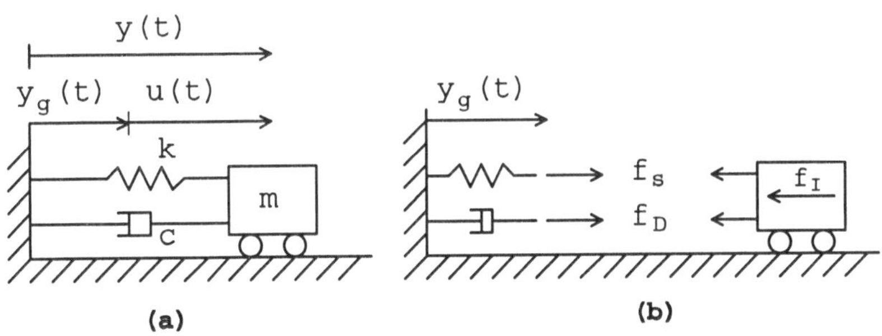

Figure 4.7.1 *(a) SDOF system subjected to harmonic ground excitation; (b) Free-body diagram.*

From the free-body diagram, equilibrium requires

$$f_I + f_D + f_s = 0 \tag{4.7.3}$$

According to Newton's second law, the relationship between the inertia force and the total acceleration $\ddot{y}$ is given by

$$f_I = m\ddot{y} \tag{4.7.4}$$

or

$$f_I = m\left(\ddot{y}_g + \ddot{u}\right) \tag{4.7.5}$$

and since

$$f_D = c\dot{u} \tag{4.7.6}$$

and

$$f_s = ku \tag{4.7.7}$$

eqn. (4.7.3) can be expressed as

$$m\ddot{u}(t) + c\dot{u}(t) + ku(t) = -mY_g\bar{\omega}^2\sin\bar{\omega}t \tag{4.7.8}$$

Comparison of eqns. (4.6.1) and (4.7.8), reveals the similarities between a ground motion and an external force excitation. As eqn. (4.6.1) indicates, the ground motion in eqn. (4.7.8) can be replaced by an external force with magnitude $P_0 = -m\bar{\omega}^2 Y_g$.

Due to the similarity of eqns. (4.6.1) and (4.7.8), all remarks for a SDOF system vibrating to a harmonic external load are also valid for the system subjected to a harmonic ground excitation.

4.8 RESPONSE TO GENERAL EXCITATION

So far we have examined the response of a SDOF system subjected to harmonic loads. However, most systems are subjected to non-harmonic loads, and the preceding discussion on harmonic loads and vibrations simply served as a means to introduce fundamental concepts of structural dynamics.

Non-harmonic excitations could be applied for either a very short period of time, which in this case are referred to as *shock excitations,* or a long time period. Examples of shock and other types of non-harmonic excitations are shown in Fig. 4.8.1. It is worth mentioning that for the majority of shock excitations:

a. the transient part of the response should be considered in the analysis, and

b. damping could be ignored since inertia and stiffness basically govern the response of the system for the time duration during which the response maxima occur.

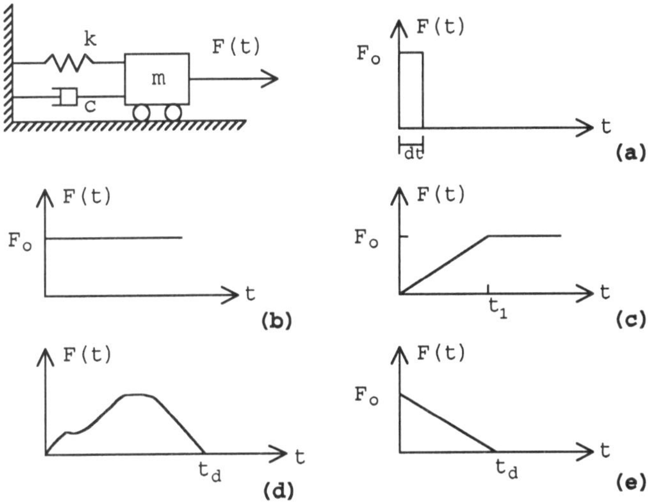

Figure 4.8.1 *General excitations: (a) Impulse load; (b) Suddenly applied load; (c) Step load with rise time t_1; (d) General arbitrary load; (e) Linearly decreasing load.*

For non-harmonic excitations, closed form solutions of even SDOF systems are difficult to obtain. Thus, the evaluation of the dynamic response of SDOF systems and models of real structures is done numerically. A method that can be applied to determine the response of linear structures is the analytical or numerical evaluation of *Duhamel's integral*:

$$y(t) = \frac{1}{m\omega_D} \int_0^t F(\tau) e^{-\xi\omega(t - \tau)} \sin\omega_D (t - \tau) d\tau \qquad (4.8.1)$$

Duhamel's integral provides the solution of the equation of motion for zero initial conditions

$$m\ddot{y} + c\dot{y} + ky = F(t) \qquad (4.8.2)$$

where F(t) is a general excitation. Evaluation of the response using Duhamel's integral is limited to simple structures. In general, current computer codes utilize methods that are based on the numerical solution of the equations of motion and not Duhamel's integral. Such methods are presented in Chapter 5.

4.9 THE CONCEPT OF RESPONSE SPECTRUM

Consider the SDOF system shown in Fig. 4.7.1(a) which is now subjected to the earthquake ground acceleration $\ddot{y}_g(t)$ shown in Fig. 4.9.1(a). The equation governing the response of the SDOF system is given by

$$m\ddot{u} + c\dot{u} + ku = -m\ddot{y}_g \qquad (4.9.1a)$$

or

$$\ddot{u} + 2\xi\omega\dot{u} + \omega^2 u = -\ddot{y}_g \qquad (4.9.1b)$$

where u is the relative displacement of the mass with respect to the ground as defined by eqn. (4.7.1).

Equation (4.9.1b) is obtained by dividing both sides of eqn. (4.9.1a) with m and substituting the expressions for ξ and ω given by eqns. (4.3.3) and (4.4.8).

Because of the irregularity of the ground acceleration, the most common approaches solve eqn. (4.9.1b) numerically. Such procedures include numerical evaluation of Duhamel's integral, or usually solutions via one of the numerical algorithms presented in Chapter 5.

For design purposes, only response maxima are usually needed. Each one of the maxima, plotted as a function of either the natural frequency or the period of the structure, constitutes a *response spectrum*. The most commonly used response spectrum is the *spectral displacement,* which is a plot of the maximum *relative* displacements S_d that SDOF systems experience for a specified ground motion, that is

$$S_d = u_{max} \qquad (4.9.2)$$

The procedure to obtain the spectral displacement for the typical earthquake motion shown in Fig. 4.9.1(a) is depicted in Figs. 4.9.1(b) through (e). Specifically, in order to obtain the spectral displacement S_{d1} corresponding to a natural period, say T_1, eqn. (4.9.1b) is solved to determine the maximum relative displacement of a SDOF system with a period T_1 and damping ratio ξ.

In order to draw Fig. 4.9.1(e), such computations are repeated many times for a series of SDOF systems with T_i $(i=1,2,3,...)$ natural periods and a damping ratio ξ.

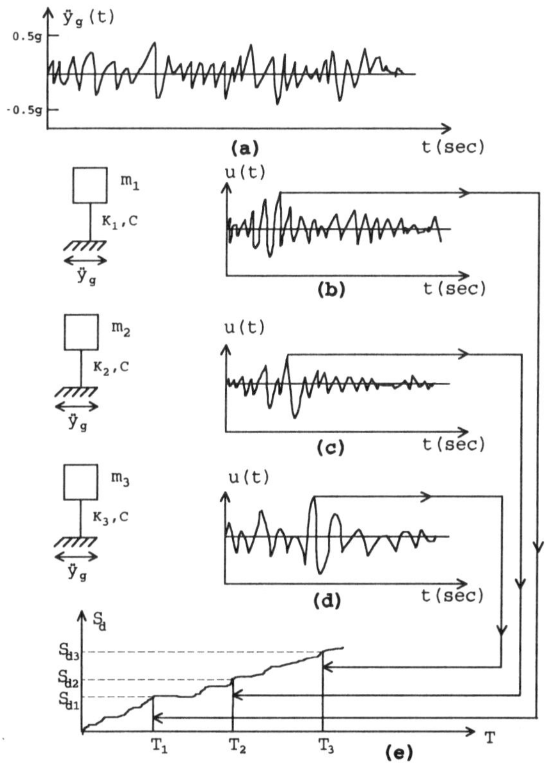

Figure 4.9.1 *(a) Earthquake motion; (b), (c), (d) Relative displacements of SDOF systems with damping ξ and natural periods T_1, T_2, and T_3; (e) Spectral displacement.*

If the ground excitation is expressed in a closed form, the solution of eqn. (4.9.1b) could also be obtained in a closed form for the spectral displacement. For example, if the ground acceleration is harmonic with a unit amplitude, the S_d will be given by

$$S_d(r,\xi) = \frac{1}{\sqrt{(1-r^2)^2 + (2\xi r)^2}} \tag{4.9.3}$$

Another convenient measure of the maximum response of a SDOF system to a specific earthquake is the relative pseudo-velocity or *spectral velocity* S_v. The spectral velocity approximates the maximum *relative* velocity of the undamped SDOF system and is defined as

$$S_v = \omega S_d \tag{4.9.4}$$

where

$$\omega = \frac{2\pi}{T} \tag{4.3.10}$$

A common alternative measure of the maximum response of a SDOF system to a given ground excitation is also the *spectral acceleration*, S_a. The spectral acceleration is the maximum total acceleration of the SDOF system, that is

$$S_a = \ddot{y}_{max} \tag{4.9.5}$$

and is related to the spectral displacement as

$$S_a = \omega^2 S_d \tag{4.9.6}$$

or

$$S_a = \left(\frac{2\pi}{T}\right)^2 S_d \tag{4.9.7}$$

The relationship between the spectral displacement, velocity, and acceleration allows their graphical representation in one plot, see Fig. 4.9.2. Figure 4.9.2 demonstrates the information provided by a typical response spectrum. Specifically, one may enter the spectra for a given natural frequency f_i and select the corresponding S_{di}, S_{vi}, and S_{ai} as indicated by the arrows. Representative design spectra can be found in various seismic design and analysis references.

Currently, there is no accurate method that can be used to develop response spectra at a specific site. For this reason, seismic codes suggest the use of *design response spectra* which represent the "average" maximum responses that can be anticipated at a site. Figure 4.9.3 shows the design spectra for SDOF systems with $\xi = 0.05$ supported by a stiff soil as recommended by the UBC-88. They have been developed based on eqns. (4.9.8) for a rocklike soil type and have been normalized with respect to the acceleration of gravity g. Before the normalized spectra of Fig. 4.9.3 are used for design, they should be

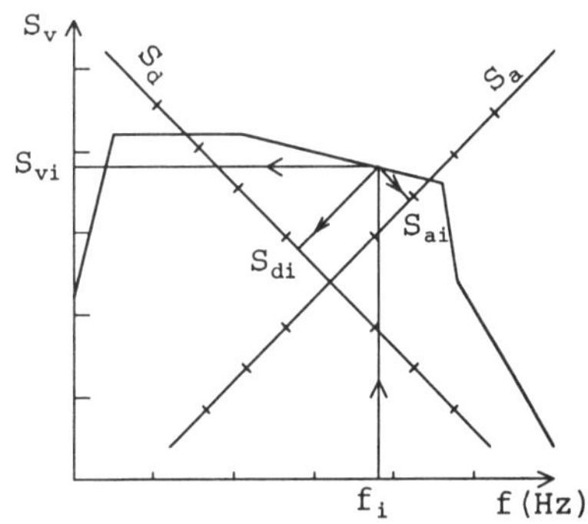

Figure 4.9.2 Use of response spectrum.

modified to account for additional considerations pertinent to the site location, the use, and type of structure. Design codes and provisions, such as the UBC, provide design spectra and procedures for the seismic design of buildings as well as mechanical and electrical equipment.

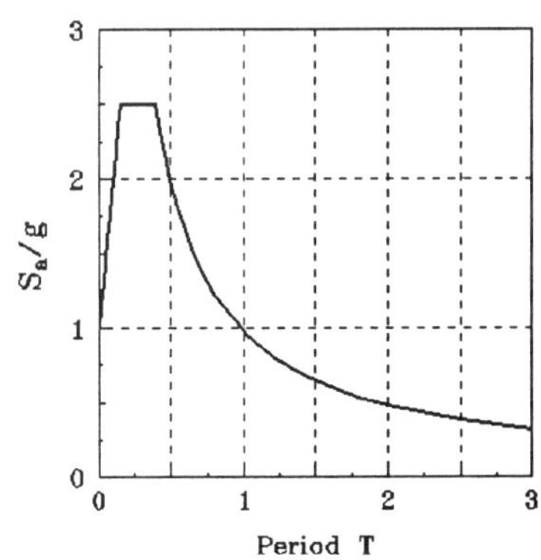

Figure 4.9.3 Normalized design spectra.

For rocklike material with a soil depth less than 200 ft the normalized spectral acceleration can be evaluated from

$$\frac{S_a}{g} = (1 + 10T) \quad for \quad 0 < T < 0.15 \text{ sec}$$

$$\frac{S_a}{g} = 2.5 \quad for \quad 0.15 < T < 0.39 \text{ sec} \tag{4.9.8}$$

$$\frac{S_a}{g} = \frac{0.975}{T} \quad for \quad 0.39 \text{ sec} < T$$

Similar formulas for various soil conditions are provided in design codes.

It should be noted that response spectra can be developed for any type of ground excitation. In fact, response spectra have been developed for numerous shock and blast excitations, e.g., Harris and Crede.

4.10 RANDOM VIBRATION ANALYSIS

4.10.1 Introduction

When all conditions under control during the dynamic excitation of a system are maintained the same and the resulting responses are always identical, the process is called *deterministic*. All types of dynamic analysis that we have discussed so far are deterministic processes. In deterministic dynamic analysis of linear systems, the properties of the system and the exciting source yield a unique response of the system through the analysis methods presented in the previous sections.

When all conditions "under control" during the dynamic excitation of the system are maintained the same and the responses continually differ from each other, the process is called *random*. For a random process only a *statistical* description of the totality of possible responses is meaningful. The randomness is the result of fluctuations in the variables of the system and the loading that are not "under control." For example, if we record the vertical displacement of one wheel of a truck that carries a load from one location to another always following a specific route and with the same velocity, the responses that we record for every trip will be different, even though all the variables "under control," that is, the truck, its load, the velocity, and the trip were always maintained the same. The differences are attributed to the variables that we either did not account for or we were unaware of that changed during each trip, e.g., small deviations from the route that the truck followed each time, inadvertent variation of the velocity during the trips.

Each one of the response records is called a *sample*, while the set of all the response records that we recorded is called the *ensemble*. Figure 4.10.1 shows a schematic representation of a sample.

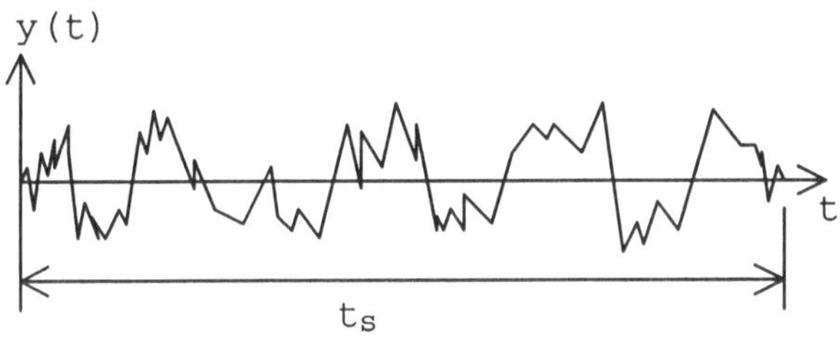

Figure 4.10.1 *Sample of random excitation.*

In deterministic analysis, the response of a SDOF system subjected to a sinusoidal excitation is characterized by its amplitude and frequency. A random vibration can often be adequately described by average amplitudes and frequency content. The average amplitudes most often used are *the mean value, the mean square value*, and the *root mean square value (RMS)*. The frequency content is described by the *power spectral density (PSD)*. Other statistical parameters can also be used to provide a more complete picture of the random vibration.

In practice, random vibration is usually restricted to *stationary* processes and in particular to a sub-class of stationary processes: the *ergotic*. In an ergotic process all properties of the process, e.g., the RMS, can be determined from a single sample adequately long in time. In our example with the truck, if we use the same driver that faces the same traffic conditions for each trip, we could expect each of the sample responses to be statistically similar to every other sample and the ergotic hypothesis could be justifiable. If y(t) describes the time variation of a sample of an ergotic excitation -in our example y(t) can be the vertical displacement of the wheel measured during one trip- then the mean value $\overline{y}$, the mean square $\overline{y^2}$, and the RMS of the random displacement are given by

$$\bar{y} = \frac{1}{t_s}\int_0^{t_s} y(t)\,dt$$
$$\overline{y^2} = \frac{1}{t_s}\int_0^{t_s} y^2(t)\,dt \qquad (4.10.1)$$
$$RMS_y = \sqrt{\overline{y^2}}$$

where t_s is the duration of the sample, see Fig. 4.10.1.

The mean and the root mean square provide measurements for the average value of the random response y(t). A measurement of how widely y(t) differs from the mean value $\overline{y}$ is given by either the *variance* σ^2_y or the *standard deviation*, σ_y

$$\sigma_y = \sqrt{\overline{y^2} - (\bar{y})^2} \qquad (4.10.2)$$

Notice that when the mean is zero, which is common in many practical cases, the RMS_y is equal to σ_y

$$RMS_y = \sqrt{\overline{y^2}} = \sigma_y \qquad (4.10.3)$$

4.10.2 Probability Distribution

Consider a sample of a random record, as shown in Fig. 4.10.2. We define as the cumulative probability of y(t), having a value $y_m < y(t) < y_n$, the ratio

$$P(y_m < y < y_n) = \frac{Dt_1 + Dt_2 + \ldots + Dt_N}{t_s} \qquad (4.10.4)$$

where t_s is the duration of the record, and Dt_i (i=1,2,...,N) are the corresponding time intervals for which $y_m < y < y_n$. In general, if we can determine the derivative of the cumulative probability p(y), then the $P(y_m < y < y_n)$ can be evaluated from

$$P(y_m < y < y_n) = \int_{y_m}^{y_n} p(y)dy \tag{4.10.5}$$

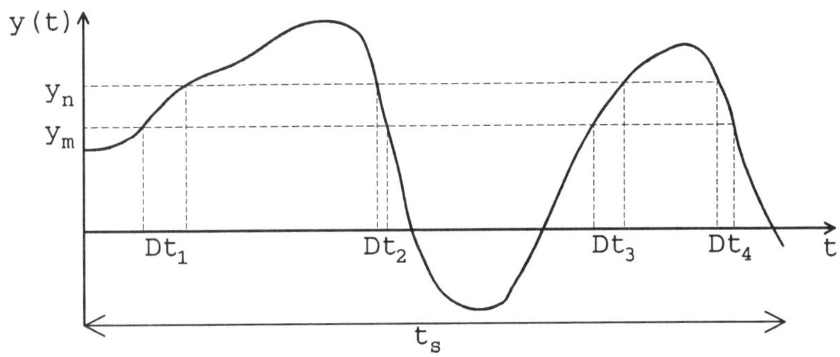

Figure 4.10.2 *Determination of probability from a sample.*

The function p(y) is called *probability density function* (PDF). The most commonly used probability density function is the *normal* or *Gaussian distribution* and is expressed by

$$p(y) = \frac{1}{\sigma_y\sqrt{2\pi}} e^{-(y - \bar{y})^2/(2\sigma_y^2)} \tag{4.10.6}$$

A graphical representation of the normal distribution for zero mean is given in Fig. 4.10.3. In view of eqn. (4.10.5), the shaded area is equal to $P(y_m < y < y_n)$.

It should be noted that calculation of probabilities as specified by eqn. (4.10.5) is provided in table form in various texts on statistics and random vibrations.

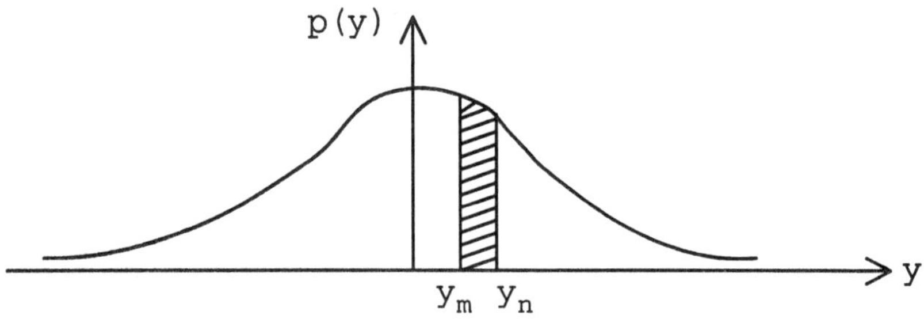

Figure 4.10.3 Normal distribution

Another commonly used probability density function is the *Rayleigh distribution*. Usually the maxima of random functions follow the Rayleigh distribution, e.g., the maximum response, say A, at a given point of a structure subjected to a support excitation. A graphical representation of the Rayleigh probability density function is given in Fig. 4.10.4, and is defined by

$$p(A)=\frac{A}{\sigma^2}e^{-A^2/(2\sigma^2)}, \qquad A>0 \tag{4.10.7}$$

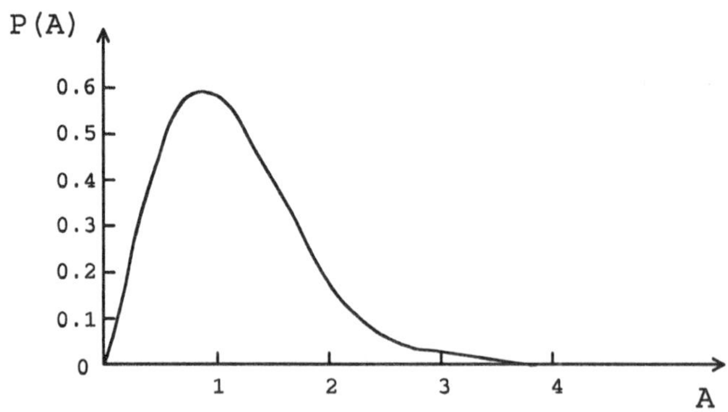

Figure 4.10.4 Rayleigh distribution.

Notice that p(A) for Rayleigh distribution is zero for negative values of A. The mean, mean square, RMS as well as the probability of A being within a specific range of values can be obtained from eqns. (4.10.1) and (4.10.5). In summary, treatment of Rayleigh distribution is similar to Gaussian.

4.10.3 Power Spectral Density and Root Mean Square Response

Consider the SDOF system shown in Fig. 4.10.5 subjected to harmonic forces $F_i(t)$, $(i=1,2,\ldots N)$ each expressed by

$$F_i(t) = F_{oi} \sin 2i\bar{f}_i t \tag{4.10.8}$$

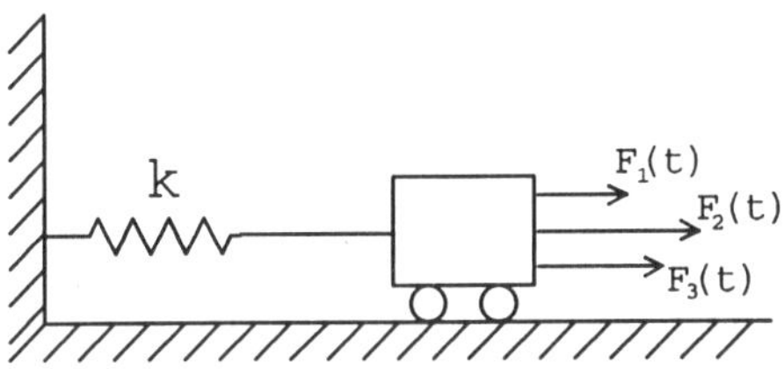

Figure 4.10.5 *SDOF system subjected to several harmonic forces.*

In view of eqns. (4.6.2), (4.6.4), and (4.6.5) the steady-state response of the system to each excitation can be expressed as

$$y_i(t) = Y_{oi} D_{mi} \sin 2i\bar{f}_i t \tag{4.10.9}$$

where D_{mi} is given by eqn. (4.6.5) for each harmonic force with amplitude F_{oi} and period $\bar{T}_i = \frac{1}{\bar{f}_i}$, and for $\xi = 0$. The mean square value of the response over a very long time interval t_s can be evaluated from eqn. (4.10.1) to give

$$\overline{y_i^2(t)} = \frac{1}{t_s} \int_0^{t_s} Y_{oi}^2 D_{mi}^2 (\sin^2 2i\bar{f}_i t)\, dt \tag{4.10.10a}$$

or

$$\overline{y_i^2(t)} = \lim_{n \to \infty} \left[\frac{n}{n\bar{T}_i/2} \int_0^{\bar{T}_i/2} Y_{oi}^2 D_{mi}^2 (\sin^2 2i\bar{f}_i t)\, dt\right] \tag{4.10.10b}$$

in which n denotes a large number of half sine waves so that $t_s = n\overline{T}_i/2$. Calculation of $\overline{y_i^2(t)}$ yields

$$\overline{y_i^2(t)} = \frac{1}{2}D_{mi}^2 Y_{oi}^2 \tag{4.10.11}$$

Similarly, the mean square of the harmonic force $F_i(t)$ is given by

$$\overline{F_i^2} = \frac{1}{t_s}\int_0^{t_s} F^2_{oi}\sin^2 2\pi \overline{f}_i t = \frac{1}{2}F_{oi}^2 \tag{4.10.12}$$

Since Y_{oi}, is the static displacement for the force F_{0i}, we obtain

$$F_{oi}^2 = k^2 Y_{oi}^2 \tag{4.10.13}$$

The total response of the system subjected to the forces $F_i(t)$ can be obtained by superimposing the individual responses $y_i(t)$. Thus, in view of eqn. (4.10.11), the mean square of the displacement response to all harmonic forces F_i is given by

$$\overline{y^2(t)} = \frac{1}{2}(D_{m1}^2 Y_{o1}^2 + D_{m2}^2 Y_{02}^2 + \ldots + D_{mN}^2 Y_{oN}^2) \tag{4.10.14}$$

The contribution of each component normalized with respect to the corresponding dynamic amplification, D_{mi}, is called *discrete spectrum*, see Fig. 4.10.6.

By substituting eqn. (4.10.13) into eqn. (4.10.14) we get

$$\overline{y^2(t)} = \left(\frac{D_{m1}}{k}\right)^2 \frac{(F_{o1}^2)}{2} + \left(\frac{D_{m2}}{k}\right)^2 \frac{(F_{02}^2)}{2} + \ldots + \left(\frac{D_{mN}}{k}\right)^2 \frac{(F_{oN}^2)}{2} \tag{4.10.15}$$

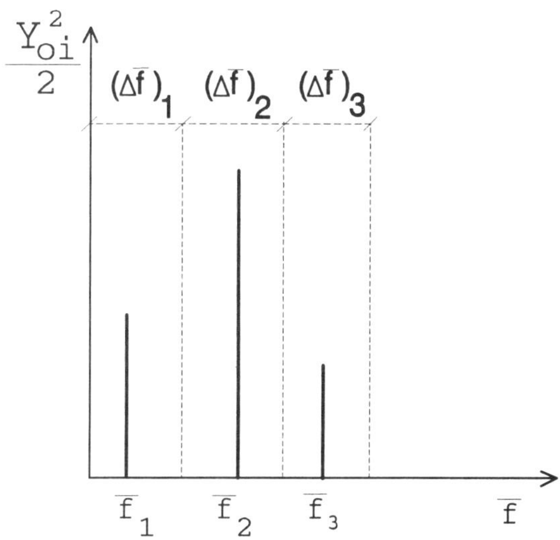

Figure 4.10.6 *Discrete spectrum.*

If many harmonic forces with different frequencies act on the system, we can extrapolate that the discrete forcing spectrum becomes a continuous function of frequency. This can be easily visualized if we define the *power spectral density (PSD)* of the displacement response $S_y(\bar{f}_i)$ as

$$\frac{1}{2}Y_{oi}^2 = S_y(\bar{f}_i)(\Delta\bar{f})_i \tag{4.10.16}$$

in which $\frac{1}{2}Y_{oi}^2$ is the contribution of the response component in the interval $(\Delta\bar{f})_i$ to the mean square of the total excitation. The term $\frac{1}{2}Y_{oi}^2$ can also be "viewed" as equal to the hatched area of the plot of S_y shown in Fig. 4.10.7.

By calling the corresponding power spectral density of the load $S_F(\bar{f}_i)$, an expression similar to eqn. (4.10.16) can be obtained, that is

$$\frac{1}{2}F_{oi}^2 = S_F(\bar{f}_i)(\Delta\bar{f})_i \tag{4.10.17}$$

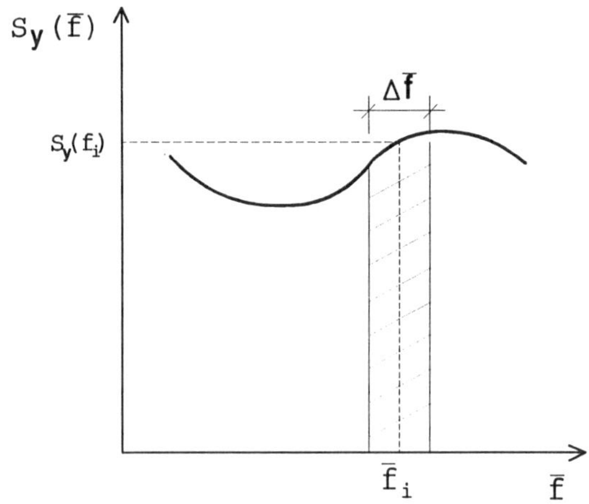

Figure 4.10.7 Power spectral density of displacement.

By combining eqns. (4.10.15) and (4.10.17) we obtain

$$\overline{y^2(t)} = \left(\frac{D_{m1}}{k}\right)^2 S_F(\bar{f}_1)(\Delta\bar{f})_1 + \left(\frac{D_{m2}}{k}\right)^2 S_F(\bar{f}_2)(\Delta\bar{f})_2 + \dots + \left(\frac{D_{mN}}{k}\right)^2 S_F(\bar{f}_N)(\Delta\bar{f})_N \tag{4.10.18}$$

In the limit, by setting $(\Delta f)_i = df$, eqn. (4.10.18) yields

$$\overline{y^2(t)} = \int_0^\infty \left(\frac{D_m}{k}\right)^2 S_F(\bar{f})d\bar{f} \tag{4.10.19}$$

Finally, the RMS_y of the excitation can be obtained from

$$RMS_y = \sqrt{\overline{y^2(t)}} \tag{4.10.1}$$

We conclude that when the power spectral density of the forcing function S_F and the ratio of the dynamic amplification factor D_m over the stiffness of the system k are known, then the root mean square of the response RMS_y can be obtained by combining eqns. (4.10.1), (4.10.19), and (4.6.5).

It should be noted that the relationship between the S_F and RMS_y, which has been derived for a series of harmonic forces F_i, can also be extended for a general forcing function. In that case, the derivation can be based on *Fourier transform* and *Parserval's theorem*. The resulting expression is identical to eqn. (4.10.19) with D_m/k replaced by the *transfer function* $H(\bar{f})$. The derivation of the transfer function as well as an introduction to Fourier transform are presented in the following section.

Many types of random excitations are Gaussian with a zero mean. Thus, calculation of their mean square response through eqn. (4.10.19) and σ_y from eqn. (4.10.3) allows evaluation of their corresponding PDF from eqn. (4.10.6). Once the PDF is determined, desired probabilities can be calculated from eqn. (4.10.5). Examples presented in Chapters 5 and 9 further elucidate random vibration analysis.

4.11 FOURIER SERIES AND FOURIER TRANSFORM

4.11.1 Introduction and Definitions

In this text, the Fourier transform is used only in random vibrations. However, the subject of Fourier transform finds extensive use in all types of dynamic analysis of linear systems as well as in modal testing and experimental stress analysis. The following is a brief overview of *Fourier series, Fourier integrals,* and *Fourier transforms*.

Consider a periodic function y(t) with period $\bar{T}$. Fourier has shown that a periodic function with a finite number of discontinuities can be expressed as the summation of an infinite number of sine and cosine terms. Such a summation is known as a Fourier series. For a periodic function, such as the one shown in Fig. 4.11.1, the Fourier series of y(t) can be written in the form

$$\begin{aligned} y(t) = a_0 + a_1\sin\bar{\omega}t + a_2\sin 2\bar{\omega}t + \ldots + a_n\sin n\bar{\omega}t + \\ + b_1\cos\bar{\omega}t + b_2\cos 2\bar{\omega}t + \ldots + b_n\cos n\bar{\omega}t + \ldots \end{aligned} \qquad (4.11.1)$$

or in the form

$$y(t) = a_0 + \sum_{n=1}^{\infty} \{a_n \sin n\bar{\omega}t + b_n \cos n\bar{\omega}t\} \tag{4.11.2}$$

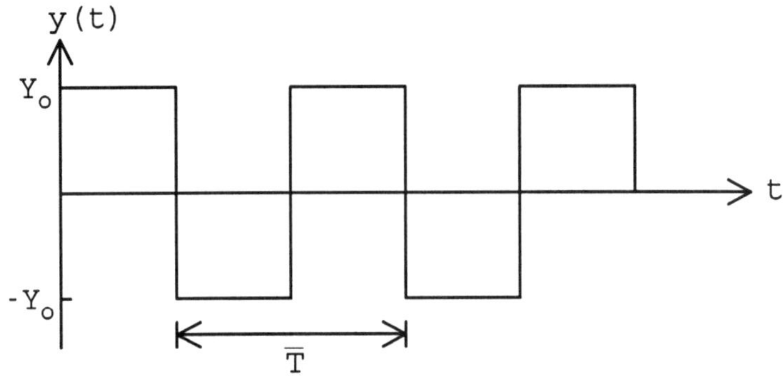

Figure 4.11.1 Periodic function.

where $\bar{\omega} = 2\pi/\bar{T}$ is the fundamental circular frequency of the periodic function. The coefficients a_0, a_n, and b_n can be determined from

$$\begin{aligned} a_0 &= \frac{1}{\bar{T}}\int_{-\bar{T}}^{\bar{T}} y(t)\,dt \\ a_n &= \frac{2}{\bar{T}}\int_{-\bar{T}}^{\bar{T}} y(t)\sin n\bar{\omega}t\,dt \\ b_n &= \frac{2}{\bar{T}}\int_{-\bar{T}}^{\bar{T}} y(t)\cos n\bar{\omega}t\,dt \end{aligned} \tag{4.11.3}$$

Next we introduce the Euler's relationships that are given by

$$\begin{aligned} \sin n\bar{\omega} &= \frac{e^{in\bar{\omega}} - e^{-in\bar{\omega}}}{2i} \\ \cos n\bar{\omega} &= \frac{e^{in\bar{\omega}} + e^{-in\bar{\omega}}}{2} \end{aligned} \tag{4.11.4}$$

where $i = \sqrt{-1}$.

Equations (4.11.4) can be used to express eqn. (4.11.2) in an exponential form

$$y(t) = \sum_{n=-\infty}^{+\infty} Y_n \exp(in\bar{\omega}t) \tag{4.11.5}$$

where the coefficients Y_n are evaluated from

$$Y_n = \frac{1}{\bar{T}} \int_{-\bar{T}/2}^{-\bar{T}/2} y(t) \exp(-in\bar{\omega}t)\, dt \tag{4.11.6}$$

If the function y(t) is not periodic, we can still apply Fourier's theory. However, in this case y(t) is expressed in the form of Fourier integral:

$$y(t) = \frac{1}{2\pi} \int_{-\infty}^{+\infty} Y(\bar{\omega}) \exp(i\bar{\omega}t)\, d\bar{\omega} \tag{4.11.7}$$

where the frequency dependent function $Y(\bar{\omega})$ is given by

$$Y(\bar{\omega}) = \int_{-\infty}^{+\infty} y(t) \exp(-i\bar{\omega}t)\, dt \tag{4.11.8}$$

The pair of functions y(t) and $Y(\bar{\omega})$ is known as the *Fourier transform pair*. The Fourier transform pair can be considered as a limiting case of the Fourier series if the non-periodic function shown in Fig. 4.11.2 is viewed as periodic with a period $\bar{T}$ extending to infinity.

The integrations indicated in eqns. (4.11.7) and (4.11.8) can be performed with an efficient numerical algorithm, the so called *Fast Fourier Transform* (FFT). The FFT is extensively used in structural dynamics since it allows determination of the frequency content of general excitations. As elaborated in Chapter 5, knowledge of the frequency content of the load is helpful in deciding on the number of modes to be determined with modal analysis.

Use of Fourier transform provides an elegant method to obtain the response of a SDOF system. This can be accomplished through the relationship of the Fourier transform of y(t), denoted as FT [y(t)], to the Fourier transform of the derivatives of y(t). The relationship can be obtained by differentiating both sides of eqn. (4.11.7) with respect to time to obtain

$$\dot{y}(t) = \frac{i\bar{\omega}}{2\pi}\int_{-\infty}^{+\infty} Y(\bar{\omega})\exp(i\bar{\omega}t)\,d\bar{\omega} \tag{4.11.9}$$

Thus, the Fourier transform of a derivative is equal to the Fourier transform of the function multiplied with $i\bar{\omega}$:

$$FT[\dot{y}(t)] = i\bar{\omega}\,FT[y(t)] \tag{4.11.10}$$

By differentiating again and using the relationship $i^2 = -1$, we obtain

$$FT[\ddot{y}(t)] = -\bar{\omega}^2\,FT[y(t)] \tag{4.11.11}$$

4.11.2 Transfer Function

Equations (4.11.10) and (4.11.11) enable us to obtain the Fourier transform of differential equations. For example, taking the Fourier transform

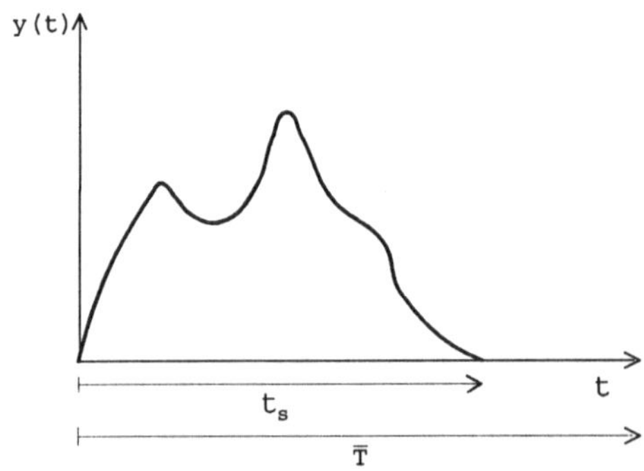

Figure 4.11.2 General forcing function.

of the equation of motion of a SDOF system with a natural frequency f

$$m\ddot{y} + c\dot{y} + ky = F(t) \tag{4.8.2}$$

and employing eqns (4.11.10) and (4.11.11) we obtain

$$(-m\bar{\omega}^2 + i\bar{\omega}c + k)\,Y(\bar{\omega}) = F(\bar{\omega}) \tag{4.11.12}$$

where $Y(\bar{\omega})$ an $F(\bar{\omega})$ are the Fourier transforms of y(t) and f(t), respectively. Solving eqn.(4.11.12) for $Y(\bar{\omega})$ yields

$$Y(\bar{\omega}) = \frac{F(\bar{\omega})}{k - m\bar{\omega}^2 + i\bar{\omega}c} \tag{4.11.13}$$

We define as transfer function, $H(\bar{\omega})$, the expression

$$H(\bar{\omega}) = \frac{Y(\bar{\omega})}{F(\bar{\omega})} = \frac{1}{k - m\bar{\omega}^2 + i\bar{\omega}c} \tag{4.11.14}$$

Thus, the transfer function is the ratio of the Fourier transform of the output to the Fourier transform of the input. In eqn. (4.11.14), by expressing the parameters in terms of the frequency ratio

$$r = \frac{\bar{\omega}}{\omega} = \frac{\bar{f}}{f} \tag{4.5.3}$$

the damping ratio

$$\xi = \frac{c}{c_{cr}} = \frac{c}{2\sqrt{mk}} \tag{4.4.8}$$

and the natural frequency of the SDOF system

$$f = \frac{\omega}{2\pi} \tag{4.3.11}$$

the transfer function can be written as

$$H(\bar{f}) = \frac{1}{k}\ \frac{1}{1-(\frac{\bar{f}}{f})^2 + i(2\xi\frac{\bar{f}}{f})} \tag{4.11.15}$$

If instead of the procedure presented in Section 4.10.3 one applies Fourier transform to relate the $S_F(\bar{f})$ of the excitation to the mean square response, the counterpart of eqn. (4.10.19) is obtained, that is,

$$\overline{y^2(t)} = \int_0^{\infty} |H(\bar{f})|^2 S_F(\bar{f}) d\bar{f} \tag{4.11.16}$$

The RMS_y can be obtained from eqn. (4.10.1).

4.11.3 Root Mean Square Response of SDOF System

If a SDOF system with a stiffness k and natural frequency f_1 is lightly damped, its transfer function is sharply peaked at resonance, see Fig. 4.11.3. If the system is also subjected to an excitation with the broad spectral density S_F shown in Fig. 4.11.3, the mean square response of the SDOF system can be approximated by

$$\overline{y^2} \cong \frac{\pi}{4\xi}\frac{f_1}{k^2}S_F(f_1) \tag{4.11.17}$$

where the $\pi/(4\xi)$ has been obtained from

$$\frac{\pi}{4\xi} = \int_0^{\infty} \frac{d(\bar{f}/f_1)}{[1-(\bar{f}/f_1)^2]^2 + (2\xi\bar{f}/f_1)^2} \tag{4.11.18}$$

and $S_F(f_1)$ is the power spectral density of the excitation at the natural frequency f_1 of the system. The calculation of the root mean square response for multi-degree-of-freedom systems is demonstrated in Chapters 5 and 9.

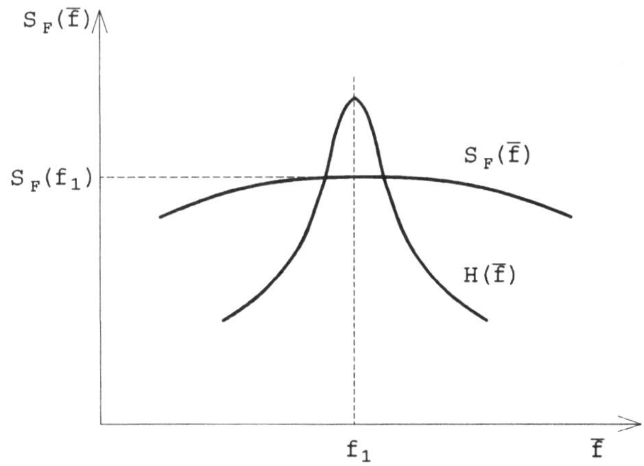

Figure 4.11.3 *Broad spectral density and transfer function of lightly damped SDOF system.*

4.12 DAMPING IN STRUCTURAL SYSTEMS

4.12.1 Introduction

Damping is present in all oscillatory systems and is a measure of a vibrating structure to dissipate energy. We are all aware of the heat developed when we bend a metal wire back and forth quickly for a number of times, and of the sound radiated away from an object given a sharp blow. Also, in many commercial applications special engineered devices are added to decrease their response by increasing damping.

In dynamic analysis, we are concerned with the effects of damping on the response. The primary influence of damping on oscillatory systems is that it reduces the response amplitude. As a result, free vibration dies out when, after initial excitation, the structure is left alone to vibrate. In forced vibrations, damping quickly eliminates the transient part of the response and decreases the response amplitude of the steady-state response. In vibrations generated from periodic loads, the loss of energy is balanced by the energy that is supplied by the excitation.

Damping is primarily important for loads of long duration such as earthquakes and forces on suspension systems. In these cases the response undergoes many reversals during which damping forces dissipate energy. On the contrary, damping can be neglected when the structure is subjected to very short duration loads, e.g., impact loads.

4.12.2 Classification of Damping

Damping can be classified into two broad categories: added and inherent. *Added damping* is generated by specially constructed damping devices attached to the structure, such as viscous and dry friction dampers. Viscous dampers are usually cylindrical systems with a closely fitted piston and filled with fluid. Dry friction dampers resist motion through friction forces developed between several surfaces pressed together under controlled forces. Examples of added damping devices include suspension systems of vehicles and base isolation of machinery and buildings. *Inherent damping* is generated from forces developed within the structure or by the surrounding media, such as molecular friction within the members, interaction with water, air and soil, and friction at the connections between members. In most structures, almost 90% of damping occurs in structural joints. The energy dissipation in a joint is a very complex process that is greatly influenced by the roughness of the material surfaces and the interface pressure.

Forces that cause dissipation of energy are called *damping forces* and they always oppose the motion of the oscillating system. Depending on the source of damping, damping forces are governed by an appropriate law which has usually been developed experimentally. The most common "types" of damping include:

Coulomb Damping: Coulomb damping forces develop at the contact area of two dry surfaces when, under the effect of dynamic forces, they move relatively to each other, as shown in Fig. 4.12.1 for a SDOF subjected to a harmonic load P(t). For example, looseness of joints can be a source of Coulomb damping in structural systems. Coulomb damping is characterized by the coefficient of kinetic friction μ that relates the damping force F_D to the force F_N acting normally to the two surfaces in contact:

$$F_D = \mu F_N \tag{4.12.1}$$

Table 4.12.1 lists representative values of μ for sliding surfaces.

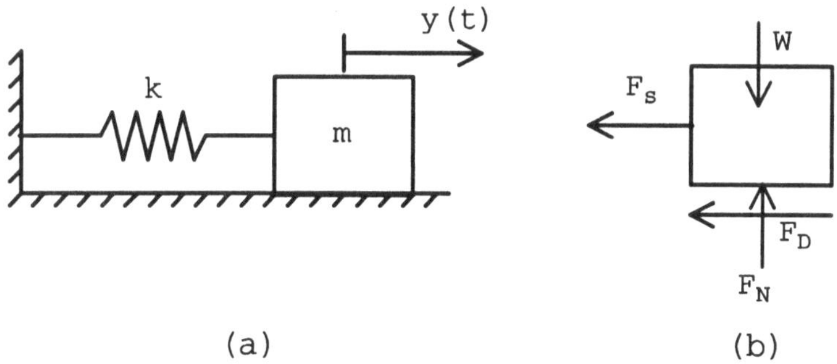

Figure 4.12.1 (a) SDOF system with Coulomb damping; (b) Free-body diagram.

The energy loss per cycle in a SDOF system with Coulomb damping U_c is given by

$$U_c = 4F_D Y_A \tag{4.12.2}$$

where Y_A is the amplitude of the excitation.

Table 4.12.1

Material	μ
Metal with Metal (Lubricated)	0.07
Steel with Steel (Unlubricated)	0.30
Rubber with Steel	1.00
Wood with wood	0.20

Structural or Hysteretic Damping: Experimental studies have shown that most structural metals, such as steel or aluminum, when cyclically stressed, develop stress-strain curves that form a narrow loop, see Fig. 4.12.2. Each area under the loading and the unloading lines is work performed by the external forces on the system. The hatched area within the loop expresses the dissipated energy during one cycle, U_h. This type of damping is called structural or hysteretic, and is a type of inherent damping. Hysteretic damping is characterized by the loss factor η. Table 4.12.2 lists the range of η for several common engineering materials.

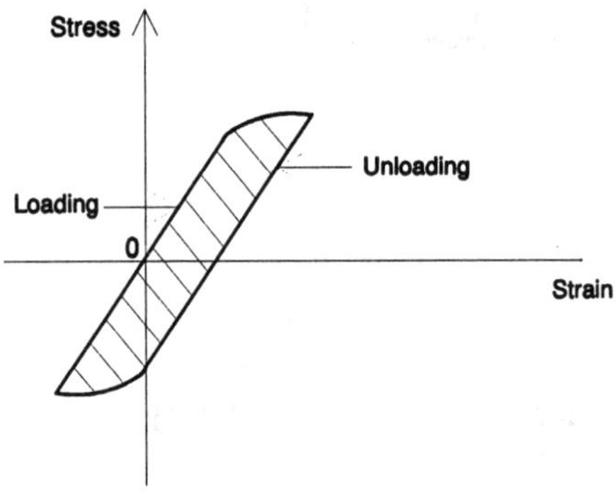

Figure 4.12.2 Hysteretic damping.

Table 4.12.2

Material	Loss Factor (η)
Aluminum - pure	0.00002 - 0.002
Cast iron	0.003 - 0.03
Concrete	0.01 - 0.06
Glass	0.0006 - 0.002
Lead	0.008 - 0.014
Rubber - natural	0.1 - 0.3
Steel	0.001 - 0.008

Experiments have shown that hysteretic damping is independent of the frequency of the exciting source and is proportional to the amplitude of vibration Y_A; thus, for a SDOF system with stiffness k subjected to a harmonic load, the U_h is given by

$$U_h = \pi \eta k Y_A^2 \tag{4.12.3}$$

Radiation Damping: This type of damping represents energy loss due to wave propagation into an infinite or semi-infinite medium. Such cases appear in soil-structure and fluid-structure interaction problems. In the former case, radiation damping expresses the energy loss through propagation of the waves in the soil, and in the latter case, the energy is lost through propagation of the waves into the fluid. Figure 4.12.3(a) shows a rigid circular machine foundation subjected to a harmonic vertical load. The foundation is placed on a deep soil deposit which allows the waves to propagate away from the foundation. Figure 4.12.3(b) shows a SDOF system that simulates the foundation-soil system.

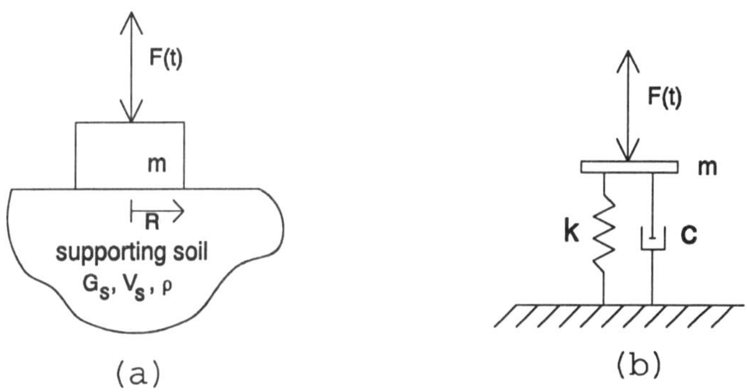

Figure 4.12.3 (a) Schematic of machine foundation; (b) Equivalent SDOF system.

The proper stiffness and damping of the SDOF system are given by

$$k = \frac{4RG_s}{1-\nu_s} \tag{4.12.4}$$

and

$$\xi = 0.425\sqrt{\frac{4\varrho R^3}{(1-\nu_s)m}} \tag{4.12.5}$$

where R is the radius of the circular foundation, m is the effective mass that includes the mass of the foundation and the machine, G_s, ν_s, and ϱ are the shear modulus, Poisson's ratio, and the mass density of the soil, respectively.

The stiffness and damping ratios for horizontal, rocking, and torsional vibrations of rigid foundations with an arbitrary shape as well as elaborate treatments of rigid and flexible foundations subjected to dynamic loads can be found in several references.

Viscous Damping: This commonly used type of damping mechanism was first introduced in Section 4.4. For viscous damping, the damping force F_D is proportional to velocity

$$F_D = c\dot{y} \tag{4.12.6}$$

where c is a constant of proportionality. The energy dissipated per cycle by the damping force in a SDOF system U_v is given by

$$U_v = 4\int_0^{Y_A} c\dot{y}\,dy \tag{4.12.7}$$

If the response is harmonic, $y(t) = Y_A \sin\bar{\omega} t$, evaluation of U_v with eqn. (4.12.7) yields

$$U_v = 4\int_0^{\bar{T}/4} cY_A^2\bar{\omega}^2\cos^2\bar{\omega}t\,dt = \pi c\bar{\omega}Y_A^2 \tag{4.12.8}$$

In view of eqns. (4.4.8) and (4.3.3), c can be expressed as

$$c = 2\xi\sqrt{km} = 2\xi\sqrt{\frac{k^2}{\omega^2}} = 2\xi\frac{k}{\omega} \tag{4.12.9}$$

By substituting c from eqn. (4.12.9) into eqn. (4.12.8) we obtain

$$U_v = 2\pi\xi kY_A^2\frac{\bar{\omega}}{\omega} \tag{4.12.10}$$

Note that $\bar{\omega}$ and ω are the circular frequency of the exciting force and the natural circular frequency of the SDOF system, respectively, and Y_A is the excitation amplitude at steady-state.

Strictly speaking, viscous damping is valid for energy dissipation caused by the laminar flow of a viscous fluid through a slot, as in a shock absorber, or around a piston in a cylinder. Nevertheless, because viscous damping can be expressed in a simple mathematical form that leads to linear equations of motion, see eqn. (4.4.1) and (5.2.1), and also the vibrational behavior of many real systems with complicated damping mechanisms is quite similar to those observed in theoretical systems with viscous damping mechanisms, the other types of damping are often expressed as equivalent viscous.

4.12.3 Equivalent Viscous Damping

Expressing the damping forces as viscous permits a computationally convenient way to handle damping. This is achieved by equating the loss of energy per cycle in a SDOF system with Coulomb U_c, hysteretic U_h, or radiation damping U_r to the dissipated energy per cycle of the SDOF system undergoing a harmonic excitation with viscous damping U_v.

For Coulomb damping, the equivalent viscous damping ratio ξ_{eq} can be obtained by equating U_c given by eqn. (4.12.2) to U_v expressed by eqn. (4.12.10):

$$2\pi k Y_A^2 \frac{\bar{\omega}}{\omega} \xi_{eq} = 4 F_D Y_A \tag{4.12.11}$$

or in a simplified form

$$\xi_{eq} = \frac{2F_D}{\pi k Y_A} \frac{\omega}{\bar{\omega}} \tag{4.12.12}$$

In view of eqn. (4.5.3), equation (4.12.12) can also be written as

$$\xi_{eq} = \frac{2F_D}{\pi k Y_A} \frac{\bar{T}}{T} \tag{4.12.13}$$

For hysteretic damping, the equivalent viscous damping can be obtained by expressing U_h , given by eqn. (4.12.3), as U_v expressed by eqn. (4.12.10):

$$2\pi k Y_A^2 \frac{\bar{\omega}}{\omega} \xi_{eq} = \pi \eta k Y_A^2 \tag{4.12.14}$$

or in a simplified form

$$\xi_{eq} = \frac{\eta}{2} \frac{\omega}{\bar{\omega}} \tag{4.12.15}$$

In view of eqn. (4.5.3), equation (4.12.15) can be written as

$$\xi_{eq} = \frac{\eta}{2} \frac{\bar{T}}{T} \tag{4.12.16}$$

Thus, the presence of another form of damping in a structural system can be expressed as *equivalent viscous damping* ξ_{eq}. In common practice, we usually express all forms of damping in the system in terms of viscous damping. In lieu of experimental data, it is assigned on the basis of former experience. For most applications, except when damping is deliberately introduced into the system, the damping ratio is less than 0.1. Table 4.12.3 provides suggested values of damping ratios ξ for several engineering systems built of common materials and vibrating in the elastic range.

Table 4.12.3

System type	Damping ratio (ξ)
Steel member	0.005 - 0.01
Welded steel system	0.01 - 0.03
Bolted steel system	0.02 - 0.07
Concrete structure	0.01 - 0.05
Timber structure	0.05 - 0.12

5

Dynamic Analysis with Finite Elements

5.1 INTRODUCTION: RECOMMENDED APPROACH

In Chapter 4 we studied the basics of the dynamic response of a SDOF system. We also discussed the physical meaning of the most important parameters that affect the response of a system subjected to dynamic loads.

The choice to perform either static or dynamic analysis, that is, whether to include or neglect the inertia effects and the time variation of the loads, is usually based on engineering judgment. Generally, when the loading is "slowly applied" the dynamic effects can be ignored, e.g., when the rise time of a ramp load is at least three times longer than the fundamental period of a basically vibrating with the first mode structural system, see Fig. 4.8.1(c). However, when the frequency of the excitation exceeds about one-third the structure's fundamental frequency, inertia effects become important and the problem should be treated as dynamic. In any case, the choice of either static or dynamic analysis should be justified, otherwise the analysis results would be meaningless.

Finite element modeling and solution of a dynamics problem is more tedious and time consuming than a statics problem. In addition to the data required for static analysis, dynamic analysis requires information about the mass distribution and damping in the system as well as the time variation of the applied loads. Besides the analytical complexities, verifying and interpreting the solution results is also more difficult to perform.

Caution and thoughtful engineering judgment is required in using a general purpose finite element program to perform dynamic analysis. Ways to avoid errors and also to get a feel for the expected results is to develop the habits of:

a. creating a very simple model, preferably a SDOF system that crudely simulates the dynamic behavior of the system, before starting a dynamic analysis;

b. following a *progressive approach* to solve the problem. With a progressive approach we mean to initially solve the problem using a FEM model with a rather coarse mesh, and proceed to the final solution by successively solving the problem using more refined models. In the successive solutions it is prudent to follow the modeling guidelines suggested in Section 3.4;

c. performing first a static analysis with a selected load when analyzing complex structures. In general, if only the fundamental mode is requested, a model which for a static analysis provides accurate results will also be adequate for dynamic analysis. The static analysis could also provide insight into the structure's behavior and help to identify modeling errors and areas where a more refined mesh would be required to improve accuracy.

In this chapter, we will elaborate on the most significant dynamic parameters of structural systems, the various types of dynamics problems, and their methods of solution. We will also provide a comparison of solution alternatives as an aid to select the most appropriate method for the application.

5.2 EQUATIONS OF MOTION

The equation that governs the dynamic response of a finite element model is given by

$$[M]\{\ddot{y}\} + [C]\{\dot{y}\} + [K]\{y\} = \{F(t)\} \qquad (5.2.1)$$

where

$[M]$	=	mass matrix
$[C]$	=	damping matrix
$[K]$	=	stiffness matrix
$\{y\}$, $\{\dot{y}\}$, $\{\ddot{y}\}$	=	vectors of nodal displacements, velocities and accelerations.
$\{F(t)\}$	=	vector of applied loads.

When a structure is subjected to a ground motion, its response is governed by

$$[M]\{\ddot{u}\} + [C]\{\dot{u}\} + [K]\{u\} = -[M]\{\ddot{y}_g(t)\} \qquad (5.2.2)$$

where

$\{u\},\{\dot{u}\},\{\ddot{u}\}$	=	vectors of nodal relative displacements, velocities, and accelerations.
$\{\ddot{y}_g(t)\}$	=	vector of ground acceleration.

In most dynamics problems the matrices [M], [C], and [K] are independent of time and the system is linear. The linearity of the system allows use of the principle of superposition. In the following, we will study linear systems only. For an in depth presentation of dynamic nonlinear problems the reader can resort to several references, e.g., Bathe, Owen and Hinton.

Equation (5.2.1), usually the most convenient way to express the equilibrium of a FEM model at time t, may also be written as

$$\{F_I\} + \{F_D\} + \{F_S\} = \{F(t)\} \qquad (5.2.3)$$

where $\{F_I\}=[M]\{\ddot{y}\}$, $\{F_D\}=[C]\{\dot{y}\}$, and $\{F_S\}=[K]\{y\}$ are the nodal inertia, damping, and elastic forces, respectively.

Equations (5.2.1) through (5.2.3) express equilibrium of the forces acting on a system. Mathematically, equation (5.2.1) represents a system of second order differential equations which, in principle, can be solved with well-established numerical methods. We will focus our discussion on the most commonly used methods, i.e., time history modal superposition, direct integration, response spectrum, frequency response, and random vibration.

This chapter starts with a discussion on the two exclusively dynamic properties of a system as implemented in finite element analysis: mass and damping. It continues with a presentation on the most widely used methods of dynamic analysis. The solution procedure for each method is demonstrated with an example. Recommendations for problems encountered in practice as well as comparisons between the methods are also presented.

5.3 MASS: LUMPED AND DISTRIBUTED

In static analysis, when we specify a non-zero value for the mass densities of the materials that comprise the system, finite element programs generate nodal loads to simulate the effects of gravity. Similarly, when we request a finite element program to perform a dynamic analysis by inputing non-zero values for the mass densities, a mass matrix is generated to simulate the inertia effects. There is, however, a large difference between the role that mass density plays in statics and dynamics. In statics, consideration of gravity loads is left to the discretion of the engineer. However, whenever in dynamics problems we fail to input the inertia loads the execution of the program is aborted.

Finite element analysis uses two numerical approaches to simulate inertia loads. The first one leads to the formation of a *distributed or consistent* mass matrix and the other one to a *lumped* mass matrix. The lumped mass matrices are diagonal matrices, i.e., all their terms except the ones along the main diagonal are zero, while the distributed mass matrices contain non-zero off-diagonal terms. The development of lumped mass matrices is demonstrated in Chapter 1.

Natural frequencies calculated with a lumped mass formulation are as good or even more accurate than if obtained with distributed mass models. Nevertheless, distributed mass formulations are superior in evaluating mode shapes, strains, and stresses, and require a less refined mesh to model the system. This deficiency of lumped mass formulations can be overcome by using a more refined mesh. Lumped mass matrices are processed by the computer much faster than consistent mass matrices. In problems with many degrees-of-freedom this advantage of lumped mass matrices is very significant in reducing processing time. Frequencies obtained by lumping the masses can be either upper or lower bounds of the system's natural frequencies, that is, they could be either greater or smaller than the natural frequencies obtained experimentally. Nevertheless, acceptable solution accuracy can be obtained if the solution trend is identified by systematically increasing the mesh.

5.4 MODAL ANALYSIS OR MODE SHAPE ANALYSIS: DEFINITIONS AND PHYSICAL INTERPRETATION

In free vibration, the externally applied loads are zero and the structure vibrates under the effect of the initial conditions. Free vibration analysis is rarely performed, since in most cases the structure is subjected to non-zero external loads. Nevertheless, solution of the free vibration problem with zero damping provides the most important dynamic properties of a structure: the natural frequencies and the mode shapes.

More specifically, when we are searching for non-zero solutions of the problem

$$[M]\{\ddot{y}\} + [K]\{y\} = \{0\} \tag{5.4.1}$$

then we are performing *modal analysis*. For an N-degree-of-fieedom system, we may assume that a possible solution is of the form

$$\{y\}_i = \{\phi\}_i \sin(\omega_i t - \alpha_i) \tag{5.4.2}$$

where $\{\phi\}_i$ is the *i-th mode shape* or *mode* with a corresponding natural circular frequency ω_i and phase angle α_i. Substituting eqn. (5.4.2) in eqn. (5.4.1) and eliminating $\sin(\omega_i t-\alpha_i)$ gives

$$([K] - \omega_i^2 [M])\{\phi\}_i = \{0\} \tag{5.4.3}$$

Equation (5.4.3) can be explicitly written as the system of N equations given by eqn. (5.4.4)

$$\begin{bmatrix} k_{11}-\omega^2 m_1 & k_{12} & \cdots & k_{1N} \\ k_{21} & k_{22}-\omega^2 m_2 & \cdots & k_{2N} \\ \cdot & \cdot & \cdots & \cdot \\ \cdot & \cdot & \cdots & \cdot \\ \cdot & \cdot & \cdots & \cdot \\ k_{N1} & k_{N2} & \cdots & k_{NN}-\omega^2 m_N \end{bmatrix} \begin{Bmatrix} \phi_1 \\ \phi_2 \\ \cdot \\ \cdot \\ \cdot \\ \phi_N \end{Bmatrix}_i = \begin{Bmatrix} 0 \\ 0 \\ \cdot \\ \cdot \\ \cdot \\ 0 \end{Bmatrix} \tag{5.4.4}$$

The formulation of eqns. (5.4.3) or (5.4.4) is an important mathematical problem known as *eigenproblem*. The eigenproblem appears in other kinds of engineering analysis besides dynamics. A characteristic example is the calculation of the buckling load in structural stability. A main feature of the eigenvalue problem is that it does not provide a unique solution of the response. More specifically, the $\{\phi\}_i$ that are calculated in modal analysis do not represent the amplitudes of the system under free vibration, but rather normalized amplitude ratios which, when properly combined, can provide the dynamic response of the system.

The objective of modal analysis is to calculate the ω_i and the corresponding $\{\phi\}_i$ that satisfy eqn. (5.4.3). A non-trivial solution, i.e., a solution for which all ω_i and $\{\phi\}_i$ are not zero, requires that the determinant of eqn. (5.4.3) is zero, that is

$$\det([K] - \omega_i^2 [M]) = 0 \tag{5.4.5}$$

Equation (5.4.5) is a polynomial equation of degree N in ω_i^2. This polynomial is known as the *characteristic equation* of the system. For each solution ω_i^2 ($i=1,2,\ldots N$) of the characteristic equation we can solve eqn. (5.4.4) for $\{\phi\}_i$. A solution ω_i^2 of the characteristic equation is called an *eigenvalue*. For each one of the eigenvalues corresponds an eigenvector, that is, for a system with N-degrees-of-freedom there are N eigenvectors and their corresponding eigenvalues. In dynamics, ω_i is called natural circular frequency and the corresponding eigenvector is called mode shape; that is, the terms *eigenvector* and *mode shape* are equivalent. The lowest natural circular frequency ω_1 is called the *fundamental circular frequency*, and its corresponding mode shape $\{\phi\}_1$ is called the *fundamental mode* of vibration.

At this point it may be helpful to present a physical interpretation of a mode shape. Equation (5.4.3) expresses equilibrium between the inertia and stiffness forces in a structure at time t. If we rewrite eqn. (5.4.3) in the form

$$[K]\{\phi\}_i = \omega_i^2 [M]\{\phi\}_i \tag{5.4.6}$$

a modal shape can be viewed as the static deflection resulting from the forces on the right-hand side of eqn. (5.4.6).

Figures 5.4.1(a) and (b) show a simply supported beam modeled with five lumped masses. The first two mode shapes of the beam are shown in Figs. 5.4.1(c) and (d). The points where the deflection of a mode shape of a beam or frame is zero are called *nodes*. In plates, shells, and membrane structures the nodes lie on nodal lines. Three-dimensional solids have nodal surfaces.

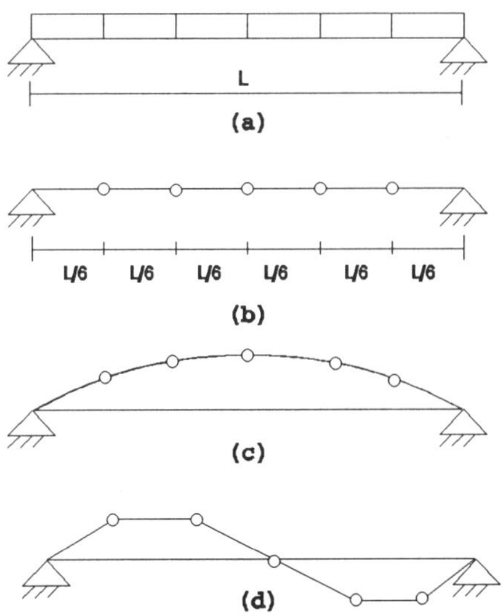

Figure 5.4.1 *(a) Six-element beam model; (b) Lumped mass model; (c) and (d) First and second mode shapes.*

The following example presents: a. the calculation of the natural frequencies and mode shapes of a structure modeled as a two-degree-of-freedom system; b. the definition and calculation of *normalized mode shapes*; and c. the *orthogonality property* of mode shapes.

EXAMPLE 5.4.1. Find the natural frequencies and mode shapes of a cantilever bar fixed at one end and free to vibrate axially, as shown in Fig. 5.4.2. Use two elements and a lumped mass formulation.

($E = 29 \times 10^6$ psi, $L = 480$ in, $A = 2$ in^2, $\varrho = 0.10$ lb-sec^2/in^4)

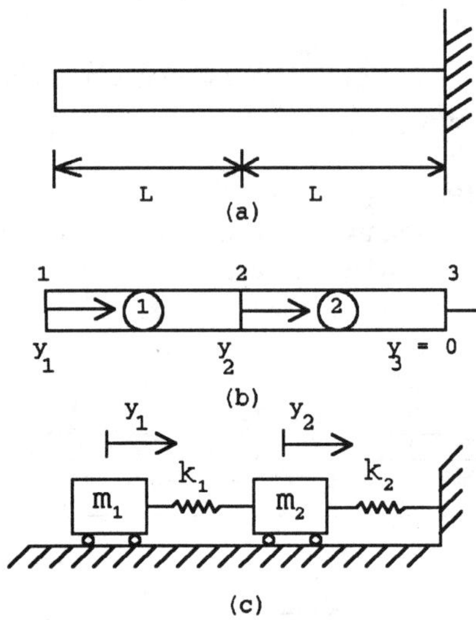

Figure 5.4.2 (a) Cantilever bar; (b) Two-element model; (c) Two-degree-of-freedom spring-mass model.

The solution procedure consists of the following four steps:

1. Formulation of the equation of motion with zero damping and zero external loads.
2. Derivation of the characteristic equation.
3. Calculation of natural frequencies and mode shapes.
4. Normalization of mode shapes.

Step 1. Formulation of the equation of motion with zero damping and zero external forces:

The equation of motion for the two-element model with zero external forces can be obtained from eqn. (1.3.23) for $F_1 = F_2 = F_3 = 0$, that is,

$$\begin{Bmatrix} 0 \\ 0 \\ 0 \end{Bmatrix} = \frac{EA}{L} \begin{bmatrix} 1 & -1 & 0 \\ -1 & 2 & -1 \\ 0 & -1 & 1 \end{bmatrix} \begin{Bmatrix} y_1 \\ y_2 \\ y_3 \end{Bmatrix} + \frac{\varrho AL}{2} \begin{bmatrix} 1 & 0 & 0 \\ 0 & 2 & 0 \\ 0 & 0 & 1 \end{bmatrix} \begin{Bmatrix} \ddot{y}_1 \\ \ddot{y}_2 \\ \ddot{y}_3 \end{Bmatrix} \tag{1.3.23a}$$

For consistency with the developments in Chapter 5, the U_i in eqn. (1.3.23) have been substituted with y_i (i = 1,2,3) in eqn. (1.3.23a).

A solution of eqn. (1.3.23a) is given by

$$\{y\} = \{Y\} \sin(\omega t - \alpha) \tag{5.4.7a}$$

Differentiating eqn. (5.4.7a) twice yields

$$\{\ddot{y}\} = -\omega^2 \{Y\} \sin(\omega t - \alpha) \tag{5.4.7b}$$

Substitution of eqns. (5.4.7a) and (5.4.7b) in eqn. (1.3.23a) and elimination of $\sin(\omega t-\alpha)$ yields

$$\begin{Bmatrix} 0 \\ 0 \\ 0 \end{Bmatrix} = \frac{EA}{L} \begin{bmatrix} 1 & -1 & 0 \\ -1 & 2 & -1 \\ 0 & -1 & 1 \end{bmatrix} \begin{Bmatrix} Y_1 \\ Y_2 \\ Y_3 \end{Bmatrix} - \omega^2 \frac{\varrho AL}{2} \begin{bmatrix} 1 & 0 & 0 \\ 0 & 2 & 0 \\ 0 & 0 & 1 \end{bmatrix} \begin{Bmatrix} Y_1 \\ Y_2 \\ Y_3 \end{Bmatrix} \tag{5.4.7c}$$

Since $Y_3=0$, equation (5.4.7c) can be reduced to

$$\left[\frac{EA}{L} \begin{bmatrix} 1 & -1 \\ -1 & 2 \end{bmatrix} - \omega^2 \frac{\varrho AL}{2} \begin{bmatrix} 1 & 0 \\ 0 & 2 \end{bmatrix} \right] \begin{Bmatrix} Y_1 \\ Y_2 \end{Bmatrix} = \begin{Bmatrix} 0 \\ 0 \end{Bmatrix} \tag{5.4.7d}$$

After performing the matrix multiplications we get:

$$\left(\frac{EA}{L} - \omega^2 \frac{\varrho AL}{2}\right) Y_1 - \frac{EA}{L} Y_2 = 0$$

$$-\frac{EA}{L} Y_1 + \left(\frac{2EA}{L} - \omega^2 \varrho AL\right) Y_2 = 0 \tag{5.4.7e}$$

Step 2. Derivation of the characteristic equation:

For a non-zero solution of eqns. (5.4.7d) or (5.4.7e) the determinant must be zero

$$\det \left| \frac{EA}{L} \begin{bmatrix} 1 & -1 \\ -1 & 2 \end{bmatrix} - \omega^2 \frac{\varrho AL}{2} \begin{bmatrix} 1 & 0 \\ 0 & 2 \end{bmatrix} \right| = 0 \tag{5.4.7f}$$

Let

$$\mu = \frac{\omega^2 (\varrho AL/2)}{EA/L} = \omega^2 \frac{\varrho L^2}{2E} \tag{5.4.7g}$$

Then, eqn. (5.4.7f) can be written as

$$\det \begin{vmatrix} 1-\mu & -1 \\ -1 & 2-2\mu \end{vmatrix} = 0 \tag{5.4.7h}$$

Expansion of the determinant leads to the *characteristic equation*

$$2\mu^2 - 4\mu + 1 = 0 \tag{5.4.7i}$$

Step 3. Calculation of natural frequencies and mode shapes:

The two roots of eqn. (5.4.7i) are $\mu_1 = 0.293$ and $\mu_2 = 1.707$. Substituting μ_1 and μ_2 in eqn. (5.4.7g), we obtain the natural circular frequencies for the fundamental and the second mode

$$\omega_1 = \frac{0.76}{L}\sqrt{\frac{E}{\varrho}} = 27.14 \quad rad/\text{sec}$$

$$\omega_2 = \frac{1.85}{L}\sqrt{\frac{E}{\varrho}} = 65.55 \quad rad/\text{sec} \tag{5.4.7j}$$

The natural circular frequencies can also be obtained through classical analytical methods that consider the beam as a continuous system. Such methods provide the "exact" ω_1 and ω_2 which for our system are given by

$$\omega_1 = \frac{\pi}{4L}\sqrt{\frac{E}{\varrho}} = 27.86 \quad rad/\text{sec}$$

$$\omega_2 = \frac{3\pi}{4L}\sqrt{\frac{E}{\varrho}} = 83.59 \quad rad/\text{sec} \tag{5.4.7k}$$

Thus, the two-element finite element analysis gave an error of 2.6% and 21.6% for the ω_1 and ω_2, respectively.

The corresponding natural frequencies obtained with the two-element model are

$$f_1 = \frac{\omega_1}{2\pi} = 4.32 \quad Hz$$

$$f_2 = \frac{\omega_2}{2\pi} = 10.44 \quad Hz \tag{5.4.7l}$$

and the natural periods

$$T_1 = \frac{1}{f_1} = 0.231 \quad \text{sec}$$
$$T_2 = \frac{1}{f_2} = 0.096 \quad \text{sec} \tag{5.4.7m}$$

In order to solve eqns. (5.4.7e) for the amplitudes Y_1 and Y_2, we observe that by setting the determinant of eqns. (5.4.7e) equal to zero, the system of the two equations is reduced to one independent equation. Thus, by substituting $\omega^2{}_1=(27.14 \text{ rad/sec})^2$ in the second of eqns. (5.4.7e) we get

$$-120833\,Y_{11} + 170949\,Y_{21} = 0 \tag{5.4.7n}$$

A second subscript in Y_1 and Y_2 has been introduced to indicate that the circular frequency ω_1 is used. Since in eqn. (5.4.7n) there are two unknowns and only one equation, we can solve for the relative value of Y_{21} over Y_{11} to obtain

$$\frac{Y_{21}}{Y_{11}} = 0.707$$

It is common to assign a unit value to one of the amplitudes; thus, we set Y_{11} = 1.00 so that

$$Y_{11} = 1.00$$
$$Y_{21} = 0.707$$

The Y_{11} and Y_{21} define the mode shape that corresponds to the first natural frequency.

Similarly, by substituting $\omega^2{}_2$ = $(65.55 \text{ rad/sec})^2$ in the second of eqns. (5.4.7e) we obtain

$$\frac{Y_{22}}{Y_{12}} = -0.707$$

We set Y_{12} = -1.00. Therefore, the Y_{22} and Y_{12} that specify the second mode shape are given by

$$Y_{22} = 0.707$$

$$Y_{12} = -1.00$$

Step 4. Normalization of mode shapes:

As we have seen so far, the amplitudes of a mode shape are relative values which could be normalized as one may choose. An especially convenient normalization for an N-degree-of-freedom system is given by

$$\phi_{ij} = \frac{Y_{ij}}{\left(\sum_{n=1}^{N} m_n \, Y_{nj}^2\right)^{1/2}} \qquad \begin{array}{l} i = 1,2,...N \\ j = 1,2,...N \end{array} \qquad (5.4.7o)$$

The components of the *normalized mode shapes* are denoted as ϕ_{ij} to distinguish them from the Y_{ij}.

For our example eqn. (5.4.7o) gives

$$\phi_{11} = \frac{Y_{11}}{\sqrt{(m_1 Y_{11}^2 + m_2 Y_{21}^2)}} = 0.102, \quad \phi_{12} = -0.102$$

$$\phi_{21} = \frac{Y_{21}}{\sqrt{(m_1 Y_{11}^2 + m_2 Y_{21}^2)}} = 0.072, \quad \phi_{22} = 0.072$$

where $m_1 = 0.5\varrho AL$ = 48.0 lb-sec²/in, and $m_2 = 2m_1$ are the lumped masses at nodes 1 and 2, respectively. The first and second normalized mode shapes can be expressed in the form

$$\{\phi\}_1 = \begin{Bmatrix} \phi_{11} \\ \phi_{21} \end{Bmatrix} = \begin{Bmatrix} 0.102 \\ 0.072 \end{Bmatrix}$$

$$\{\phi\}_2 = \begin{Bmatrix} \phi_{12} \\ \phi_{22} \end{Bmatrix} = \begin{Bmatrix} -0.102 \\ 0.072 \end{Bmatrix} \qquad (5.4.7p)$$

A graphical representation of the two mode shapes is shown in Fig. 5.4.3.

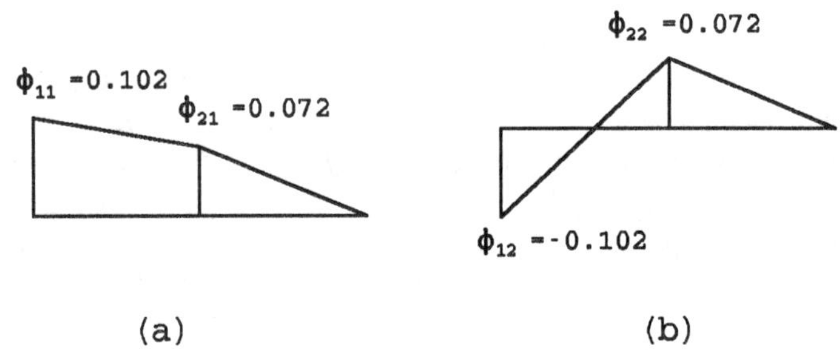

Figure 5.4.3 *Mode shapes: (a) Fundamental; (b) Second.*

For an N-degree-of-freedom system, all mode shapes can also be expressed in the form of the *modal matrix* $[\Phi]$

$$[\Phi] = \begin{bmatrix} \phi_{11} & \phi_{12} & \cdots & \phi_{1N} \\ \phi_{21} & \phi_{22} & \cdots & \phi_{2N} \\ . & . & . & . \\ \phi_{N1} & \phi_{N2} & \cdots & \phi_{NN} \end{bmatrix} \qquad (5.4.7q)$$

In our case eqn. (5.4.7q) becomes

$$[\Phi] = \begin{bmatrix} 0.102 & -0.102 \\ 0.072 & 0.072 \end{bmatrix} \qquad (5.4.7r)$$

The most important characteristic of mode shapes is that they satisfy the orthogonality conditions. For an N-degree-of-freedom system the orthogonality conditions are given by

$$\begin{aligned} \{\phi\}_i^T[M]\{\phi\}_j &= 0 \quad \text{for} \quad i \neq j \\ \{\phi\}_i^T[M]\{\phi\}_j &= 1 \quad \text{for} \quad i = j \end{aligned} \tag{5.4.7s}$$

and

$$\begin{aligned} \{\phi\}_i^T[K]\{\phi\}_j &= 0 \quad \text{for} \quad i \neq j \\ \{\phi\}_i^T[K]\{\phi\}_j &= \omega_i^2 \quad \text{for} \quad i = j \end{aligned} \tag{5.4.7t}$$

where i,j = 1, 2, ..., N, and the $\{\phi\}_i^T$ are the rows of the transpose of the modal matrix $[\Phi]^T$.

In order to verify the orthogonality condition for our example, set i=1 and j=1 in eqn. (5.4.7s), to get

$$\lfloor 0.102 \quad 0.072 \rfloor \begin{bmatrix} 48 & 0 \\ 0 & 96 \end{bmatrix} \begin{Bmatrix} 0.102 \\ 0.072 \end{Bmatrix} = 1$$

and for i=1 and j=2, to obtain

$$\lfloor 0.102 \quad 0.072 \rfloor \begin{bmatrix} 48 & 0 \\ 0 & 96 \end{bmatrix} \begin{Bmatrix} -0.102 \\ 0.072 \end{Bmatrix} = 0$$

We will refer again to the orthogonality property when we discuss the time history modal superposition, frequency response analysis, response spectrum, and random vibration analysis.

Practical Considerations for Modal Analysis

Regardless of what type of analysis is selected to solve a linear dynamics problem, modal analysis should be the first step. Natural frequencies and mode shapes are required for time history modal superposition, frequency response, response spectrum, and random vibration analysis. Even for time history direct integration, that does not require modal analysis, prior calculation of a few modes is recommended, since knowledge of the natural frequencies can help to select the proper time step.

Modal analysis calculates the lowest natural frequencies best, and provides less accuracy in evaluating higher natural frequencies and mode shapes. As a result, there should be sufficient justification to calculate higher modes and include them in subsequent dynamic analysis, e.g., time history modal superposition, response spectrum, and random vibration analysis. As a general rule, the number of modes to request in modal analysis depend on the natural frequencies of the system as related to the frequency content and the spatial variation of the applied loads. The loads with frequencies that are close to the natural frequencies of the system affect the response due to large values of the dynamic magnification factor as elaborated in Section 4.6.2 for a SDOF system. The spatial variation of the loads has an effect on which modes will be mostly excited. For example, a concentrated load acting at the center of a simply supported beam excites only the odd modes, see Section 8.2. The following classification provides additional in-sight in deciding on the proper number of modes that should be obtained with modal analysis.

In general, dynamics problems can be classified into *wave propagation* and *structural dynamics* problems. A typical example of a wave propagation problem is the study of the waves generated in a plate excited by dynamic loads. The evaluation of the plate's response, after the waves have propagated in the whole plate have reflected and refracted on its boundaries, is a structural dynamics problem. The primary difference between wave propagation and structural dynamics is that in wave propagation problems a larger number of mode shapes are excited in the system. In structural dynamics, only the lower and the intermediate modes must usually be determined.

Modal analysis also provides useful insight into the dynamic behavior of a structure. Examination of the mode shapes can help to identify modeling errors such as incorrect boundary conditions and nodal connectivity. Inadvertent omission of restraining nodal displacements or rotations can be identified through rigid body modes of the whole or parts of the model. Additional remarks and recommendations to perform modal analysis are provided in Section 7.1.

5.5 MODAL ANALYSIS WITH LOAD STIFFENING

The presence of an axial load in a beam affects its flexural response. The degree that compressive static loads influence the flexural behavior leading to *buckling* is greatly dependent upon the way the beam is connected to the rest of the structural system and its slenderness ratio L/r, where r is given by

$$r = \sqrt{\frac{I}{A}} \tag{5.5.1}$$

with L denoting the unsupported length of the beam, I the moment of inertia, and A the cross-sectional area. Extensive information on buckling caused by static loads is available in the literature, e.g., Chajes.

As for the static case, the presence of axial loads also affects the dynamic flexural response of slender beams. More specifically, compressive loads decrease the flexural stiffness of the beam, while the reverse occurs when the axial loads are tensile. Figures 5.5.1(a) and (b) show the fundamental flexural mode of the same beam under a tensile and a compressive load P, respectively. The fundamental natural frequency of the beam in tension is greater than the fundamental frequency of the beam in compression. In both cases the magnitude of the axial load is kept constant with time.

Both tensile and compressive axial loads can have a significant effect on the flexural dynamic response of slender beams. Similarly, tensile and compressive loads can significantly affect the response of structures that include slender plate and shell components. In practice, however, compressive loads present more interest, since they may lead to failure of the system caused by excessive flexural vibrations. It should be noted that compressive loads primarily affect the fundamental frequency with a lesser effect on higher natural frequencies. The effect is substantial when the axial load is close to the buckling load. Thus, in the presence of large compressive loads, it is prudent to perform a buckling analysis in order to calculate the critical load before a dynamic analysis is carried out. An example in Section 7.7 demonstrates the significance that axial loads play on flexural natural frequencies and mode shapes of beams.

Finally, it should be noted that the flexural response of systems comprised of either beams having small slenderness ratios or thick plates and shells is practically not affected by the presence of axial loads.

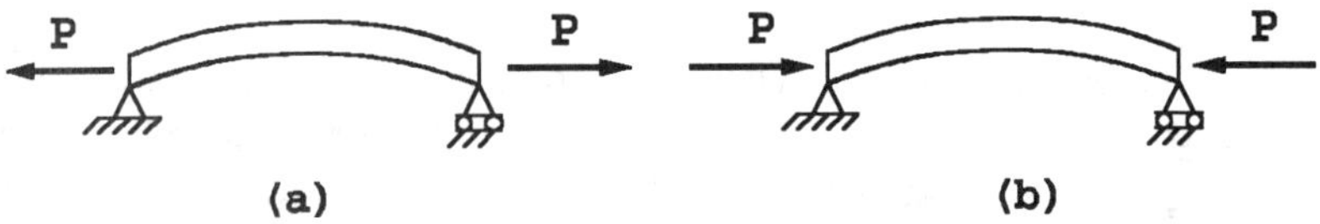

Figure 5.5.1 Fundamental flexural mode of beam subjected to a constant axial load: (a) Tensile; (b) Compressive.

5.6 STURM FREQUENCY CHECK

In Example 5.4.1 the natural frequencies were obtained from the characteristic equation (5.4.7i). Finite element programs employ efficient algorithms to calculate natural frequencies and mode shapes. These algorithms usually include procedures to examine if any natural frequencies were missed in modal analysis. Such a procedure is the Sturm frequency check.

The computational effort to perform the Sturm frequency check is considerable and should be used only when it appears to be necessary. A Sturm frequency check is suggested in the following three cases:

a. when modal analysis is performed to a relatively refined model of the structural system for the first time.
b. when the natural frequencies of the system are closely spaced.
c. when the calculated frequencies appear to be higher than anticipated.

5.7 CUT-OFF FREQUENCY

As we will discuss in more detail in Section 5.11, in many applications only a few low modes contribute significantly to the response. Thus, in many cases calculation of higher modes is unnecessary. The highest frequency in the analysis is called the cut-off frequency f_c. In finite element programs, specification of f_c is used to terminate the computation of all natural frequencies that are greater than f_c. Only those modes whose frequencies are less than the f_c are then used in subsequent dynamic analysis.

The following suggestions can be of help for an effective use of f_c in modal analysis of large systems:

a. When modal analysis results will be used in a subsequent dynamic analysis of a structure subjected to a seismic excitation, a cut-off frequency of 50 Hz is usually sufficient.

b. When modal analysis results will be used to calculate the response of machinery parts, then a cut-off frequency of as high as 250 Hz may be selected.

5.8 RIGID BODY MODES

The natural frequencies of a system can either be positive or zero. The frequencies that are zero correspond to rigid body modes, i.e., modes in which none of the elastic elements is deformed. A representative example is an airplane body and wings modeled as a flexible beam with three lumped masses. The mass of the fuselage is denoted as m_f, and m_w represents the mass of one wing. The first three modes of the three lumped mass model are shown in Fig. 5.8.1. Notice that the first two modes are rigid body modes with frequencies equal to zero.

Three-dimensional models that are deliberately unsupported exhibit up to six rigid body modes. Any additional zero natural frequencies either correspond to parts of the model that respond as mechanisms or have been introduced inadvertently by not specifying sufficient boundary conditions to restrain rigid body displacements. In the latter case, errors in providing inappropriate connectivity between structural parts can be detected by examining the corresponding mode shapes. Rigid body modes can also be present in structures supported on soft springs, see Fig. 5.8.2.

An example of a structure with rigid body modes is presented in Chapter 7.

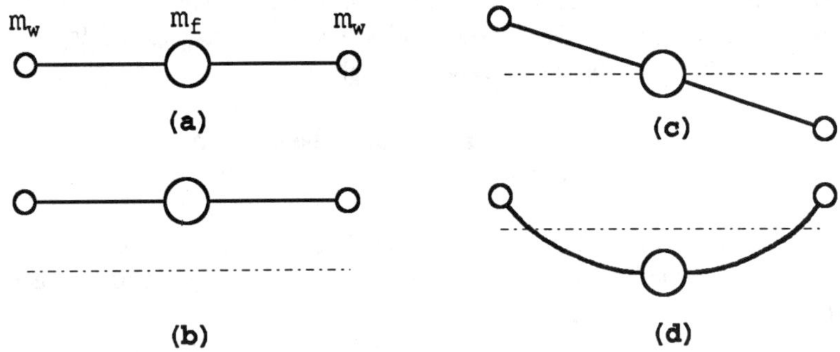

Figure 5.8.1 (a) Simple airplane model; Three lowest modes: (b) Fundamental mode (rigid body translation); (c) Second mode (rigid body rotation); (d) Third mode (flexural deformation).

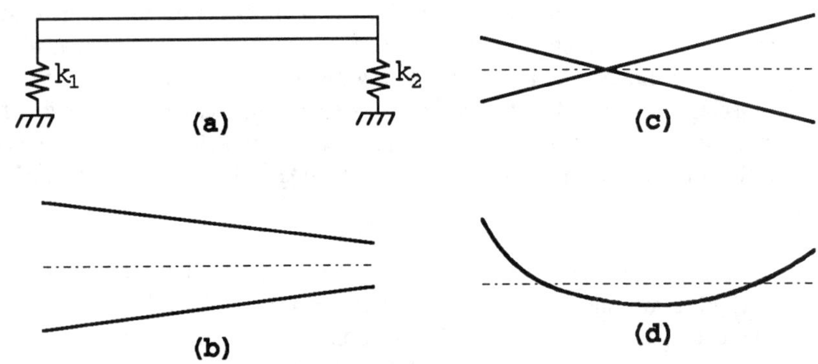

Figure 5.8.2 (a) Beam on unequal "soft" springs; Three lowest modes: (b) First mode (rigid body translation); (c) Second mode (rigid body rotation); (d) Third mode (flexural deformation).

5.9 HOW TO INCORPORATE DAMPING IN FEM

In modal analysis damping is neglected because it is usually small and practically does not affect the calculation of natural frequencies and mode shapes, see Section 4.4. Also, damping can be neglected in systems with viscous damping ratios less than 0.02, provided the excitation is harmonic with frequency ratios that are not in the range of 0.5 to 1.45, see Section 4.6.2. Nevertheless, damping is one of the primary concerns in dynamic analysis and is accounted for with the damping matrix [C]. In general, the matrix [C] cannot be constructed from element damping matrices, as is the case for the mass and stiffness matrices. The most effective computational procedure to account for damping is to express [C] either as an algebraic combination of the mass and stiffness matrices of the system or in terms of damping ratios and natural circular frequencies. These procedures are elucidated in Sections 5.10 and 5.12.

In response spectrum analysis, damping is incorporated into the response spectra. As presented in Section 4.9, one of the parameters in selecting the proper spectra is the amount of damping anticipated in the system. In time history modal superposition, frequency response, and random vibration, damping is usually expressed in the form of equivalent viscous, while Rayleigh damping is usually used in time history direct integration. Particular details about consideration of damping for each type of dynamic analysis are given in Chapters 5, 8, and 9.

Regarding the significance that accurate consideration of damping has on the solution accuracy, the rule is that the longer the duration of the dynamic loads the greater the effect of damping on the system's response. Thus, for blast or shock loadings damping plays very little role on the response and as a result it is usually neglected. For seismic loads, damping can significantly reduce the response and therefore it must be included in the analysis. Damping should be accurately specified for the modes that have the largest contribution to the response. These modes can be identified by comparing the natural frequencies of the structure to the frequencies of the harmonic terms of the Fourier analysis of the dynamic loads. Thus, if a Fourier series expansion or a Fourier transform of the loads shows that only frequencies up to f_s are included in the loading, the modal analysis should accurately calculate modes up to $f_c=4f_s$. If the frequency content of the load is not determined through a Fourier transform or a series expansion, then an estimate of the f_c can be obtained from the suggestions given in Section 5.7.

Finally, if discrete damping elements are used, these elements must be assigned appropriate damping coefficients. Use of discrete elements arises in vehicle dynamics, machine foundations, and seismic analysis when soil modeling is necessary.

5.10 TIME HISTORY MODAL SUPERPOSITION

Time history modal superposition is probably the most commonly used method to calculate the response of systems of finite extent subjected to loads with a known time variation. The method, strictly applicable to linear systems, requires prior modal analysis. For systems with infinite or semi-infinite extent which are usually encountered in soil-structure and fluid-soil-structure interaction the method is no longer applicable since the notion of "mode shape" looses its meaning. However, even such systems could be modeled with properly truncated regions of the infinite or semi-infinite media, e.g., soil and fluid, and analyzed with time history modal superposition. The procedure to obtain a system's response with time history modal superposition is described below.

As we saw in Section 5.2, eqn. (5.2.1) represents a system of N coupled ordinary differential equations. Being coupled implies that they cannot be solved independently from each other. This computational difficulty is circumvented by modal analysis in conjunction with the orthogonality property of the mode shapes. We will demonstrate that the response of a structural system governed by eqn. (5.2.1) can be obtained as the superposition of the solution of N independent equations. Each one of the independent equations is the equation of motion of a SDOF system which can be solved with a step-by-step integration scheme. The procedure of obtaining the response of a multi-degree-of-freedom system to a dynamic load as the superposition of the responses of N single-degree-of-freedom systems is called time history modal superposition. More specifically the method employs integration schemes to combine mode shapes and calculate the response at specified time intervals for the desired time duration.

Damping is assumed to be viscous, that is

$$\begin{aligned} \{\phi\}_i^T [C] \{\phi\}_j &= 2\omega_i \xi_i \quad \textit{for } i=j \\ \{\phi\}_i^T [C] \{\phi\}_j &= 0 \quad \textit{for } i \neq j \end{aligned} \tag{5.10.1}$$

where ξ_i is the modal damping ratio of the i-th mode. As elaborated in Example 5.10.1, this assumption also allows modal decomposition for damping. For a discussion on the selection of ξ_i refer to Section 4.12. Note that for every mode, ξ_i can be different; however, common practice is to assign the same value of ξ_i to all modes.

There are two implications in using eqn. (5.10.1) in time history: First, the total damping in the structure is accounted for as the sum of the individual damping assigned to each mode. Second, when solving the decoupled equations of motion as presented in Example 5.10.1, it is not necessary to form the damping matrix [C] but only the stiffness and mass matrices [K] and [M].

The following example demonstrates the time history modal superposition by determining the response of the cantilever bar of Example 5.4.1 subjected to a suddenly applied load. For every step of the solution applied to the two-degree-of-freedom system we also provide the corresponding step for an N-degree-of-freedom system.

EXAMPLE 5.10.1. Determine the response of the cantilever bar of Example 5.4.1 subjected to a suddenly applied (step) load at the free end as shown in Fig. 5.10.1. Assume zero initial conditions at t = 0 and assign 10% damping ($\xi_1 = \xi_2 = \xi = 0.1$) to each mode. The amplitude of the step load is F_0 = 1000 lb. Use the two-element model of Example 5.4.1.

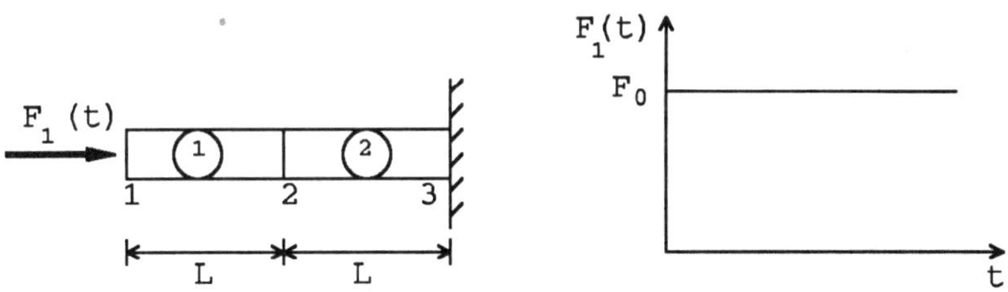

Figure 5.10.1 Cantilever subjected to a suddenly applied load.

Italics are used to outline the procedure for the cantilever beam modeled as an N-degree-of-freedom system.

The analytical solution consists of the following four steps:

1. Formulation of the equation of motion.
2. Modal analysis.
3. Uncoupling of the equations of motion.
4. Combination of the modal responses.

Step 1. Formulation of the equation of motion:

After applying the boundary conditions, i.e., $U_3=0$ and the nodal forces $F_2(t)=0$ and $F_1(t)$, the eqn. (1.3.24) for the cantilever becomes

$$\frac{\varrho AL}{2}\begin{bmatrix}1 & 0\\ 0 & 2\end{bmatrix}\begin{Bmatrix}\ddot{y}_1\\ \ddot{y}_2\end{Bmatrix}+\begin{bmatrix}C_{11} & C_{12}\\ C_{21} & C_{22}\end{bmatrix}\begin{Bmatrix}\dot{y}_1\\ \dot{y}_2\end{Bmatrix}+\frac{EA}{L}\begin{bmatrix}1 & -1\\ -1 & 2\end{bmatrix}\begin{Bmatrix}y_1\\ y_2\end{Bmatrix}=\begin{Bmatrix}F_1(t)\\ 0\end{Bmatrix} \tag{5.10.2a}$$

Notice that in eqn. (5.10.2a) we have also added viscous damping and U_i has been substituted with y_i ($i = 1,2$).

Step 2. Modal analysis:

This step has been presented in Example 5.4.1. The analysis provided the modal matrix $[\Phi]$ and the natural circular frequencies given by eqns. (5.4.7r) and (5.4.7j), respectively.

Step 3. Uncoupling of the equations of motion:

To uncouple eqns. (5.10.2a), we introduce the transformation

$$\{y_i(t)\} = [\Phi]\,\{z_i(t)\} \tag{5.10.2b}$$

for which differentiation with respect to time yields

$$\{\dot{y}_i(t)\} = [\Phi]\,\{\dot{z}_i(t)\} \tag{5.10.2c}$$

and

$$\{\ddot{y}_i(t)\} = [\Phi]\,\{\ddot{z}_i(t)\} \tag{5.10.2d}$$

where $z_i(t)$, $\dot{z}_i(t)$ and $\ddot{z}_i(t)$ are time varying functions.

For the two-degree-of-freedom system, eqn. (5.10.2b) is given by

$$\begin{Bmatrix} y_1(t) \\ y_2(t) \end{Bmatrix} = \begin{Bmatrix} 0.102 & -0.102 \\ 0.072 & 0.072 \end{Bmatrix} \begin{Bmatrix} z_1(t) \\ z_2(t) \end{Bmatrix} \qquad \textit{(5.10.2e)}$$

If we substitute eqns. (5.10.2b), (c), and (d) in equation (5.10.2a), then premultiply eqn. (5.10.2a) with $[\Phi]^T$ and use the orthogonality property expressed by eqn. (5.4.7s), eqn. (5.4.7t), and eqn. (5.10.1), we get

$$\begin{bmatrix} 1 & 0 \\ 0 & 1 \end{bmatrix} \begin{Bmatrix} \ddot{z}_1 \\ \ddot{z}_2 \end{Bmatrix} + \begin{bmatrix} 2\xi_1\omega_1 & 0 \\ 0 & 2\xi_2\omega_2 \end{bmatrix} \begin{Bmatrix} \dot{z}_1 \\ \dot{z}_2 \end{Bmatrix} + \begin{bmatrix} \omega_1^2 & 0 \\ 0 & \omega_2^2 \end{bmatrix} \begin{Bmatrix} z_1 \\ z_2 \end{Bmatrix}$$

$$= \begin{bmatrix} 0.102 & 0.072 \\ -0.102 & 0.072 \end{bmatrix} \begin{Bmatrix} F_1(t) \\ 0 \end{Bmatrix} \qquad \textit{(5.10.2f)}$$

For an N-degree-of-freedom system eqn. (5.10.2f) takes the form

$$\begin{bmatrix} 1 & 0 & ...0 \\ 0 & 1 & ...0 \\ . & . & . \\ 0 & 0 & 1 \end{bmatrix} \begin{Bmatrix} \ddot{z}_1 \\ \ddot{z}_2 \\ . \\ \ddot{z}_N \end{Bmatrix} + \begin{bmatrix} 2\xi_1\omega_1 & 0 & ...0 \\ 0 & 2\xi_2\omega & ...0 \\ . & . & . \\ 0 & 0 & 2\xi_N\omega_N \end{bmatrix} \begin{Bmatrix} \dot{z}_1 \\ \dot{z}_2 \\ . \\ \dot{z}_N \end{Bmatrix} + \begin{bmatrix} \omega_1^2 & 0 & ..0 \\ 0 & \omega_2^2 & ..0 \\ . & . & . \\ 0 & 0 & \omega_N^2 \end{bmatrix} \begin{Bmatrix} z_1 \\ z_2 \\ . \\ z_N \end{Bmatrix}$$

$$= \begin{bmatrix} \phi_{11} & \phi_{21} & ...\phi_{N1} \\ \phi_{12} & \phi_{22} & ...\phi_{N2} \\ . & . & . \\ \phi_{1N} & \phi_{2N} & ...\phi_{NN} \end{bmatrix} \begin{Bmatrix} F_1(t) \\ F_2(t) \\ . \\ F_N(t) \end{Bmatrix} \qquad \textit{(5.10.2g)}$$

Notice that the matrices on the right-hand side of eqns. (5.10.2f) and (5.10.2g) are the $[\Phi]^T$. By performing the matrix multiplications indicated in eqn. (5.10.2f) we obtain the uncoupled system of equations

$$\begin{aligned} \ddot{z}_1 + 2\xi\omega_1\dot{z}_1 + \omega_1^2 z_1 &= 0.102\,F_1(t) \\ \ddot{z}_2 + 2\xi\omega_2\dot{z}_2 + \omega_2^2 z_2 &= -0.102\,F_1(t) \end{aligned} \qquad (5.10.2h)$$

Reducing the coupled system of equations of motion, eqn. (5.10.2a), to a system of uncoupled equations, eqns. (5.10.2h), is probably the most significant step of time history modal superposition.

For an N-degree-of-freedom system the uncoupled equations of motion are given by

$$\ddot{z}_i + 2\xi\omega_i\dot{z}_i + \omega_i^2 z_i = \phi_{1i}F_1 + \phi_{2i}F_2 + \ldots + \phi_{Ni}F_N \qquad (i = 1,2,3,\ldots,N)$$

or in a concise form

$$\ddot{z}_i + 2\xi_i\omega_i\dot{z}_i + \omega_i^2 z_1 = \sum_{r=1}^{N} \phi_{ri}F_r \qquad (5.10.2i)$$

If each force can be expressed in terms of the same function of time g(t) so that $F_r = f_r\, g(t)$, then eqn. (5.10.2i) takes the form

$$\sum_{r=1}^{N} \phi_{ri}F_r = g(t)\Gamma_i \qquad (5.10.2j)$$

where

$$\Gamma_i = \sum_{r=1}^{N} \phi_{ri}\, f_r$$

The term Γ_i is known as the modal participation factor.

If we assume that the system is initially at rest, the solution of eqns. (5.10.2h) is obtained from eqn. (4.8.1) which takes the following form:

$$z_1(t) = \frac{102}{\omega_{D1}} \int_0^t e^{-\xi\omega_1(t-\tau)} \sin\omega_{D1}(t-\tau)d\tau$$

$$z_2(t) = -\frac{102}{\omega_{D2}} \int_0^t e^{-\xi\omega_2(t-\tau)} \sin\omega_{D2}(t-\tau)d\tau \qquad (5.10.2k)$$

where ω_{D1} and ω_{D2} are defined in eqn. (4.4.9), that is

$$\omega_{D1} = \omega_1\sqrt{(1-\xi^2)}$$

$$\omega_{D2} = \omega_2\sqrt{(1-\xi^2)} \qquad (4.4.9)$$

For this simple case, evaluation of the above integrals can be expressed in a closed form

$$z_1(t) = \frac{102}{\omega_1^2}\left\{1 - e^{-\xi\omega_1 t}\left[\cos\omega_{D1}t + \left(\frac{\xi\omega_1}{\omega_{D1}}\right)\sin\omega_{D1}t\right]\right\}$$

$$z_2(t) = -\frac{102}{\omega_2^2}\left\{1 - e^{-\xi\omega_2 t}\left[\cos\omega_{D2}t + \left(\frac{\xi\omega_2}{\omega_{D2}}\right)\sin\omega_{D2}t\right]\right\} \qquad (5.10.2l)$$

Step 4. Combination of the modal responses:

Finally, the response of the cantilever is obtained by substituting the $z_1(t)$ and $z_2(t)$ given by eqn. (5.10.2l) in eqn. (5.10.2e)

$$\begin{Bmatrix} y_1(t) \\ y_2(t) \end{Bmatrix} = \begin{Bmatrix} 0.102 \\ 0.072 \end{Bmatrix} z_1(t) + \begin{Bmatrix} -0.102 \\ 0.072 \end{Bmatrix} z_2(t) \qquad (5.10.2m)$$

Clearly, eqn. (5.10.2m) indicates that the response is obtained as a superposition of the modal responses. *Modal response* is defined as the product of the the i-th mode shape with the corresponding $z_i(t)$. If we perform the multiplications indicated in eqn. (5.10.2m), the solution can also be written in the form

$$\begin{aligned} y_1(t) &= 0.102 z_1(t) - 0.102 z_2(t) \\ y_2(t) &= 0.072 z_1(t) + 0.072 z_2(t) \end{aligned} \tag{5.10.2n}$$

For an N-degree-of-freedom system the counterpart of eqn. (5.10.2m) is given by

$$\begin{Bmatrix} y_1 \\ y_2 \\ \cdot \\ \cdot \\ y_N \end{Bmatrix} = \begin{Bmatrix} \phi_{11} \\ \phi_{21} \\ \cdot \\ \cdot \\ \phi_{N1} \end{Bmatrix} z_1(t) + \begin{Bmatrix} \phi_{12} \\ \phi_{22} \\ \cdot \\ \cdot \\ \phi_{N2} \end{Bmatrix} z_2(t) + \ldots + \begin{Bmatrix} \phi_{1N} \\ \phi_{2N} \\ \cdot \\ \cdot \\ \phi_{NN} \end{Bmatrix} z_N(t) \tag{5.10.2o}$$

5.11 PRACTICAL CONSIDERATIONS FOR TIME HISTORY MODAL SUPERPOSITION

Finite element programs use efficient algorithms to perform the "uncoupling of the equations of motion" and the "superposition of the independent solutions" described in Example 5.10.1. The response is not provided in a closed form but rather at user defined time increments (time steps) of duration Dt. Such efficient algorithms include the Wilson's θ and the Newmark's β methods. Both methods are *unconditionally stable*, which implies that they provide the response of the system regardless of the selected time step Dt. However, selection of an arbitrary value for Dt does not warrant that the obtained response is accurate. The selection of Dt is very critical, since it affects both the computational efficiency and the solution accuracy. A large value for Dt will provide inaccurate results, while a very small Dt will unnecessarily increase the computational time and cost, without a justifiable improvement in solution accuracy. A recommendation for the selection of a proper Dt is given in the following.

A five-step procedure can be used to obtain the response of a system subjected to dynamic loads with time history modal superposition. Even though the first step can be omitted, its execution greatly expedites the solution, since it can be of help in developing the proper model and determining the appropriate natural frequencies and mode shapes through modal analysis.

1. Determine the frequency content of the dynamic loads. For a periodic loading this can be easily accomplished with a Fourier series expansion. When the load is transient the frequency content can be obtained with a Fast Fourier Transform, see Section 4.11.1. For several common types of dynamic loads, approximate upper limits of the frequency content have been presented in Section 5.7.

2. Perform modal analysis on a finite element model that can accurately calculate natural frequencies up to f_c. For a discussion on the critical frequency f_c and selection of the number of modes to request in modal analysis refer to Sections 5.4, 5.7, and 5.9.

3. Select a time step Dt that minimizes the computational inaccuracies inherent in the integration scheme used in time history modal superposition. As a general rule, a time step $Dt<0.05T_c$, where $T_c=1/f_c$, provides accurate results. In most cases, however, a $Dt<0.1T_c$ is sufficient.

4. Select the appropriate values for the damping ratio ξ according to the recommendations given in Section 4.12.

5. Define the duration that the response should be evaluated. In selecting the duration of the analysis, we must consider the type and duration of the load as well as the amount of damping in the system. For lightly damped systems ($0.0<\xi<0.05$), the analysis could be performed for about 10 T_1, where T_1 is the fundamental period of the system. For more heavily damped systems, the analysis could be performed for about $5T_1$ to $8T_1$, see Fig. 4.6.3. For impact loads we ususally ignore damping and calculate the response for a duration slightly greater than the duration of the load. A discussion on the effect of loading duration on the response is presented in Section 4.6.3. The discussion in Section 4.6.3 refers to harmonic loads. The conclusions, however, apply as a conservative extension to general excitations.

Examples in Chapter 8 illustrate various aspects of the five-step procedure.

5.12 TIME HISTORY DIRECT INTEGRATION

An alternative method to obtain the response of a system to dynamic loads is the time history direct integration. This method uses time step-by-step integration algorithms to solve the system of the coupled equations of motion, i.e., eqns. (5.2.1). Such algorithms include Wilson's θ and Newmark's β methods that calculate the response at specified time intervals. As mentioned in the previous section, both methods are unconditionally stable, which means that whatever the selection of the time interval Dt, the solution scheme will provide an answer. However, this answer may not necessarily be correct.

As in modal superposition, an accurate solution is obtained with a $Dt<0.05T_c$. For most applications, however, we can obtain a sufficiently accurate solution with a Dt that satisfies the relationship

$$Dt \leq 0.1\,T_c \tag{5.12.1}$$

where T_c is the smallest natural period that should be accounted for in the analysis. Suggestions for selecting the proper f_c and the corresponding T_c are presented in Sections 5.4, 5.7, and 5.9. A larger value of Dt could still provide a sufficiently accurate solution, but contribution of modes with natural frequencies greater than T_c will be *filtered*, i.e., their contribution to the response will be significantly decreased.

One of the main differences between time history modal superposition and direct integration is the way damping is treated by the two methods. In principle, direct integration does not require the damping in the system to be viscous; the damping matrix does not have to satisfy any orthogonality properties that allow decoupling of eqn. (5.2.1). In practice, however, the damping matrix is constructed by assigning selected values to the damping ratios ξ_i. In direct integration *Rayleigh damping* is used. For Rayleigh damping, the damping matrix is given by

$$[C] = \alpha[M] + \beta[K] \tag{5.12.2}$$

The constants α and β are calculated from the system of equations

$$\alpha + \beta \omega_i^2 = 2\omega_i \xi_i \tag{5.12.3}$$

where ω_i is obtained through modal analysis, and ξ_i are damping ratios specified by the analyst. From eqn. (5.12.3), we can easily prove that if we assign two damping ratios ξ_i and ξ_j corresponding to the natural circular frequencies ω_i and ω_j, the α and β can be determined from

$$\begin{aligned} \alpha &= \frac{2\omega_i\omega_j}{\omega_j^2 - \omega_i^2}(\omega_j\xi_i - \omega_i\xi_j) \\ \beta &= \frac{2}{\omega_j^2 - \omega_i^2}(\omega_j\xi_j - \omega_i\xi_i) \end{aligned} \tag{5.12.4}$$

Rayleigh damping, being a linear combination of the mass and stiffness matrices, fulfills the orthogonality conditions (5.4.7s) and (5.4.7t), which in turn imply that Rayleigh damping allows decoupling of the equations of motion. It is important to note that if $[C]=\alpha[M]$, $(\beta=0)$, the higher modes of the structure will be assigned very little damping, while for $[C]=\beta[K]$, $(\alpha=0)$, the higher modes will be heavily damped. Thus, by giving appropriate values to α and β, we can filter or retain the effect of the higher modes.

A numerical example demonstrating the procedure to calculate α and β is also presented in Section 8.4.1. As mentioned earlier, time history direct integration does not require prior modal analysis. Nevertheless, the fact that Rayleigh damping is expressed in terms of damping ratios and natural circular frequencies implies that, in damped systems, modal analysis must precede time history direct integration, except if the natural frequencies and the damping are already known.

5.13 TIME HISTORY: MODAL SUPERPOSITION VERSUS DIRECT INTEGRATION

In dynamic analysis using time history, the time at which each one of the forces or the ground excitation start to act on the system must be specified. Since the part of the response caused by the initial conditions is usually diminished very rapidly due to damping, common practice sets the initial conditions, i.e., displacements, velocities, and accelerations equal to zero.

From the discussion in Sections 5.10 through 5.12, it should be clear now that there are more similarities and fewer differences between modal superposition and direct integration. For the same time step Dt, direct integration provides identical results to modal superposition for which all significant natural frequencies and mode shapes are included in the analysis. The remaining part of this section emphasizes the differences between the two methods.

The main advantage of direct integration over modal superposition is that direct integration can be used for both linear and nonlinear dynamics problems. Direct integration should be preferred when many modes must be included and the response is required for a short duration of time, such as in shock vibration problems. Direct integration provides more choices in accounting for damping. In principle, the damping matrix [C] does not have to satisfy the orthogonality conditions in direct integration analysis. However, use of Rayleigh damping implies modal orthogonality.

Contrary to direct integration, modal analysis must precede modal superposition. This is an advantage of modal superposition, since modal analysis allows detection of modeling errors as well as selection of the proper Dt. A second advantage of modal superposition is that, once a modal analysis of the system is performed, several modal superposition analyses can be performed for the same system subjected to different loads with a relatively small additional computational effort. Modal superposition is preferred over direct integration when the duration of the loads is long and a relatively small number of modes contributes to the response.

5.14 FREQUENCY RESPONSE ANALYSIS

If the externally applied loads or the ground excitation are harmonic, the preferred method of analysis is usually frequency response analysis. As demonstrated in Example 5.14.1 and later in Chapter 9, frequency response analysis provides the steady-state response. The contribution of the transient part to the system's response cannot be captured with frequency response analysis. Thus, strictly speaking, the results obtained with frequency response analysis are not conservative for design. If we want to determine both the transient and the steady-state part of the response, then we should choose either time history modal superposition or direct integration. When, however, the duration of the transient part of the response is either very short or not of interest, as is the case for most machine vibration problems, then use of frequency response analysis is appropriate.

As demonstrated in Example 5.14.1, frequency response analysis is very similar to time history modal superposition. In fact, frequency response analysis can be viewed as modal superposition that is limited to the calculation of the steady-state response of systems subjected to harmonic loads. It also requires prior modal analysis of the structure. The following example illustrates the method as applied to an undamped multi-degree-of-freedom system subjected to a harmonic load. It should be noted, however, that if the system is subjected to more than one harmonic load, the maximum responses due to the individual harmonic loads will most likely not occur at the same time. In this case, the steady-state response amplitudes should be determined with a modal combination method, such as the square root of the sum of the squares (SRSS), see Section 5.16.

EXAMPLE 5.14.1 Determine the steady-state response of the plane frame shown in Fig. 5.14.1. The second story is subjected to the harmonic load F_2 = 15,000 sin 15.7t lb. The frame is a typical model of a two-story building with the girders representing the floor slabs considered as rigid. The steel columns are rigidly connected to the ground and to the girders, thus, allowing only horizontal motion at the top of the columns. The total mass of each floor is m_1 = 116.46 lb-sec²/in and m_2 = 38.82 lb-sec²/in. The Young's modulus of the columns is E = 30x10⁶ psi. Zero damping is assigned to the system. As shown in Fig. 5.14.1, the lower columns are W10X49 with I_{z1} = 272 in⁴, A_1 = 14.4 in², and L_1 = 17 ft, and the columns of the upper story are W10X26 with I_{z2} = 144 in⁴, A_2 = 7.61 in², and L_2 = 12 ft.

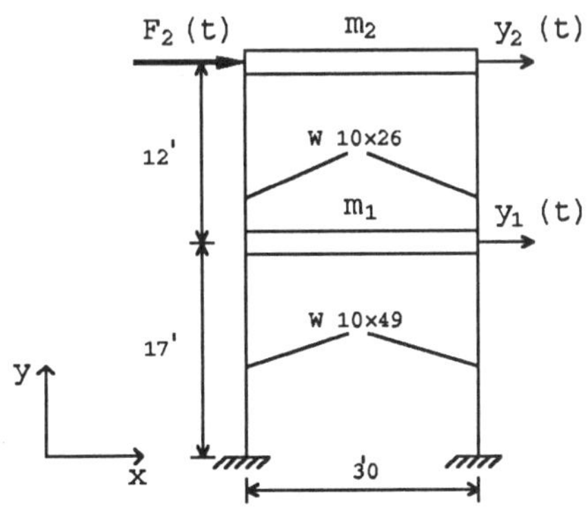

***Figure 5.14.1** Plane frame subjected to concentrated load.*

As in time history modal superposition, the analytical solution consists of four steps:

1. Formulation of the equations of motion.
2. Modal analysis.
3. Uncoupling of the equations of motion.
4. Combination of the modal responses.

Step 1. Formulation of the equations of motion:

The frame can be modeled as the two-degree-of-freedom system shown in Fig. 5.14.2(a). The stiffness for the two columns of each floor is evaluated from

$$k_i = \frac{12E(2I_{zi})}{L_i^3} \qquad (i=1,2) \tag{5.14.1a}$$

giving $k_1 = 23{,}068$ lb/in and $k_2 = 34{,}722$ lb/in.

The free-body diagram shown in Fig. 5.14.2(b) indicates the elastic and inertia forces that develop on each floor with displacements $y_1(t)$ and $y_2(t)$.

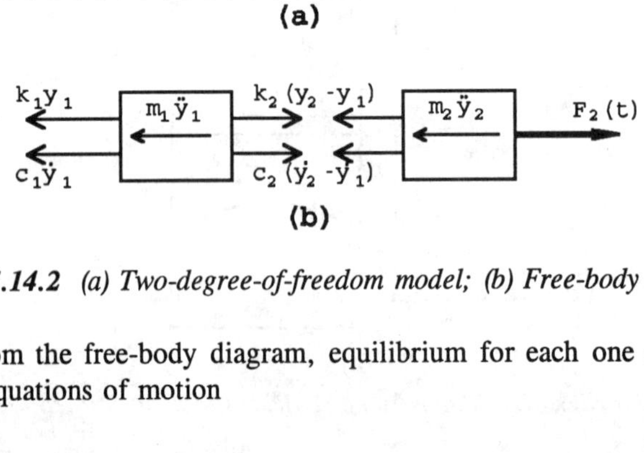

Figure 5.14.2 *(a) Two-degree-of-freedom model; (b) Free-body diagram.*

From the free-body diagram, equilibrium for each one of the floors yields the equations of motion

$$m_1\ddot{y}_1 + (k_1 + k_2)y_1 - k_2 y_2 = 0$$
$$m_2\ddot{y}_2 - k_2 y_1 + k_2 y_2 = F_2(t) \qquad \textit{(5.14.1b)}$$

or in a matrix form

$$\begin{bmatrix} m_1 & 0 \\ 0 & m_2 \end{bmatrix} \begin{Bmatrix} \ddot{y}_1 \\ \ddot{y}_2 \end{Bmatrix} + \begin{bmatrix} k+k_2 & -k_2 \\ -k_2 & k_2 \end{bmatrix} \begin{Bmatrix} y_1 \\ y_2 \end{Bmatrix} = \begin{Bmatrix} 0 \\ F_2(t) \end{Bmatrix} \qquad \textit{(5.14.1c)}$$

where $F_2(t)$ = 15,000 sin 15.7t lb.

Step 2: Modal analysis:

Modal analysis involves solution of the system

$$\begin{bmatrix} m_1 & 0 \\ 0 & m_2 \end{bmatrix} \begin{Bmatrix} \ddot{y}_1 \\ \ddot{y}_2 \end{Bmatrix} + \begin{bmatrix} k_1+k_2 & -k_2 \\ -k_2 & k_2 \end{bmatrix} \begin{Bmatrix} y_1 \\ y_2 \end{Bmatrix} = \begin{Bmatrix} 0 \\ 0 \end{Bmatrix} \qquad \textit{(5.14.1d)}$$

Substitution of the numerical values in eqn. (5.14.1d) yields

$$\begin{bmatrix} 116.46 & 0 \\ 0 & 38.82 \end{bmatrix} \begin{Bmatrix} \ddot{y}_1 \\ \ddot{y}_2 \end{Bmatrix} + \begin{bmatrix} 57{,}790 & -34{,}722 \\ -34{,}722 & 34{,}722 \end{bmatrix} \begin{Bmatrix} y_1 \\ y_2 \end{Bmatrix} = \begin{Bmatrix} 0 \\ 0 \end{Bmatrix} \qquad \textit{(5.14.1e)}$$

As described in Section 5.4, performing modal analysis provides the natural frequencies and the normalized mode shapes

$$f_1 = 1.895\ Hz \qquad f_2 = 5.624\ Hz$$

$$\{\phi\}_1 = \begin{Bmatrix} \phi_{11} \\ \phi_{21} \end{Bmatrix} = \begin{Bmatrix} 0.0764 \\ 0.0908 \end{Bmatrix} \qquad \{\phi\}_2 = \begin{Bmatrix} \phi_{12} \\ \phi_{22} \end{Bmatrix} = \begin{Bmatrix} -0.0524 \\ 0.1323 \end{Bmatrix} \qquad \textit{(5.14.1f)}$$

By assembling the mode shapes we form the modal matrix

$$[\Phi] = \begin{bmatrix} \phi_{11} & \phi_{12} \\ \phi_{21} & \phi_{22} \end{bmatrix} = \begin{bmatrix} 0.0764 & -0.0524 \\ 0.0908 & 0.1323 \end{bmatrix} \tag{5.14.1g}$$

Step 3: Uncoupling of the equations of motion:

As in modal superposition, in order to uncouple the equations of motion given by eqns. (5.14.1c), we introduce the transformation

$$\begin{Bmatrix} y_1(t) \\ y_2(t) \end{Bmatrix} = [\,\Phi\,] \begin{Bmatrix} z_1(t) \\ z_2(t) \end{Bmatrix} \tag{5.14.1h}$$

Substitution of eqn. (5.14.1h) in eqn. (5.14.1c), premultiplication of the resulting equation with $[\Phi]^T$, and application of the orthogonality properties given by eqns. (5.4.7s) and eqns.(5.4.7t), yields

$$\begin{aligned} \ddot{z}_1 + \omega_1^2 z_1 &= \phi_{21} F_2(t) \\ \ddot{z}_2 + \omega_2^2 z_2 &= \phi_{22} F_2(t) \end{aligned} \tag{5.14.1i}$$

Note that eqns. (5.14.1i) can also be obtained from eqn. (5.10.2i) for $\xi_i = 0$, (i=1,2).

Substituting the numerical values in eqn. (5.14.1i) leads to

$$\begin{aligned} \ddot{z}_1 + 141.77 z_1 &= 1362.0 \sin 15.7t \\ \ddot{z}_2 + 1248.67 z_2 &= 1984.5 \sin 15.7t \end{aligned} \tag{5.14.1j}$$

According to eqn. (4.5.4), the steady-state responses corresponding to the total responses $z_1(t)$ and $z_2(t)$ are given by

$$z_{s1} = \frac{1362}{141.77} \frac{\sin 15.7t}{1 - r_1^2}$$

$$z_{s2} = \frac{1984.5}{1248.67} \frac{\sin 15.7t}{1 - r_2^2} \qquad (5.14.1k)$$

The frequency ratios can be evaluated from eqn. (4.5.3) to give

$$r_1 = \frac{\bar{\omega}}{\omega_1} = \frac{15.7}{\sqrt{141.77}} = 1.32 \qquad r_2 = \frac{\bar{\omega}}{\omega_2} = \frac{15.7}{\sqrt{1248.67}} = 0.44 \qquad (5.14.1l)$$

Substituting eqns. (5.14.1l) in eqns. (5.14.1k) yields

$$z_{s1} = -12.94 \ \sin 15.7t$$

$$z_{s2} = 1.98 \ \sin 15.7t \qquad (5.14.1m)$$

Step 4. Combination of the modal responses:

Finally, substituting eqns. (5.14.1m) in eqn. (5.14.1h), the steady-state response of the frame is obtained

$$y_{s1}(t) = \phi_{11} z_{s1}(t) + \phi_{12} z_{s2}(t)$$

$$y_{s2}(t) = \phi_{21} z_{s1}(t) + \phi_{22} z_{s2}(t) \qquad (5.14.1n)$$

Notice that the subscript s has been introduced to indicate that eqns. (5.14.1n) provide the steady-state response. After substituting the numerical values, eqns. (5.14.1n) become

$$y_{s1}(t) = -1.092 \ \sin 15.7t$$

$$y_{s2}(t) = -0.913 \ \sin 15.7t \qquad (5.14.1o)$$

Since the response of the system is steady-state, the minus sign in the above equations can be omitted. For this system that has zero damping, the response is *in-phase* with the load, i.e., there is a zero phase-angle between the load $F_2(t)$ and the responses $y_{si}(t)$ $(i=1,2)$. As will be demonstrated in Sections 9.1 through 9.1.4, damping introduces an *out-of-phase* response term as well. Example 5.14.1 is also elaborated in Chapter 9 for various values of damping.

5.15 RESPONSE SPECTRUM ANALYSIS

Response spectrum analysis is commonly used to determine the response of structures subjected to ground motions. A *response spectrum* is a plot of the maximum response, i.e., maximum acceleration, velocity, or displacement to a specified dynamic input acting on all possible SDOF systems. The concept of response spectrum and its development is presented in Section 4.9.

Response spectrum analysis presents many similarities with time history modal superposition. They both require modal analysis and determine the response through modal combination. However, response spectrum analysis does not provide the response as a function of time. Specifically, it does not determine the response at selected time intervals but in the form of maximum response amplitudes. These amplitudes are obtained through statistically based modal combination algorithms. Such algorithms include the square root of the sum of the squares (SRSS), the formulas recommended by the Nuclear Regulatory Commission (NRC), and the complete quadratic combination (CQC).

The most commonly used combination method is the SRSS. The SRSS usually provides good maximum response estimates when the structure has well separated natural modes. The NRC and the CQC procedures should always be used when modal analysis reveals that the structure has several closely spaced natural frequencies. A discussion and suggestions on how to use modal combination methods are given in Section 5.16.

The differences and similarities between response spectrum analysis and time history modal superposition are demonstrated through the following example. Before going through Example 5.15.1, it is suggested to review Section 4.9 and steps 1 and 2 of Example 5.14.1.

EXAMPLE 5.15.1. Determine the maximum relative displacements of the frame shown in Fig. 5.15.1(a). The frame is subjected to the design earthquake specified by eqns. (4.9.8). Recall that use of the design spectra implies a damping ratio $\xi = 0.05$ for all modes.

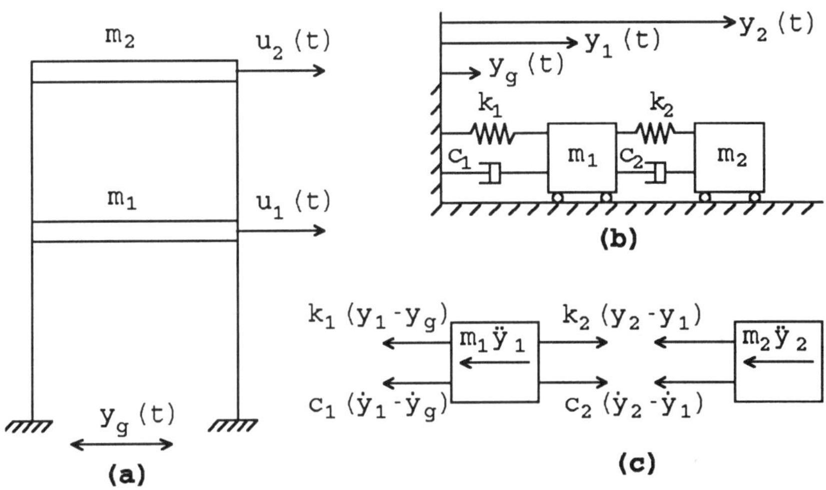

Figure 5.15.1 (a) Two-story frame; (b) Two-degree-of-freedom system; (c) Free-body diagram.

The analytical solution consists of four steps:

1. Formulation of the equations of motion.
2. Modal analysis.
3. Uncoupling of the equations of motion.
4. Use of design spectra to calculate the maximum relative displacements.

Step 1. Formulation of the equations of motion:

As discussed in Example 5.14.1, the frame can be modeled as the two-degree-of-freedom system shown in Fig. 5.15.1(b). Considering the free-body diagram of Fig. 5.15.1(c), the equation of motion is given by

$$m_1\ddot{y}_1+c_1(\dot{y}_1-\dot{y}_g)+k_1(y_1-y_g)-k_2(y_2-y_1)-c_2(\dot{y}_2-\dot{y}_1)=0$$
$$m_2\ddot{y}_2+k_2(y_2-y_1)+c_2(\dot{y}_2-\dot{y}_1)=0 \tag{5.15.1a}$$

As demonstrated in Section 4.7 for a SDOF system, the relative displacements of the floors u_1 and u_2 can be expressed in terms of the absolute displacements y_1 and y_2 and the ground motion y_g as given by

$$u_1=y_1-y_g$$
$$u_2=y_2-y_g \tag{5.15.1b}$$

In view of eqn. (5.15.1b), equation (5.15.1a) can be written in the matrix form

$$\begin{bmatrix} m_1 & 0 \\ 0 & m_2 \end{bmatrix}\begin{Bmatrix} \ddot{u}_1 \\ \ddot{u}_2 \end{Bmatrix}+\begin{bmatrix} c_1+c_2 & -c_2 \\ -c_2 & c_2 \end{bmatrix}\begin{Bmatrix} \dot{u}_1 \\ \dot{u}_2 \end{Bmatrix}+\begin{bmatrix} k_1+k_2 & -k_2 \\ -k_2 & k_2 \end{bmatrix}\begin{Bmatrix} u_1 \\ u_2 \end{Bmatrix}=-\begin{Bmatrix} m_1 \\ m_2 \end{Bmatrix}\ddot{y}_g \tag{5.15.1c}$$

In order to derive eqn. (5.15.1c) the relationship $u_2-u_1 = y_2-y_1$ that can be easily obtained from eqns. (5.15.1b) has also been used.

Step 2. Modal analysis:

Modal analysis involves solution of the system

$$\begin{bmatrix} m_1 & 0 \\ 0 & m_2 \end{bmatrix}\begin{Bmatrix} \ddot{u}_1 \\ \ddot{u}_2 \end{Bmatrix}+\begin{bmatrix} k_1+k_2 & -k_2 \\ -k_2 & k_2 \end{bmatrix}\begin{Bmatrix} u_1 \\ u_2 \end{Bmatrix}=\begin{Bmatrix} 0 \\ 0 \end{Bmatrix} \tag{5.15.1d}$$

Substitution of the numerical values in eqn. (5.15.1d) results in

$$\begin{bmatrix} 116.46 & 0 \\ 0 & 38.82 \end{bmatrix}\begin{Bmatrix} \ddot{u}_1 \\ \ddot{u}_2 \end{Bmatrix}+\begin{bmatrix} 57{,}790 & -34{,}722 \\ -34{,}722 & 34{,}722 \end{bmatrix}\begin{Bmatrix} u_1 \\ u_2 \end{Bmatrix}=\begin{Bmatrix} 0 \\ 0 \end{Bmatrix} \tag{5.15.1e}$$

The natural frequencies and mode shapes have been obtained in step 2 of Example 5.14.1.

Step 3. Uncoupling of the equations of motion:

As in time history modal superposition, to uncouple the equations of motion we introduce the transformation

$$\begin{Bmatrix} u_1(t) \\ u_2(t) \end{Bmatrix} = [\Phi] \begin{Bmatrix} z_1(t) \\ z_2(t) \end{Bmatrix} \tag{5.15.1f}$$

Substitution of eqn. (5.15.1f) in eqn. (5.15.1c), pre-multiplication of the resulting equation with $[\Phi]^T$, application of the orthogonality properties , i.e., eqns. (5.4.7s) and (5.4.7t), and finally application of the orthogonality property to the damping matrix as specified by eqn. (5.10.1) leads to

$$\begin{aligned} \ddot{z}_1 + 2\xi\omega_1\dot{z}_1 + \omega_1^2 z_1 &= -(m_1\phi_{11} + m_2\phi_{21})\ddot{y}_g \\ \ddot{z}_2 + 2\xi\omega_2\dot{z}_2 + \omega_2^2 z_2 &= -(m_1\phi_{12} + m_2\phi_{22})\ddot{y}_g \end{aligned} \tag{5.15.1g}$$

By comparing eqns. (5.15.1g) to eqns. (5.10.2i) and (5.10.2j), we observe that $\ddot{y}_g = g(t)$ and the participation factors Γ_1 and Γ_2 are given by

$$\begin{aligned} \Gamma_1 &= -(m_1\phi_{11} + m_2\phi_{21}) = -12.42 \\ \Gamma_2 &= -(m_1\phi_{12} + m_2\phi_{22}) = 0.97 \end{aligned} \tag{5.15.1h}$$

By substituting in eqn. (5.15.1g) $\xi = 0.05$, the circular natural frequencies $\omega_1 = (2\pi)(1.895) = 11.9$ rad/sec and $\omega_2 = (2\pi)(5.624) = 35.34$ rad/sec, we obtain

$$\begin{aligned} \ddot{z}_1 + 1.19\dot{z}_1 + 141.77z_1 &= -12.42\ddot{y}_g \\ \ddot{z}_2 + 3.53\dot{z}_2 + 1248.67z_2 &= 0.97\ddot{y}_g \end{aligned} \tag{5.15.1i}$$

We introduce now the following transformation:

$$z_1(t)=\Gamma_1 v_1(t)$$
$$z_2(t)=\Gamma_2 v_2(t) \qquad (5.15.1j)$$

Substitution of eqn. (5.15.1j) in eqn. (5.15.1g) yields

$$\ddot{v}_1+2\xi\omega_1\dot{v}_1+\omega_1^2 v_1=\ddot{y}_g$$
$$\ddot{v}_2+2\xi\omega_2\dot{v}_2+\omega_2^2 v_2=\ddot{y}_g \qquad (5.15.1k)$$

Substituting the numerical values in eqn. (5.15.1k) gives

$$\ddot{v}_1+1.19\dot{v}_1+141.77v_1=\ddot{y}_g$$
$$\ddot{v}_2+3.53\dot{v}_2+1248.67v_2=\ddot{y}_g \qquad (5.15.1l)$$

Next, we substitute eqn. (5.15.1j) in eqn. (5.15.1f) and perform the matrix multiplication, to obtain

$$u_1(t)=\Gamma_1\phi_{11}v_1(t)+\Gamma_2\phi_{12}v_2(t)$$
$$u_2(t)=\Gamma_1\phi_{21}v_1(t)+\Gamma_2\phi_{22}v_2(t) \qquad (5.15.1m)$$

or

$$u_1(t)=0.949v_1(t)+0.051v_2(t)$$
$$u_2(t)=1.128v_1(t)-0.128v_2(t) \qquad (5.15.1n)$$

Step 4. Use of design spectra to calculate the maximum relative displacements:

Notice that each one of eqns. (5.15.1k) resembles eqn. (4.9.1b). Thus, the spectral accelerations S_{a1} and S_{a2} can be obtained from eqn. (4.9.8) for the corresponding natural periods T_1 = 1/1.895 = 0.53 sec and T_2 = 1/5.624 = 0.178 sec, respectively

$$S_{al} = \frac{0.975}{0.53} g = (1.84)(386.4) = 710.83 \ in/\sec^2$$

$$S_{a2} = 2.50 \ g = (2.50)(386.4) = 966.00 \ in/\sec^2 \tag{5.15.1o}$$

As mentioned in Section 5.9, the damping ratio is accounted for in the spectra. The corresponding spectral displacements S_{d1}, and S_{d2}, are calculated in terms of S_{a1} and S_{a2} from eqn. (4.9.6). Finally, application of the SRSS to eqn. (5.15.1m) with $v_{1max} = S_{d1}$ and $v_{2max} = S_{d2}$, see eqn. (5.16.1), yields the maximum relative displacements

$$u_{1max} = \sqrt{(\Gamma_1 \phi_{11} \frac{S_{al}}{\omega_1^2})^2 + (\Gamma_2 \phi_{12} \frac{S_{a2}}{\omega_2^2})^2}$$

$$u_{2max} = \sqrt{(\Gamma_1 \phi_{21} \frac{S_{al}}{\omega_1^2})^2 + (\Gamma_2 \phi_{22} \frac{S_{a2}}{\omega_2^2})^2} \tag{5.15.1p}$$

Substitution of the numerical values for Γ_i, ϕ_{ij}, ω_i^2, and S_{ai} (i,j = 1,2) in eqn. (5.15.1p) provides the maximum relative displacements

$$u_{1max} = 4.76 \ \ in$$

$$u_{2max} = 5.66 \ \ in \tag{5.15.1q}$$

Using a multi-degree-of-freedom finite element model, Example 5.15.1 is also presented in Section 9.2.5.

5.16 MODAL COMBINATION METHODS

Modal combination methods are used to calculate the response in response spectrum, frequency response, and random vibration analysis. Be reminded that the term "response" used in this text refers to displacements, velocities, accelerations, member forces, and stresses.

The most widely used modal combination methods are the square root of the sum of the squares (SRSS), the U.S. Nuclear Regulatory Commission

(NRC) methods, and the complete quadratic combination (CQC). The methods are based on random vibration concepts. They approximate the response by combining the effects of several modes while accounting for the fact that all modal maxima do not occur at the same time.

It must be noted that when the modal member forces are combined, they no longer satisfy equilibrium. Also the combined modal displacements are not consistent with member forces. However, both displacement-member force compatibility and equilibrium are satisfied on an individual mode basis. It is also important to note that the results obtained by modal combination methods are not always conservative. However, under proper conditions, each method provides results of acceptable accuracy.

In order to describe the three modal combination methods we define as U_{ids} the response to i-th mode, for a ground motion along the d-th direction, and for the component of the spectrum along the global s-th axis. For example, U_{1xx} is the fundamental modal response for the ground motion along the x-axis and the response spectrum component along the x-axis. The subscribes vary as i = 1,2,....,m, where m is the number of modes, d = x,y,z, and s = x,y,z, with x,y,z defining the global coordinate system.

Square Root of the Sum of the Squares (SRSS): According to SRSS the response amplitude U is given by

$$U=\sqrt{U_1^2+U_2^2+...+U_m^2} \tag{5.16.1}$$

where

$$U_i=U_{ixx}+U_{iyy}+U_{izz} \tag{5.16.2}$$

We define as *cluster factor* (CF) the ratio

$$\frac{f_i-f_{i-1}}{f_{i-1}}\leq CF \tag{5.16.3}$$

Two consecutive modes i and (i-1) are considered as closely spaced if their $CF \leq 0.1$ The SRSS introduces substantial numerical errors when closely spaced modes are present. Further details to lessen such errors are presented in Chapter 9. Closely spaced modes usually occur in structures in which torsional effects are significant, e.g., multi-story frames with rigid slabs, and also in systems with almost identical members, e.g., continuous beams, plate, and shell structures with almost identical spans. The adequacy of SRSS also decreases with damping, since for increasing damping modal coupling effects become more important. Alternatives that account for modal coupling are the NRC methods, and the CQC.

NRC methods: The NRC has proposed several methods to combine modal responses that could include closely spaced modes. We present here two of these methods. Specifically, the combined response is calculated from either eqn. (5.16.4a), known as the NRC guide 1.92 procedure, or eqn. (5.16.4b), known as the NRC modified procedure.

$$U = \sqrt{U_{xx}^2 + U_{yy}^2 + U_{zz}^2} \qquad (5.16.4a)$$

or

$$U = U_{xx} + U_{yy} + U_{zz} \qquad (5.16.4b)$$

where

$$\begin{aligned} U_{xx} &= \sqrt{\sum U_{ixx}^2 + \sum (U_{jxx} + U_{kxx})^2} \\ U_{yy} &= \sqrt{\sum U_{iyy}^2 + \sum (U_{jyy} + U_{kyy})^2} \\ U_{zz} &= \sqrt{\sum U_{izz}^2 + \sum (U_{jzz} + U_{kzz})^2} \end{aligned} \qquad (5.16.5)$$

The first summation in eqns. (5.16.5) refers to the modes that are not closely spaced. The second summation is performed for $j \neq k$ (j,k=1,2,...,m) and for the closely spaced modes, i.e., for consecutive modes with CF less than a selected value, e.g., CF=0.1.

CQC: This is the most widely accepted method for modal combination. According to the CQC the combined response is given by

$$U = \sqrt{U_{xx}^2 + U_{yy}^2 + U_{zz}^2} \tag{5.16.6}$$

where

$$U_{xx}^2 = \sum_{i=1}^{N} \sum_{j=1}^{N} U_{ixx} U_{jxx} \varepsilon_{ij} \tag{5.16.7}$$

The U_{yy} and U_{zz} are calculated from eqn. (5.16.7) for the subscript x substituted with y and z, respectively. If the damping ratio ξ is the same for all modes, the *cross-modal coefficient* ϵ_{ij} is obtained from

$$\varepsilon_{ij} = \frac{8\xi^2 (1 + r) r^{2/3}}{(1 - r^2)^2 + 4\xi^2 r (1 + r)^2} \tag{5.16.8}$$

where the frequency ratio r is defined in eqn. (4.5.3). Notice that for $r = 1$, ϵ_{ij} is equal to 1, regardless of the value of ξ. Furthermore, if the modal periods are well spaced, i.e., CF > 0.1, the cross-modal parameter is practically equal to zero for $i \neq j$, the CQC becomes identical to SRSS.

In conclusion, the NRC methods and CQC are always acceptable. In general, use of SRSS is acceptable for CF greater than 0.1.

5.17 RANDOM VIBRATION ANALYSIS

When the exciting forces or ground motions are statistically defined, the response can be determined with random vibration analysis. For SDOF systems the method is discussed in Section 4.10. In most applications, random vibration analysis involves the power spectral density of the excitation. The power spectral density can be determined electronically with instruments called frequency analyzers or spectral density analyzers. Once the power spectral density is known, the mean square displacement response can be calculated from

$$\overline{y^2(t)} = \int_0^{+\infty} |H(\bar{f})|^2 S_F(\bar{f})\, d\bar{f} \qquad \textbf{(4.11.16)}$$

where $S_F(\bar{f})$ is the power spectral density of the excitation, and $H(\bar{f})$ is the transfer function as defined in Sections 4.10.3 and 4.11.2, respectively.

Knowing the mean square value of the response and using standard probability functions, such as the normal distribution, we can predict the system's response in probabilistic terms. The following example elucidates the procedure. Using a finite element model, this example is also presented in Chapter 9.

EXAMPLE 5.17.1. Consider the plane frame shown in Fig. 5.15.1(a). The same frame was also used in Examples 5.14.1 and 5.15.1. The foundation is subjected to the random acceleration $\ddot{y}_g(t)$ with the constant power spectral density $S_F(\bar{f})$ = 0.20 g²/Hz shown in Fig. 5.17.1. Calculate the probability that the horizontal relative response at the top of the frame will exceed 9 in. Assume that the mean value of the random excitation is zero and the damping ratio ξ=0.05 for all modes.

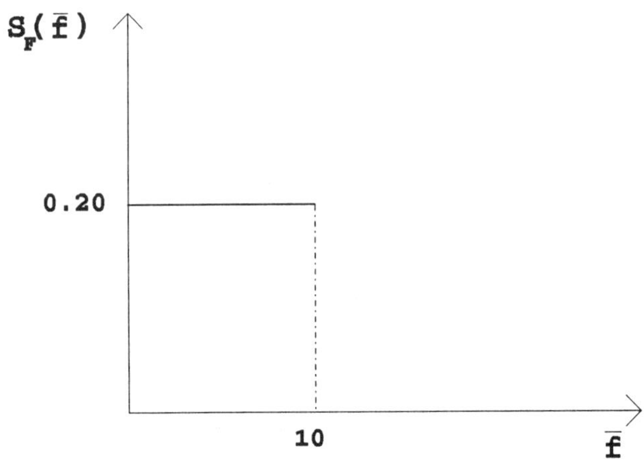

Figure 5.17.1 Power spectral density of ground excitation.

The solution procedure consists of five steps:

1. Formulation of the equations of motion.
2. Modal analysis.
3. Uncoupling of the equations of motion.
4. Calculation of the root mean square displacement response.
5. Calculation of requested probability.

The first three steps of the solution procedure are identical to the ones presented in Example 5.15.1. Thus, eqns. (5.15.1a) through (5.15.1n) are also valid for this example. For convenience, we repeat here the eqns. (5.15.1k) through (5.15.1n), and then continue with steps 4 and 5:

$$\begin{aligned} \ddot{v}_1+2\xi\omega_1\dot{v}_1+\omega_1^2v_1&=\ddot{y}_g \\ \ddot{v}_2+2\xi\omega_2\dot{v}_2+\omega_2^2v_2&=\ddot{y}_g \end{aligned} \tag{5.15.1k}$$

$$\begin{aligned} \ddot{v}_1+1.19\dot{v}_1+141.77v_1&=\ddot{y}_g \\ \ddot{v}_2+3.53\dot{v}_2+1248.67v_2&=\ddot{y}_g \end{aligned} \tag{5.15.1l}$$

or

$$\begin{aligned} u_1(t)&=\Gamma_1\phi_{11}v_1(t)+\Gamma_2\phi_{12}v_2(t) \\ u_2(t)&=\Gamma_1\phi_{21}v_1(t)+\Gamma_2\phi_{22}v_2(t) \end{aligned} \tag{5.15.1m}$$

$$\begin{aligned} u_1(t)&=0.949v_1(t)+0.051v_2(t) \\ u_2(t)&=1.128v_1(t)-0.128v_2(t) \end{aligned} \tag{5.15.1n}$$

Step 4. Calculation of the root mean square displacement response:

Equations (5.15.1k) or (5.15.1l) can be viewed as the equations of motion of two SDOF systems with natural frequencies f_1 and f_2 and stiffnesses

$k_1 = \omega_1^2$ and $k_2 = \omega_2^2$, respectively. Since both systems are subjected to a broad power spectral density excitation, their mean square responses can be approximated through eqn. (4.11.17). Thus, the $\overline{v_1^2}$ and the $\overline{v_2^2}$ are given by

$$\overline{v_1^2} = \frac{\pi}{4\xi}\frac{f_1}{k_1^2}S_F(f_1) \tag{5.17.1a}$$

and

$$\overline{v_2^2} = \frac{\pi}{4\xi}\frac{f_2}{k_2^2}S_F(f_2) \tag{5.17.1b}$$

By substituting $f_1 = 1.895$ Hz, $f_2 = 5.624$ Hz, $k_1 = 141.77$, $k_2 = 1248.67$, and $S_F(f_1) = S_F(f_2) = (0.2)(386.4)^2$ in eqns. (5.17.1a) and (5.17.1b), we obtain

$$\begin{aligned} \overline{v_1^2} &= 44.22 \\ \overline{v_2^2} &= 1.692 \end{aligned} \tag{5.17.1c}$$

Next, we apply the SRSS, see Section 5.16, to determine the RMS of the response

$$\begin{aligned} \sqrt{\overline{u_1^2}} &= [(\Gamma_1\phi_{11})^2\overline{v_1^2} + (\Gamma_2\phi_{12})^2\overline{v_2^2}]^{\frac{1}{2}} \\ \sqrt{\overline{u_2^2}} &= [(\Gamma_1\phi_{21})^2\overline{v_1^2} + (\Gamma_2\phi_{22})^2\overline{v_2^2}]^{\frac{1}{2}} \end{aligned} \tag{5.17.1d}$$

Recall from eqn. (5.15.1h) that Γ_1=-12.42 and Γ_2=0.97, and that the mode shapes are given by eqn. (5.14.1f). Substituting the numerical values in eqn. (5.17.1d), we obtain the RMS for the response of the top story of the frame

$$\sqrt{\overline{u_2^2}} = \sqrt{7.50^2 + 0.167^2} = 7.50 \quad in \tag{5.17.1e}$$

Step 5. Calculation of requested probability:

We can now calculate the probability that the horizontal relative response at the top exceeds 9 in. The calculations will be based on the assumption that the random response is Gaussian. Then, for zero mean of the response, i.e., $\overline{u}_2 = 0$ and in view of eqn. (4.10.2) and (4.10.3), the root mean square displacement response is equal to the standard deviation

$$\sqrt{\overline{u_2^2}} = \sigma_2 \tag{5.17.1f}$$

As mentioned in Section 4.10.2, the probability for a Gaussian distribution can be calculated from eqns. (4.10.5) and (4.10.6). In evaluating the requested probability, the following two properties for a Gaussian random function y(t) and a given constant *a* are utilized:

$$P(|y|>a) = 2P(y>a)$$

and

$$P(|y|>a) = P(|\frac{y-\overline{y}}{\sigma}|>\frac{a-\overline{y}}{\sigma})$$

By making use of probability properties, and the symmetry of Gaussian distribution we obtain

$$\begin{gathered} P(|u_2(t)|>9) = P(|\frac{u_2(t)-\overline{u}_2}{\sigma_2}|>\frac{9-\overline{u}_2}{\sigma_2}) = \\ P(|\frac{u_2(t)}{7.50}|>\frac{9}{7.50}) = 2P(\frac{u_2(t)}{7.50}>1.2) = 2(0.115) = 0.23 \end{gathered} \tag{5.17.1g}$$

Thus, the probability that the horizontal relative displacement will exceed 9 in is 23%.

Similar calculations with the RMS of the stresses can be used to determine the probability that the stresses in the columns would exceed the allowable or the yield stresses. As a consequence, we could identify the range of load intensity for linear analysis.

6

Static Analysis: Numerical Examples

6.1 INTRODUCTION

In Chapter 3, we presented a general discussion on modeling fundamentals and recommendations on evaluating the analysis results. The examples discussed in this chapter are fairly simple. They illustrate the concepts set forth earlier, and present a systematic procedure to create accurate models, to check the applied external loads, to evaluate the obtained displacements and stresses, and finally to decide whether the analysis should be repeated with a more refined model.

The examples have been selected to introduce modeling aspects that an analyst should consider and take advantage of in static analysis. Their solution is also demonstrated with Algor as an attempt to familiarize the user with basic commands and modeling options. It should be noted, however, that the procedure and suggestions are valid for most commercial finite element programs.

In the examples presented in Chapters 6, 7, 8, and 9, the translational and rotational degrees-of-freedom are denoted as T and R, respectively. The restrained degrees-of-freedom are specified with subscripts. For example, assigning T_xR_{xyz} at a node denotes that the X-translational and the X-, Y-, and Z-rotational degrees-of-freedom are restrained.

6.2 ANALYSIS OF SHORT BEAM

Consider the beam with a small thickness t subjected to the uniformly distributed load q shown in Fig. 6.2.1. The beam is allowed to deform only in the YZ plane. Only points A and B are fully restrained, i.e., their degrees-of-freedom are $T_{xyz}R_{xyz}$. Referring to Fig. 6.2.1, the dimensions of the beam, the elastic properties, and the magnitude of the applied load are as follows: L = 12 in, c = 3 in, t = 1 in, E = 3 x 10^7 psi, $\nu = 0.30$, q = 1000 lb/in^2. The weight density of the beam is γ = 0.2836 lb/in^3.

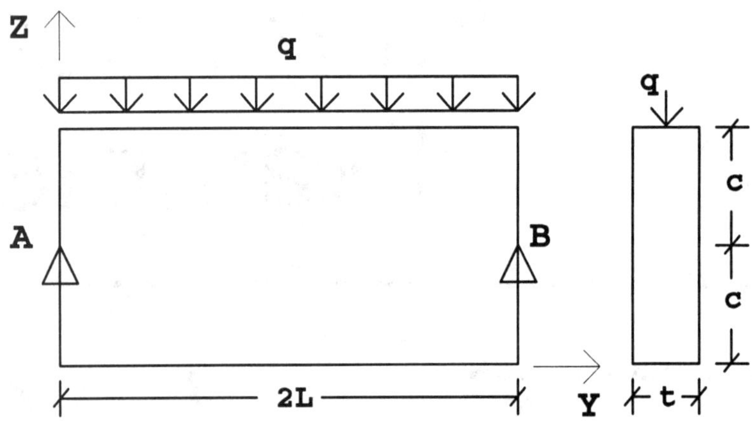

Figure 6.2.1 Short beam subjected to uniform load.

Since the beam thickness is small in comparison with the other dimensions and the loading is applied in the plane of the beam, we can analyze the system as a plane stress problem, see Sections 2.4 and 3.2. For this introductory example, the mesh refinement is not based on a rigorous application of the guidelines suggested in Chapter 3. In our first attempt to develop the model, we use a uniform mesh with quadrilateral plane stress elements. Figure 6.2.2 shows the model and the node numbering. Since the aspect ratio is (24/8)/(6/2) = 1, the mesh should be adequate for an accurate stress analysis, see Section 3.4.

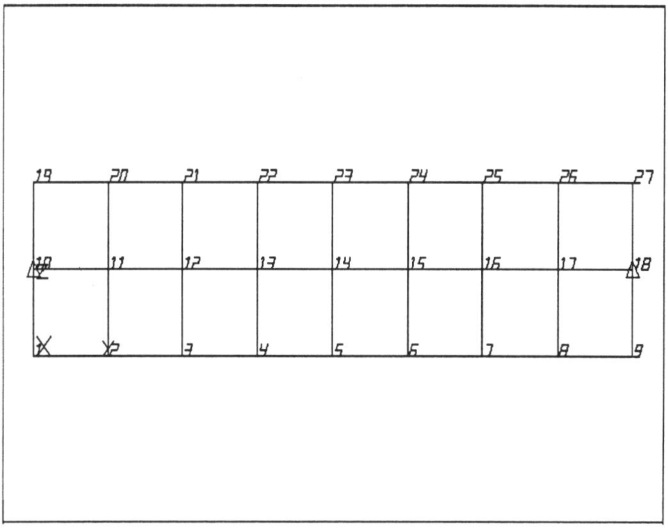

Figure 6.2.2 Sixteen-element model of short beam.

Before proceeding to the analysis, it is good practice to calculate the number of degrees-of-freedom of the model. Since this is a plane stress problem, there are no rotational degrees-of-freedom about any axis and no translational degrees-of-freedom along the X-axis. The nodes at A and B are restrained in all directions to simulate fixed-end boundary conditions. In order to calculate the degrees-of-freedom of the model, we first determine the total possible degrees-of-freedom of all nodes, which is 27 x 6 = 162. The total degrees-of-freedom of our model is the difference between the total possible degrees-of-freedom (27 x 6), minus the eliminated rotational and translational degrees-of-freedom (27 x 4), minus the translational degrees-of-freedom of the two mid-nodes at the boundaries (2 x 2), that is

$$162 - (27X4) - (2X2) = 50$$

It should be noted that since the model has one axis of symmetry, half the structure could have been analyzed with the proper boundary conditions on the axis of symmetry. If we use the eight-element model shown in Fig. 6.2.3, then the total degrees-of-freedom would be the difference of the total possible degrees-of-freedom (15 x 6 = 90), minus the eliminated rotational and translational degrees-of-freedom (15 x 4), minus the translational degrees-of-freedom of the mid-node at the boundary (1 x 2), minus the Y-translational degrees-of-freedom of the 3 nodes at the axis of symmetry (3 x 1), that is

$$90 - (15X4) - (1X2) - (3X1) = 25$$

The reduction of the total degrees-of-freedom clearly indicates the advantage of using symmetry whenever possible. Use of symmetry leads to a smaller model that is easier to develop, to identify possible modeling errors, to examine the results, and is faster to process.

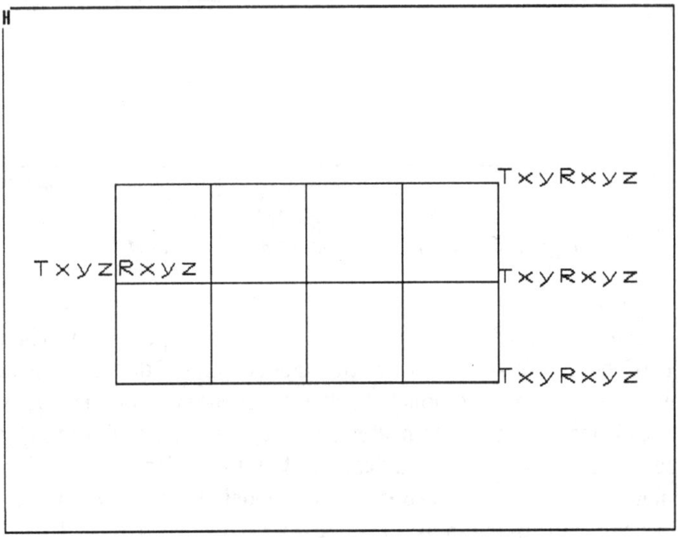

Figure 6.2.3 Half-model of short beam.

6.3 ANALYSIS WITH ALGOR

Even though this section is specifically written for users of Algor, the general procedure outlined and several of the comments made are applicable to most finite element software. The processors as well as the pre- and postprocessors refer to the Hyper version, and are applicable with minor modifications to all the other versions. In the following sections, either capital letters or quotes will be used to designate processors, options, and commands of the Algor software.

Solution with Algor involves use of the preprocessors SuperDraw (SD2H) and DECODS, the processor SSAP0H which performs static analysis of linear elastic systems with small displacements and strains, and the postprocessor SuperView (SVIEWH). Since a detailed description of the commands and features of the pre-, post-, and processors are provided in ViziCad and the Algor Processor Reference Manual, the following discussion focusses on illustrating the solution procedure and attempts to elucidate the use of certain commands and features of SD2H, DECODS, and SVIEWH.

The model created in SD2H is shown in Fig. 6.3.1. There are several ways to create the mesh shown in this figure as described in ViziCad. In any case, however, the model must be drawn in the YZ-plane as required by Algor for all two-dimensional problems, i.e., plane stress, plane strain, and axisymmetric. Besides the mesh, the boundary conditions are also defined in SD2H, see Fig. 6.3.1. Next, we define the COLOR, GROUP, and LAYER for each line in the model. For the appropriate use of these parameters refer to ViziCad. We specify the COLOR of all lines in the model as green (No.1), except the lines at z = 6 where the uniform load is applied. The COLOR assigned to lines z = 6 is red (No.2). The color of GROUP for all lines in the model is also green. Note that GROUP is used to assign material properties in DECODS. Thus, by assigning only one color for the GROUP, we can assign only one material property in DECODS. Generation of the model in SD2H may lead to create overlapping lines. In order to assure that there are no duplicate lines in the model, it is good practice to use the sequence of commands: CONSTRUCT;CLEAN;DUPLICATE, which deletes all duplicate lines. It should be noted that two lines sharing exactly the same end nodes are considered as duplicate when they are assigned the same color, group, and layer. Execution of SD2H creates a binary file "file.esd." The file name given to the model of this example is "EX62.esd". This file is further processed with DECODS. One can access DECODS by starting from the main menu of SD2H and following the command sequence: TRANSFER;STRESS.

The model created in SD2H specifies only the geometry and not the type of finite elements to be used. Also, it does not define the material properties of the structure and the applied uniform loading. All these quantities are defined in DECODS through several options. Specifically, in ELEMENTS we select TYPE;2-Dim, and in INFO we select "plane stress" and "incompatible." The justification for our selection is given in Section 3.2, where a discussion on incompatible plane stress elements is presented. Within the ELEMENTS menu, the weight density, Young's modulus of elasticity, Poisson's ratio, the thermal expansion coefficient, and the shear modulus G are defined

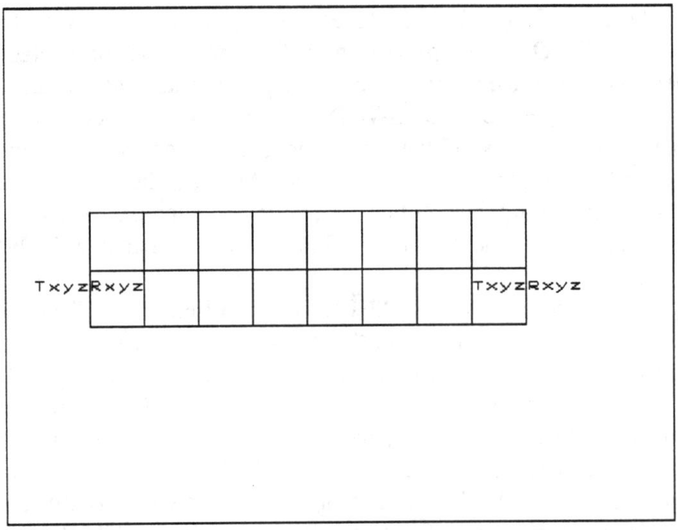

Figure 6.3.1 Sixteen-element model created in SD2H.

in GROUP. In COLOR we specify the thickness, the magnitude of the uniform load, and the orientation of the elements. In this example two colors have been specified in SD2H: we have assigned color No. 1 to the elements between the lines z = 0 and z = 3 and color No. 2 to the rest of the elements. Referring to Fig. 6.3.2 the elements 1 through 8 are assigned color No. 1 and the elements 9 through 16 are designated color No. 2. All elements have the same thickness t = 1 in. However, the color No.1 elements are assigned a pressure with magnitude equal to zero, while a pressure q = 1000 psi is specified to the color No. 2 elements. Note that the color of an element is the highest color assigned to one of its sides. In ANALYSIS we select "static" and leave the defaults unchanged in the "stress analysis" menu. The uniform load is activated within the GLOBAL;LOAD CASE menu. The selection of LOAD CASE leads to the "load case multipliers" where we set 1 in column "A(Press)" for the first load case (LC). The gravity load along the Z-axis could be included in the analysis by setting 1 in column "B(Accel)" for the first LC, and the default Az = -1 in the ANALYSIS;STATIC;STRESS ANALYSIS menu. Table 6.3.1 lists the data inputed in DECODS.

Table 6.3.1

Elements	Type	Information		Group []	Color[]
	4) 2-Dim	Type	Incompat.		
		Plane S.	Incompat.		

Group						
Name	Lib	Density	Young's	Poisson	Alpha	G
steel	yes	0.2836	3E7	0.3	6.5E-6	

Color						
Color	Thickness	Tref	Pressure	Code	IPy	IPz
1	1					
2	1		1000			

Analysis: Static: Stress Analysis			
Grav	Ax	Ay	Az
386.4	0	0	-1

Global: Load case [ ]				
Load Case	A(press)	B(Accel)	C(Disp)	D(Therm)
1	1	0	0	0

To complete the data processing with DECODS we select "All" and then "Run" within the DECODE menu. Execution of DECODS creates the following binary files: "EX62.ems", "EX62.esg", and "EX62.sst." It also creates the "EX62. " which is an ASCII file with no extension. The "EX62.sst" and "EX62. " are of particular interest to the analyst. The "EX62.sst" is accessed by SVIEWH, while the "EX62. " is the input file of the processor pertinent to the analysis type, in our case the SSAP0H. Before we elaborate on the use of the "EX62.sst" and "EX62. " files, it is worth discussing the use of certain options in the DECODE menu of DECODS, i.e., "BC + Force, Material, Intersect lines, Invalid lines, and Invalid regions." Activation of "BC

+ Force" or "Material" is of particular interest when the model has already been decoded once. Specifically, when "BC + Force" or "Material" is selected, DECODS operates only on changes related to boundary conditions and applied forces or material properties. Since decoding of all the other quantities will not be repeated, the processing time will be reduced. The savings could be substantial when repeating the analysis of large models for different boundary conditions and material properties.

There are several instances in which we may inadvertently create geometric shapes in SD2H that cannot be converted to finite elements when processed with DECODS. A typical example is a model that contains non-intersecting lines. In this case, executing DECODS could lead to loss of elements that would appear as voids in the finite element model. A means of recovering lost elements is through activating the options of "Intersecting lines," "Invalid lines," or "Invalid regions" of DECODS in conjunction with CONSTRUCT;CLEAN;VERTICES of SD2H, as described in ViziCad.

The ASCII "EX62. " created by DECODS is given in Table 6.3.2. This is probably the most significant file for a number of reasons. First, this is the only file used by the processor. Thus, if we retain "EX62. " we can always process it to obtain the displacements and stresses of the structure. Second, the "EX62. " contains all the model data written in an ASCII format. Therefore, by using the Algor Processor Reference Manual we can verify and -if needed- modify the "EX62. " with a text editor. We can then run the "EX62. " again to obtain the system's displacements and stresses. This process is the most expedient way to repeat the analysis, after we have incorporated any desired changes to the model. Obviously, by making the desirable changes in the "EX62. " with a text editor, we avoid the lengthier process of incorporating the changes in SD2H and DECODS before running the processor again. The user should be reminded, however, that if he/she chooses to make any changes in the model by modifying the "EX62. " file, these changes will not be reflected in the files created by SD2H and DECODS. To an experienced user, changing the model through editing the "EX62. " can lead to substantial savings in preprocessing time.

After running DECODS and before processing the "EX62. " with SSAP0H, it is good practice to check the model. The "EX62.sst" accessed with SVIEWH can serve as a means of identifying possible errors or omissions in the model. At this stage we cannot examine the displacements and the stresses with SVIEWH, since the model has not been processed yet. However, there is a number of checks that can be performed in order to examine whether the model simulates the structure and the loads are applied as intended. For example, we

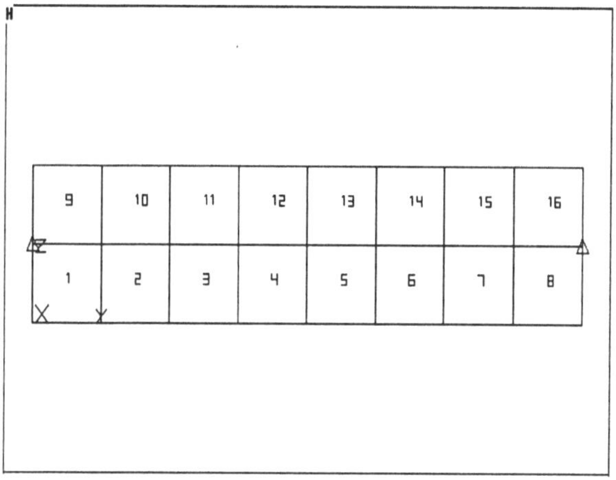

Figure 6.3.2 *Element numbers of sixteen-element mesh.*

can investigate whether any elements are lost, identify nodes with similar degrees-of-freedom, examine the numbering of elements and nodes. Lost elements create void areas in the model that can be easily identified when the model is relatively simple. For rather involved models and always starting from the main menu in SVIEWH, sequential execution of the commands: F5;SHRINK;F10;REDRAW, STRESS-DI;LIGHT, and STRESS-DI;DO-DITHER depicts the model in a way that makes it easier to locate any lost elements. Access of INQUIRE provides several options to examine the nodal degrees-of-freedom, the properties of finite elements, and the applied loads. In many instances, it is necessary to know the element and node numbering. In this case, the command F5 can be used to view the model with its node and element numbers. Figure 6.3.2 has been obtained by accessing F5, activating the "Ele num," and executing F10;REDRAW. A similar figure with the element node numbers can be obtained if we activate the "Node num."

Once we carefully review our model, we can process the "EX62. " file with SSAP0H. Execution of SSAP0H generates several binary files, e.g., "EX62.do," "EX62.nso," as well as the ASCII "EX62.l" and "EX62.s" files. The "EX62.do" and the "EX62.nso" are accessed by SVIEWH when we request information regarding the displacements and stresses. The "EX62.l" and "EX62.s" provide information about the input data of the model as well as the nodal displacements and element stresses. For a complete list of the files created by the pre- and the processors refer to the Algor Processor Reference Manual. We will examine the use of the output files in the next section where the analysis results are critically evaluated.

Table 6.3.2

Prepared by DECODS 2.10-S

```
27    1        1      0  0 0    0   0     0     0 000    0     0     0 0 386.4
 1  1  0  0  1  1  1  0.000000E+00 0.000000E+00 0.000000E+00  0    0. 0 0
 2  1  0  0  1  1  1  0.000000E+00 3.000000E+00 0.000000E+00  0    0.
 3  1  0  0  1  1  1  0.000000E+00 6.000000E+00 0.000000E+00  0    0.
 4  1  0  0  1  1  1  0.000000E+00 9.000000E+00 0.000000E+00  0    0.
 5  1  0  0  1  1  1  0.000000E+00 1.200000E+01 0.000000E+00  0    0.
 6  1  0  0  1  1  1  0.000000E+00 1.500000E+01 0.000000E+00  0    0.
 7  1  0  0  1  1  1  0.000000E+00 1.800000E+01 0.000000E+00  0    0.
 8  1  0  0  1  1  1  0.000000E+00 2.100000E+01 0.000000E+00  0    0.
 9  1  0  0  1  1  1  0.000000E+00 2.400000E+01 0.000000E+00  0    0.
10  1  1  1  1  1  1  0.000000E+00 0.000000E+00 3.000000E+00  0    0.
11  1  0  0  1  1  1  0.000000E+00 3.000000E+00 3.000000E+00  0    0.
12  1  0  0  1  1  1  0.000000E+00 6.000000E+00 3.000000E+00  0    0.
13  1  0  0  1  1  1  0.000000E+00 9.000000E+00 3.000000E+00  0    0.
14  1  0  0  1  1  1  0.000000E+00 1.200000E+01 3.000000E+00  0    0.
15  1  0  0  1  1  1  0.000000E+00 1.500000E+01 3.000000E+00  0    0.
16  1  0  0  1  1  1  0.000000E+00 1.800000E+01 3.000000E+00  0    0.
17  1  0  0  1  1  1  0.000000E+00 2.100000E+01 3.000000E+00  0    0.
18  1  1  1  1  1  1  0.000000E+00 2.400000E+01 3.000000E+00  0    0.
19  1  0  0  1  1  1  0.000000E+00 0.000000E+00 6.000000E+00  0    0.
20  1  0  0  1  1  1  0.000000E+00 3.000000E+00 6.000000E+00  0    0.
21  1  0  0  1  1  1  0.000000E+00 6.000000E+00 6.000000E+00  0    0.
22  1  0  0  1  1  1  0.000000E+00 9.000000E+00 6.000000E+00  0    0.
23  1  0  0  1  1  1  0.000000E+00 1.200000E+01 6.000000E+00  0    0.
24  1  0  0  1  1  1  0.000000E+00 1.500000E+01 6.000000E+00  0    0.
25  1  0  0  1  1  1  0.000000E+00 1.800000E+01 6.000000E+00  0    0.
26  1  0  0  1  1  1  0.000000E+00 2.100000E+01 6.000000E+00  0    0.
27  1  0  0  1  1  1  0.000000E+00 2.400000E+01 6.000000E+00  0    0.
 4  16     1     1     2    0
 1    1     0.2836      7.34E-04         0.
 0. 3.E+07 3.E+07      3.E+07        0.3           0.3           0.3 1.154E+07
 6.5E-06    6.5E-06     6.5E-06
        0.          1.          0.          0.          0.
        0.          0.          0.          0.         -1.
        0.          0.          0.          0.          0.
        1.          0.          0.          0.          0.
 1     1     2    11     10     1         0.           0.  0  0          1.
 2     2     3    12     11     1         0.           0.  0  0          1.
 3     3     4    13     12     1         0.           0.  0  0          1.
 4     4     5    14     13     1         0.           0.  0  0          1.
 5     5     6    15     14     1         0.           0.  0  0          1.
 6     6     7    16     15     1         0.           0.  0  0          1.
 7     7     8    17     16     1         0.           0.  0  0          1.
 8     8     9    18     17     1         0.           0.  0  0          1.
 9    20    19    10     11     1         0.        1000.  0  0          1.
10    21    20    11     12     1         0.        1000.  0  0          1.
11    22    21    12     13     1         0.        1000.  0  0          1.
12    23    22    13     14     1         0.        1000.  0  0          1.
13    24    23    14     15     1         0.        1000.  0  0          1.
14    25    24    15     16     1         0.        1000.  0  0          1.
15    26    25    16     17     1         0.        1000.  0  0          1.
16    27    26    17     18     1         0.        1000.  0  0          1.
 0     0
 1.    0.         0.          0.
```

6.4 INTERPRETATION OF GRAPHICAL RESULTS

The engineer is the one responsible for the accuracy of the results and not the finite element program. The program is merely a tool used to facilitate the analysis and perform the computations. Thus, the engineer carries the responsibility of evaluating the reliability of the finite element program, the development of sufficiently accurate models as well as the interpretation of the results. The importance of the latter responsibility cannot be overemphasized enough, since wrong interpretation of the results could lead to serious and possibly dangerous engineering errors.

The difficulty in interpreting the results depends on the complexity of the system and on the type(s) of elements used. Results involving volume and area elements are more difficult to interpret than results from beam and truss element models. The amount of the solution output depends on the size of the model and the type of the analysis. It is easier to examine the output for a small model of a system than the voluminous results for a refined model of the same system. The advantages of starting our analysis with small models and progressively moving to more elaborate ones have been discussed in Chapter 3.

The output is reported in graphical and numerical forms. The binary files with the extensions ".sst," ".nso," ".do," and ".ro" contain graphical results which can be viewed with SVIEWH. The ASCII files with the extensions ".l," ".s," and ".rl" can be accessed with a text editor.

The recommended approach to evaluate the results is to examine first the graphical output in order to verify that the displacements and stresses look reasonable, and second the numerical output. Sections 6.4.1 through 6.4.6 discuss various options and commands of SVIEWH that can be used to examine the graphical output. The evaluation of the numerical output is presented in Sections 6.5 through 6.5.3.

6.4.1 Displacement Results

Once the "EX62. " is processed with SSAP0H, we proceed to examine the deformed structure. This is accomplished through SVIEWH and by following the sequence of commands: DISPLACED;CAL SCAL;DISPLACED. In many instances viewing the model by activating both the DISPLACED and the WITH UND allows seeing the relationship between the deformed and the undeformed shapes. Viewing simultaneously the deformed and the undeformed shapes can help us to identify possible modeling errors. However, activating both options in large models could lead to plots that are too involved to evaluate the relationship between the undeformed and the deformed configurations. In this case, it may be preferable to view the deformed or portions of the deformed model, separately. In order to examine portions of a structure we may start from the main menu in SVIEWH and follow the sequence: OPTIONS;HIDE ELE;SELECT-E. The last command leads to the SELECT menu. After selecting the elements that we wish to view, we return to the HIDE ELE menu and activate the HIDE USEL to view only the selected elements. We can examine the displacements and stresses of only this part of the structure through the options provided in the DISPLACED and STRESS-DI menus. If we wish to view the whole structure again, we can disengage the HIDE USEL command by starting from the main menu and following the sequence: OPTIONS;HIDE ELE;UNHIDE.

The deformed shape of the short beam is shown in Fig. 6.4.1.1. It is consistent with what we expect from the physical behavior of the system. The anticipated deformation shape is further confirmed if we inquire the displacements at symmetrical nodes. That is, we verify that symmetric nodes have equal displacements parallel to the axis of symmetry and opposite displacements along an axis normal to the axis of symmetry. This test is a

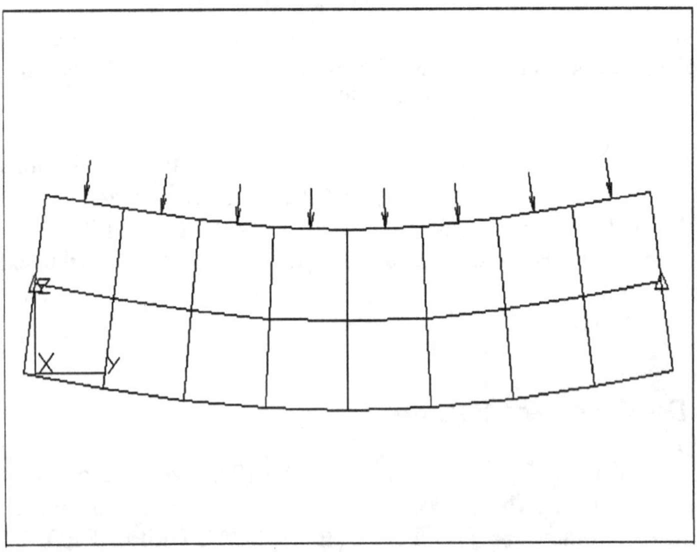

Figure 6.4.1.1 Deformed shape of short beam.

indication on whether the loads and the boundary conditions have been applied as symmetric. The magnitude of the displacements can serve as an additional check of the elastic properties specified in our model and the magnitude of the applied loads.

6.4.2 Stress Results

Once we make sure that the displacements look reasonable, we proceed to evaluate the reaction forces and stresses. Before discussing the specifics, we will briefly present the various choices to view the stress results. The SVIEWH provides two alternatives to display the stresses: a dithered plot, and a contour plot. Either alternative provides only one stress component. Usually, the most useful stresses to examine are the von Mises, the Tresca, the principal stresses, and the components along the directions of interest.

A dithered plot is a graphical display of the stress variation in the form of colored areas. Activation of SOLID-DI in the GENERAL option within the STRESS-DI menu provides a dithered plot of the stress variation that assigns solid colors to regions in the model based on their stress intensity. If the SOLID-DI option is not selected, then a smooth transition of the colors depicting the stress variation is drawn, thus, portraying a continuous variation of stresses in the model, see Fig. 6.7.2.3. Dithered plots provide a vivid and impressive presentation of stresses. Contour plots also provide an easy way to portray, examine, and understand stress results, see Fig. 6.4.3.1.

The recommended procedure to evaluate stress results is to examine:

a. the critical stresses in order to identify areas that have yielded. Such critical stresses include the von Mises and Tresca for ductile materials, pertinent critical stresses for composite materials, e.g., Tsai-Wu, and the maximum and minimum principal stresses for brittle materials. As elaborated in Section 6.4.3, the presence of yield stresses could invalidate the linear analysis results;

b. the reactions, as an additional verification of the model;

c. the stresses in the directions of interest; and

d. the precision to assess the accuracy of the results, and identify regions of the mesh that should be further refined.

The four-step sequence is described in Sections 6.4.3 through 6.4.6.

6.4.3 Von Mises, Tresca, and Principal Stresses

The von Mises stresses are simply the values obtained by carrying out the calculations on the left-hand side of eqn. (2.9.2) for every node. These stresses should not be used for design. They should be simply compared with the yield stress of the structural material under uniaxial tension, σ_y. If the von Mises stress at a node has exceeded σ_y, the part of the structure around this node has yielded.

Identifying areas that have yielded has several implications to our analysis and design. Strictly speaking, since the applied loading has developed stresses that have exceeded the linear range, linear analysis is no longer valid. If we are interested in calculating the stresses around the yielded region, we should perform a nonlinear analysis using an appropriate finite element program, e.g., AccuPak of Algor. If the yielded area is rather small and, in addition, it is not critical in maintaining the stability of the structure, then we may assert that the displacements and stresses away from the yielded region are acceptable. In order to obtain a dithered plot of the von Mises stresses we execute the sequence: STRESS-DI;POST;VON MISES followed by STRESS-DI;DO DITHER. In SVIEWH, we can easily locate any yielded areas through the sequence: STRESS-DI;AUX POST;THRESHOLD, setting the THRESHOLD value equal to σ_y and then performing a DO-DITHER in the STRESS-DI menu.

Figure 6.4.3.1 shows the contour plot of the von Mises stresses for the short beam. It has been obtained through the sequence: STRESS-DI;AUX POST;L CONTOUR;COLORS;NUMBER which was set equal to 4, followed by ESC;DISPLAY. Further editing in SD2H gave the final form of Fig. 6.4.3.1. Each contour line is assigned a color representing a constant value of stress. In this model four different colors have been used, each representing a stress increment of about 2226 psi. The contour plot has been scaled in increments based on the maximum and minimum stress values in the model. One may obtain an estimate of the stress variation between two successive contour lines by linear interpolation. The part of the model that has the largest number of contour lines is the most highly stressed. In our example these areas are near the top and the bottom fibers of the beam. In Fig. 6.4.3.1 the von Mises stress at the top contour line is about 11619.9 psi which, for the yield stress for uniaxial tension of steel $\sigma_y = 30000$ psi, indicates that no region in the structure has yielded.

If instead of the von Mises we request the Tresca*2 stresses, the stresses reported at a node are the values obtained by carrying out the calculations on the left-hand side of the following equation:

$$\sigma_1 - \sigma_3 = \sigma_y \qquad (6.4.3.1)$$

where σ_1 and σ_3 are the principal nodal stresses. As defined in Section 2.9, the difference (σ_1 - σ_3) is the Tresca*2 stress. Recall that the Tresca stresses are used to locate the areas in the model which are stressed beyond the elastic limit. Note that by calculating the Tresca*2 stresses, Algor allows direct comparison with σ_y and immediate identification of the structural regions that have yielded. When we identify any yielded regions and members, then we may choose one of the analysis alternatives for the von Mises stresses discussed earlier in this section.

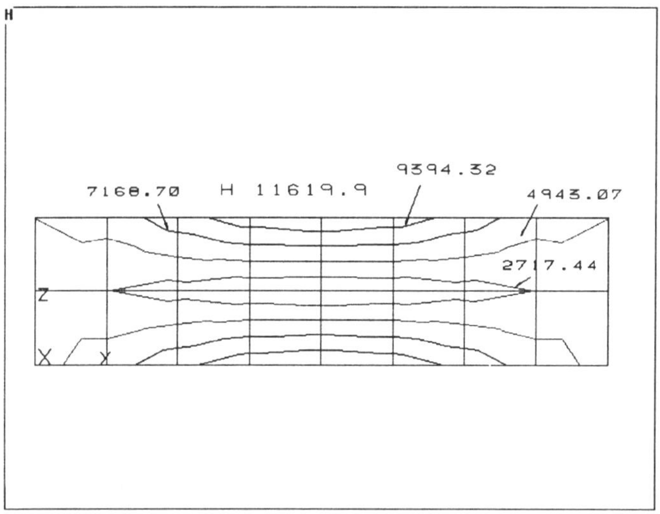

Figure 6.4.3.1 Contour plot of von Mises stresses.

6.4.4 Reaction Forces

Examination of the reaction forces can serve several purposes. It can be used as a check on whether we have applied the loads and boundary conditions correctly, and also visualize how equilibrium of the whole system is satisfied. It can also help us re-evaluate and modify the system's configuration, supports, and linkages.

There are two ways to determine the reactions:

a. The first one requires use of boundary elements at the supports. If we are interested in the components of the reaction forces along the global coordinate system, then we can use rigid boundary elements. However, if we want to calculate the reactions along any axes that are not parallel to the global coordinate system, then we must use elastic boundary elements, see Section 3.8.

b. The second method accesses the "EX62.rl" and "EX62.ro" files. The "EX62.rl" provides a listing of the nodal forces, reactions, and residual forces, and as an ASCII file it can be examined with a text editor. The size of this file is rather voluminous even for a moderate size model. The "EX62.ro" is binary. It contains the same data with the "EX62.rl" but is accessed with the SVIEWH. Specifically, in order to view the contour plot of the nodal reaction forces along the Z-axis in SVIEWH we can follow the sequence of commands: STRESS-DI;POST;REACT VEC;FRC REACT;VECTOR;Z-DIR, which should be succeeded by the sequence: STRESS-DI;AUX POST;L CONTOUR;DISPLAY. In order to obtain a dithered plot we must follow the sequence: STRESS-DI;DO DITHER.

6.4.5 Components of the S-tensor

Stress evaluation is a primary task in static and dynamic analysis. Wrong interpretation of stresses is probably one of the most serious errors; therefore, it requires caution and adequate effort.

As many finite element programs, Algor does not report element nodal stresses with respect to the global Cartesian system used to specify the geometry of the structure. It defines the stresses in the S-TENSOR sub-menu of POST-DI with respect to *local* or *element* coordinate systems. Specifically, Algor calculates the stresses S_{ij} (i, j = 1,2 3) at Gaussian points in each element and then, through interpolation, it determines the S_{ij} at the nodes. By providing the stresses at the nodes, it reports the stresses in a more useful way than by providing them inside the elements. Thus, in order to correctly interpret the nodal stresses we must know the orientation of the local coordinate systems.

Figure 6.4.5.1 shows the element numbering and the orientation of each element indicated by the skewed lines. The skewed or *orientation lines* have been drawn through the sequence of commands: OPTIONS;ELEM OPT;ELAST;ORIENT. The ORIENT option prompts for specification of a color. Once a color is selected, then F10;R draws a figure of the model that includes the orientation lines. They connect the center of each element to the point on the ij edge nearer to the i-node. For example, for element 5 the nodes i, j, k, and l are shown in Fig. 6.4.5.2. Once the element i,j,k,l node numbering is known, the local axes of the element can be defined. Specifically, the local 1-axis is determined by a line drawn from the midpoint of the il-side to the midpoint of the jk-side. The local 2-axis is normal to the 1-axis. Figure 6.4.5.2 displays the local 1- and 2- axes of element 5. The procedure to identify the element orientation is also applicable to quadrilateral plate/shell elements. A discussion on how to identify the local element axes for one-, two-, and three-dimensional elements is given in ViziCad.

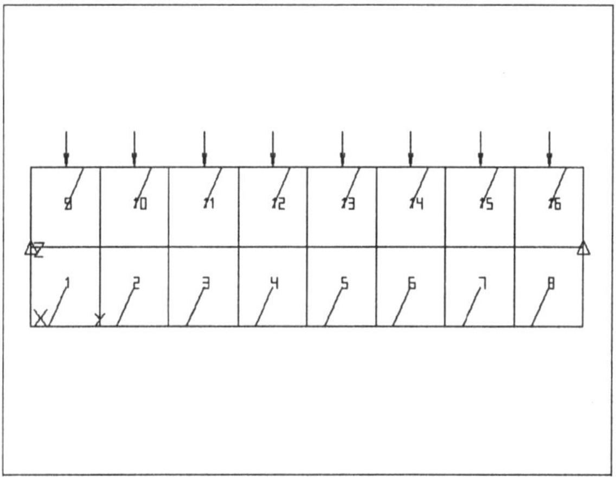

Figure 6.4.5.1 *Orientation lines.*

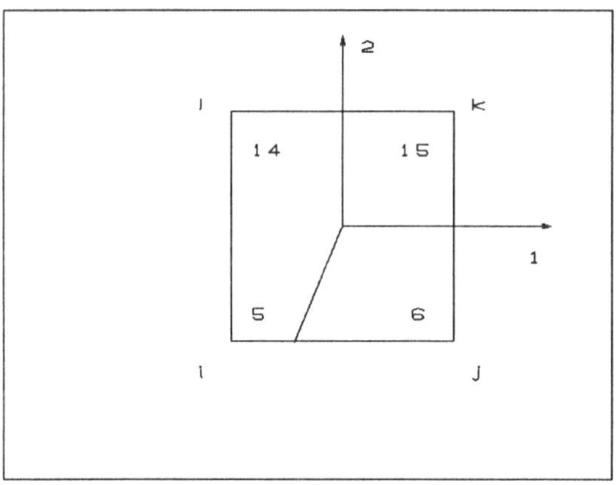

Figure 6.4.5.2 *Element orientation and local axes.*

6.4.6 Precision

The PRECISION allows a user to assess the accuracy of the mesh in evaluating the stresses, and also to identify the parts of a structure that must be modeled with a more refined mesh in order to increase the accuracy of the results.

As specified in ViziCad, the precision at each node is evaluated from

$$precision = \frac{0.5\,(\max\ val - \min\ val)}{Global\ von\ Mises\ \max} \qquad (6.4.6.1)$$

The nominator indicates the difference between the maximum and minimum von Mises stresses of the elements that share the node, while the denominator is the maximum von Mises stress in the model. Figure 6.4.6.1 shows the contour plot of precision for the short beam. Since the model is symmetric, the precision plot is also symmetric. A dithered plot of precision can be obtained through the sequence: STRESS-DI;POST;PRECISION, followed by STRESS-DI;DO DITHER.

For most applications, the mesh can be considered as adequate for the parts of the model where precision is less than 0.1. Nevertheless for areas with high precision which are away from the regions of interest, a finer mesh is usually not necessary, since according to Saint Venant's principle, further refinement would not significantly affect the areas of interest. The precision should be examined in conjunction with critical stresses, such as the von Mises, that is, regions with low precision (which implies good accuracy) and high stresses may still need to be further refined in order to obtain accurate results. Finally, further mesh refinement could be avoided in areas with high precision and low critical stresses.

6.5 INTERPRETATION OF NUMERICAL RESULTS

Graphical representation of the results provides an indication of the solution accuracy and the adequacy of the model to simulate the system. However, the real measure of the solution accuracy lies on the numerical results of the analysis.

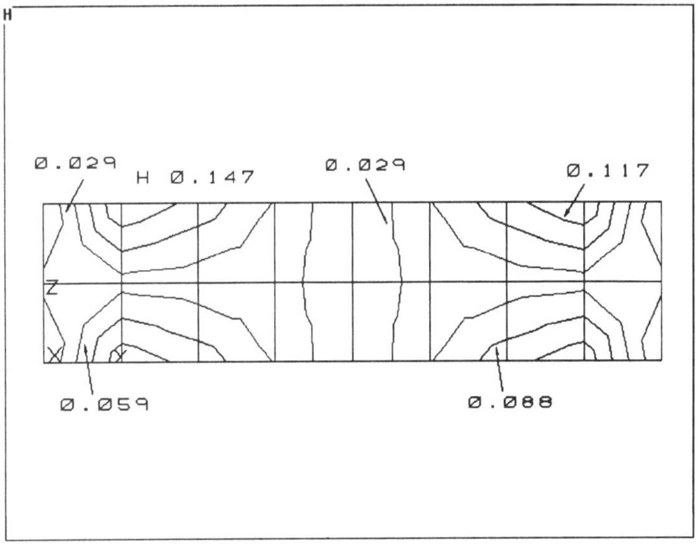

Figure 6.4.6.1 Contour plot of precision.

A coarse mesh could render reasonable deformations and stresses. Nevertheless, it might not provide an acceptable solution accuracy. Examination of the numerical results can help us assess the solution accuracy, and obtain useful information about the model. Numerical results of static analysis with Algor are provided in the files with extensions ".l", ".s", and ".rl." Listings of these files can serve as additional documentation of the calculations that accompany an analysis or a design report.

6.5.1 Displacement Output

A printout of the "EX62.l" is given in Table 6.5.1.1. The first part of the listing contains geometric, material, loading, and connectivity input data. Close examination of this part may reveal inadvertent modeling errors.

The second part contains information concerning data processing, such as bandwidth minimization, equation parameters, hard disk file size information for the processor, element load multipliers, and stiffness matrix parameters. Bandwidth minimization, equation parameters, and hard disk file size inform us about the size of the model, thus, assisting us to estimate the computational and hardware requirements for large size models in order to achieve the desired computational accuracy. By checking the element load multipliers we can verify the loading input. The stiffness matrix parameters provide information about the stiffness variation within the model. Specifically the "maximum/minimum" ratio indicates the presence of parts in the model that are simulated as either very soft or very stiff. A large maximum/minimum ratio might either create numerical instability with a considerable effect on the solution accuracy or could be detected by the processor as insufficient boundary conditions. A discussion on how to deal with soft and stiff regions in a model is presented in Section 3.7.

The last part of "EX62.1" provides the nodal displacements for each load case. Notice that the vertical displacement at node 5 is -0.0088 in.

Table 6.5.1.1

```
1**** Algor (c) Linear Stress Analysis - SSAP0 Rel. Ver. 11.02-3H

INPUT  FILE.............ex62
------------------------------------------------

     Prepared by DECODS 2.10-S

1**** CONTROL INFORMATION

     number of node points          (NUMNP)  =   27
     number of element types        (NELTYP) =    1
     number of load cases           (LL) =        1
     number of frequencies          (NF) =        0
     geometric stiffness flag       (GEOSTF) =    0
     analysis type code             (NDYN) =      0
     solution mode                  (MODEX) =     0
     equations per block            (KEQB) =      0
     weight and c.g. flag           (IWTCG) =     0
     bandwidth minimization flag    (MINBND) =    0
     gravitational constant         (GRAV) =      3.8640E+02

     bandwidth minimization specified
```

1**** TWO-DIMENSIONAL SOLID ELEMENTS

plane stress analysis

number of elements = 16
number of materials = 1
maximum temperatures per material = 1
analysis code = 2
axisymmetric.....0
plane strain.....1
plane stress.....2
incompatible displacement modes = 0
include.........0
suppress.........1

1**** MATERIAL PROPERTIES

material i.d. number = 1
number of temperatures = 1
weight density = 2.8360E-01
mass density = 7.3400E-04
beta angle = 0.0000E+00

TEMPERATURE	E(N)	E(S)	E(T)	NU(NS)	NU(NT)	NU(ST)	G(NS)
.0	3.000E+07	3.000E+07	3.000E+07	.300	.300	.300	1.154E+07

ALPHA(N)	ALPHA(S)	ALPHA(T)
6.500E-06	6.500E-06	6.500E-06

1**** ELEMENT LOAD MULTIPLIERS

	CASE A	CASE B	CASE C	CASE D
TEMP	0.000E+00	0.000E+00	0.000E+00	1.000E+00
PRES	1.000E+00	0.000E+00	0.000E+00	0.000E+00
X-DIR	0.000E+00	0.000E+00	0.000E+00	0.000E+00
Y-DIR	0.000E+00	0.000E+00	0.000E+00	0.000E+00
Z-DIR	0.000E+00	-1.000E+00	0.000E+00	0.000E+00

1**** ELEMENT CONNECTIVITY DATA

ELEM NO.	NODE I	NODE J	NODE K	NODE L	MAT'L INDEX	REFERENCE TEMP	I-J FACE PRESSURE	OP	THICKNESS
1	1	2	11	10	1	0.000E+00	0.000E+00	20	1.000E+00
2	2	3	12	11	1	0.000E+00	0.000E+00	20	1.000E+00
3	3	4	13	12	1	0.000E+00	0.000E+00	20	1.000E+00
4	4	5	14	13	1	0.000E+00	0.000E+00	20	1.000E+00
5	5	6	15	14	1	0.000E+00	0.000E+00	20	1.000E+00
6	6	7	16	15	1	0.000E+00	0.000E+00	20	1.000E+00
7	7	8	17	16	1	0.000E+00	0.000E+00	20	1.000E+00
8	8	9	18	17	1	0.000E+00	0.000E+00	20	1.000E+00
9	20	19	10	11	1	0.000E+00	1.000E+03	20	1.000E+00
10	21	20	11	12	1	0.000E+00	1.000E+03	20	1.000E+00
11	22	21	12	13	1	0.000E+00	1.000E+03	20	1.000E+00
12	23	22	13	14	1	0.000E+00	1.000E+03	20	1.000E+00
13	24	23	14	15	1	0.000E+00	1.000E+03	20	1.000E+00
14	25	24	15	16	1	0.000E+00	1.000E+03	20	1.000E+00
15	26	25	16	17	1	0.000E+00	1.000E+03	20	1.000E+00
16	27	26	17	18	1	0.000E+00	1.000E+03	20	1.000E+00

1**** ELEMENT LOAD MULTIPLIERS

load case	case A	case B	case C	case D
1	1.000E+00	0.000E+00	0.000E+00	0.000E+00

1**** STIFFNESS MATRIX PARAMETERS

minimum non-zero diagonal element =	1.3627E+07
maximum diagonal element =	5.4507E+07
maximum/minimum =	4.0000E+00
average diagonal element =	3.2704E+07
density of the matrix =	4.6714E+01

1**** STATIC ANALYSIS

LOAD CASE = 1

Displacements/Rotations(degrees) of nodes

NODE number	X- translation	Y- translation	Z- translation	X- rotation	Y- rotation	Z- rotation
1	0.0000E+00	-3.0966E-03	-2.8693E-04	0.0000E+00	0.0000E+00	0.0000E+00
2	0.0000E+00	-2.8763E-03	-3.4764E-03	0.0000E+00	0.0000E+00	0.0000E+00
3	0.0000E+00	-2.1562E-03	-6.3283E-03	0.0000E+00	0.0000E+00	0.0000E+00
4	0.0000E+00	-1.1559E-03	-8.1406E-03	0.0000E+00	0.0000E+00	0.0000E+00
5	0.0000E+00	0.0000E+00	-8.7810E-03	0.0000E+00	0.0000E+00	0.0000E+00
6	0.0000E+00	1.1559E-03	-8.1406E-03	0.0000E+00	0.0000E+00	0.0000E+00
7	0.0000E+00	2.1562E-03	-6.3283E-03	0.0000E+00	0.0000E+00	0.0000E+00
8	0.0000E+00	2.8763E-03	-3.4764E-03	0.0000E+00	0.0000E+00	0.0000E+00
9	0.0000E+00	3.0966E-03	-2.8693E-04	0.0000E+00	0.0000E+00	0.0000E+00
10	0.0000E+00	0.0000E+00	0.0000E+00	0.0000E+00	0.0000E+00	0.0000E+00
11	0.0000E+00	-1.2523E-05	-3.6573E-03	0.0000E+00	0.0000E+00	0.0000E+00
12	0.0000E+00	-9.9525E-06	-6.4604E-03	0.0000E+00	0.0000E+00	0.0000E+00
13	0.0000E+00	-4.5194E-06	-8.3370E-03	0.0000E+00	0.0000E+00	0.0000E+00
14	0.0000E+00	0.0000E+00	-8.9780E-03	0.0000E+00	0.0000E+00	0.0000E+00
15	0.0000E+00	4.5194E-06	-8.3370E-03	0.0000E+00	0.0000E+00	0.0000E+00
16	0.0000E+00	9.9525E-06	-6.4604E-03	0.0000E+00	0.0000E+00	0.0000E+00
17	0.0000E+00	1.2523E-05	-3.6573E-03	0.0000E+00	0.0000E+00	0.0000E+00
18	0.0000E+00	0.0000E+00	0.0000E+00	0.0000E+00	0.0000E+00	0.0000E+00
19	0.0000E+00	3.0234E-03	-3.9925E-04	0.0000E+00	0.0000E+00	0.0000E+00
20	0.0000E+00	2.8489E-03	-3.5601E-03	0.0000E+00	0.0000E+00	0.0000E+00
21	0.0000E+00	2.1392E-03	-6.4226E-03	0.0000E+00	0.0000E+00	0.0000E+00
22	0.0000E+00	1.1465E-03	-8.2346E-03	0.0000E+00	0.0000E+00	0.0000E+00
23	0.0000E+00	0.0000E+00	-8.8746E-03	0.0000E+00	0.0000E+00	0.0000E+00
24	0.0000E+00	-1.1465E-03	-8.2346E-03	0.0000E+00	0.0000E+00	0.0000E+00
25	0.0000E+00	-2.1392E-03	-6.4226E-03	0.0000E+00	0.0000E+00	0.0000E+00
26	0.0000E+00	-2.8489E-03	-3.5601E-03	0.0000E+00	0.0000E+00	0.0000E+00
27	0.0000E+00	-3.0234E-03	-3.9925E-04	0.0000E+00	0.0000E+00	0.0000E+00

6.5.2 Stress Output

Selected excerpts from the "EX62.s" are given in Table 6.5.2.1. The first part of this file lists input data pertinent to the number and type of elements in the model as well as the analysis type.

The second part lists the components of the S-tensor as well as the maximum and minimum principal stresses for each element for all load cases. As in SVIEWH, the stresses are given with respect to the element systems. For every element, each node is named as i, j, k, l, see Fig. 6.5.2.1 which depicts the four elements that are close to the left support of the model with their assigned element and node numbers as well as their orientation lines. The corresponding i, j, k, and l to the nodes of each element are assigned according to the orientation described in Section 6.4.5.

The information about the S-tensor given in "EX62.s" is more descriptive than the graphical stresses variation seen with SVIEWH. Specifically, SVIEWH provides average stress components at each node, while the "EX62.s" lists the stress components of all elements at each node. Thus, from the "EX62.s" we can calculate the average nodal stresses shown in

SVIEWH. For example, SVIEWH gives an σ_{22} = -449.5 psi at node 11. This value is the average of the σ_{22} stresses at the nodes k, l, l, k, of the elements 1, 2, 9, and 10, respectively, see Fig. 6.5.2.1, and is given below:

$$-449.5 = \frac{1}{4}(-1414 - 772.7 + 514.9 - 126.3) \qquad (6.5.2.1)$$

The nodal stresses on the right-hand side of eqn. (6.5.2.1) are underlined in Table 6.5.2.1.

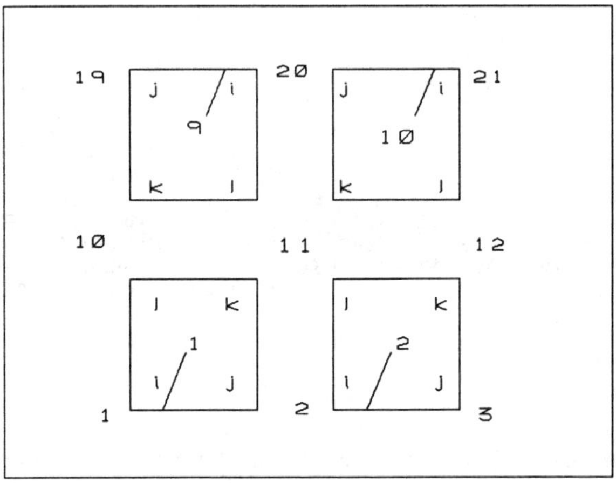

Figure 6.5.2.1 Orientation lines, node and element numbers.

Table 6.5.2.1

1**** Algor (c) FEA Stress Processor MKNSO, Ver 11.06-3H

INPUT FILE............ex62

**** 2-D Elasticity elements:

Number of elements = 16
Number of materials = 1
Maximum temperature pts =1
Analysis code = 2
0 : axisymmetric

1 : plane strain
2 : plane stress
Incompatible modes = 0
0 : included
1 : not included

**** Nodal stresses for 2-D elasticity elements:

El.#	LCND		Sigma-11	Sigma-22	Sigma-33	Tau-12	Sigma-Max	Sigma-Min	Sigma-Int
1	1	I	2.481E+03	3.264E+03	0.000E+00	-1.705E+03	4.622E+03	0.000E+00	1.123E+03
1	1	J	2.481E+03	-1.414E+03	0.000E+00	-1.705E+03	3.122E+03	-2.055E+03	0.000E+00
1	1	K	1.523E+02	-1.414E+03	0.000E+00	-1.705E+03	1.245E+03	-2.507E+03	0.000E+00
1	1	L	1.523E+02	3.264E+03	0.000E+00	-1.705E+03	4.017E+03	-5.999E+02	0.000E+00
2	1	I	7.042E+03	-7.727E+02	0.000E+00	-1.240E+03	7.234E+03	-9.649E+02	0.000E+00
2	1	J	7.042E+03	-2.848E+02	0.000E+00	-1.240E+03	7.246E+03	-4.891E+02	4.547E-13
2	1	K	-1.329E+02	-2.848E+02	0.000E+00	-1.240E+03	1.034E+03	-1.451E+03	0.000E+00
2	1	L	-1.329E+02	-7.727E+02	0.000E+00	-1.240E+03	8.280E+02	-1.734E+03	0.000E+00
5	1	I	1.148E+04	-2.520E+02	0.000E+00	2.501E+02	1.149E+04	-2.574E+02	-9.095E-13
5	1	J	1.148E+04	-2.462E+02	0.000E+00	2.501E+02	1.149E+04	-2.516E+02	-9.095E-13
5	1	K	-2.955E+01	-2.462E+02	0.000E+00	2.501E+02	1.347E+02	-4.105E+02	0.000E+00
5	1	L	-2.955E+01	-2.520E+02	0.000E+00	2.501E+02	1.329E+02	-4.145E+02	0.000E+00
9	1	I	-2.335E+03	5.149E+02	0.000E+00	-1.795E+03	1.382E+03	-3.202E+03	0.000E+00
9	1	J	-2.335E+03	-4.450E+03	0.000E+00	-1.795E+03	0.000E+00	-5.476E+03	-1.309E+03
9	1	K	-7.155E+02	-4.450E+03	0.000E+00	-1.795E+03	7.400E+00	-5.173E+03	0.000E+00
9	1	L	-7.155E+02	5.149E+02	0.000E+00	-1.795E+03	1.797E+03	-1.998E+03	0.000E+00
10	1	I	-7.224E+03	-7.210E+02	0.000E+00	-1.260E+03	0.000E+00	-7.460E+03	-4.855E+02
10	1	J	-7.224E+03	-1.263E+02	0.000E+00	-1.260E+03	9.061E+01	-7.441E+03	0.000E+00
10	1	K	-1.014E+02	-1.263E+02	0.000E+00	-1.260E+03	1.146E+03	-1.374E+03	0.000E+00
10	1	L	-1.014E+02	-7.210E+02	0.000E+00	-1.260E+03	8.861E+02	-1.708E+03	0.000E+00

6.5.3 Reaction Force Output

An excerpt from the printout of "EX62.rl" is given in Table 6.5.3.1. The first part of "EX62.rl" is not listed, since it contains information on the input data already presented in Tables 6.5.1.1 and 6.5.2.1. The ASCII file "EX62.rl" and the binary file "EX62.ro" are generated by the reaction force processor MKRFOH.

Table 6.5.3.1 lists the "element level nodal reactions" as well as the "global level nodal reactions, applied, and residual forces." The "element level nodal reactions" include the components of the reactions at each node. The information of the "EX62.rl" is more detailed than the graphical representation of nodal reactions seen with SVIEWH. Specifically, SVIEWH provides average nodal reaction forces, while "Ex61.rl" lists the reactions for all element nodes. For example, the vertical reaction obtained with SVIEWH at node 10, shown in Fig. 6.5.3.1, is R_{10Z} = 12016.0 lbs. The R_{10Z} is related to the vertical reaction forces at node 10 of elements 1 and 9, see the underlined forces in Table 6.5.3.1, as given by

$$-5114.7 - 6885.3 = -12000.0 \qquad (6.5.3.1)$$

The list of the "global level nodal reactions, applied forces, and residual forces" contains information that can also be accessed in SVIEWH. The residual forces are defined as the algebraic summation of the reactions and the applied forces. Obviously, such listing of the reaction forces at the supports is an alternative to the rigid boundary elements that provide the reaction forces at the boundaries. A discussion on the use of boundary elements is presented in Chapter 3.

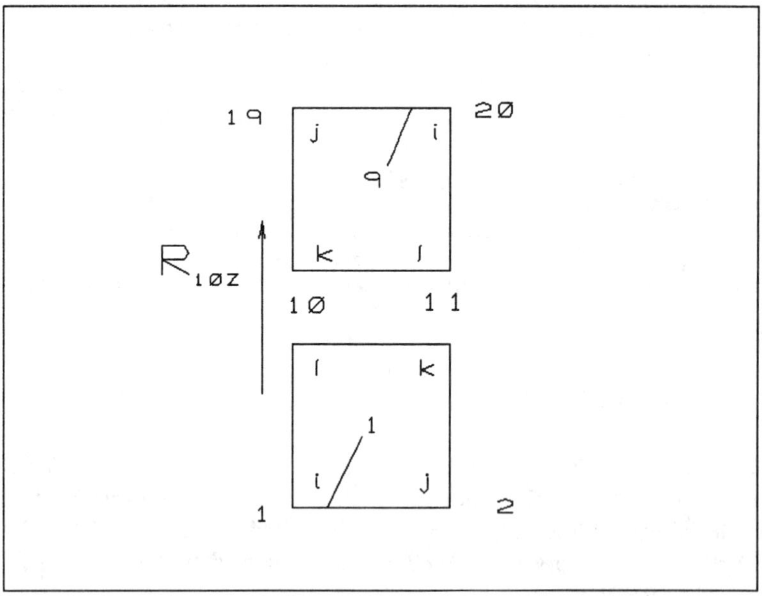

Figure 6.5.3.1 Vertical reaction at node 10.

Table 6.5.3.1

1**** Algor (c) Static FEA reaction processor MKRFO version 1.00-3H

INPUT FILE............ex62

Prepared by DECODS 2.10-S

1**** ELEMENT LEVEL NODAL REACTIONS

Group-number	Element-type	Element-number	Loadcase
1	4	1	1

NODE	FORCE-x	FORCE-y	FORCE-z	MOMENT-x	MOMENT-y	MOMENT
1	0.0000E+00	-1.0914E-11	0.0000E+00	0.0000E+00	0.0000E+00	0.0000E+
APPLD.	0.0000E+00	0.0000E+00	0.0000E+00	0.0000E+00	0.0000E+00	0.0000E+
2	0.0000E+00	-5.1147E+03	2.7755E+03	0.0000E+00	0.0000E+00	0.0000E+
APPLD.	0.0000E+00	0.0000E+00	0.0000E+00	0.0000E+00	0.0000E+00	0.0000E+
11	0.0000E+00	1.1644E+03	2.3392E+03	0.0000E+00	0.0000E+00	0.0000E+
APPLD.	0.0000E+00	0.0000E+00	0.0000E+00	0.0000E+00	0.0000E+00	0.0000E+
10	0.0000E+00	3.9503E+03	-5.1147E+03	0.0000E+00	0.0000E+00	0.0000E+
APPLD.	0.0000E+00	0.0000E+00	0.0000E+00	0.0000E+00	0.0000E+00	0.0000E+

Group-number	Element-type	Element-number	Loadcase
1	4	2	1

NODE	FORCE-x	FORCE-y	FORCE-z	MOMENT-x	MOMENT-y	MOMENT-z

Group-number	Element-type	Element-number	Loadcase
1	4	9	1

NODE	FORCE-x	FORCE-y	FORCE-z	MOMENT-x	MOMENT-y	MOMENT-
20	0.0000E+00	5.3853E+03	4.4028E+03	0.0000E+00	0.0000E+00	0.0000E+
APPLD.	0.0000E+00	0.0000E+00	-1.5000E+03	0.0000E+00	0.0000E+00	0.0000E+
19	0.0000E+00	3.6380E-12	1.5000E+03	0.0000E+00	0.0000E+00	0.0000E+
APPLD.	0.0000E+00	0.0000E+00	-1.5000E+03	0.0000E+00	0.0000E+00	0.0000E+
10	0.0000E+00	-4.5756E+03	-6.8853E+03	0.0000E+00	0.0000E+00	0.0000E+
APPLD.	0.0000E+00	0.0000E+00	0.0000E+00	0.0000E+00	0.0000E+00	0.0000E+
11	0.0000E+00	-8.0969E+02	9.8252E+02	0.0000E+00	0.0000E+00	0.0000E+
APPLD.	0.0000E+00	0.0000E+00	0.0000E+00	0.0000E+00	0.0000E+00	0.0000E+

1**** GLOBAL LEVEL NODAL REACTIONS(R), APPLIED FORCES(F), & RESIDUAL(R+F)

NODE	LCASE	RF+	FORCE-x	FORCE-y	FORCE-z	MOMENT-x	MOMENT-y	MOMENT
10	1	R	0.00E+00	-6.25E+02	-1.20E+04	0.00E+00	0.00E+00	0.00E+0
10	1	F	0.00E+00	0.00E+00	0.00E+00	0.00E+00	0.00E+00	0.00E+0
10	1	+	0.00E+00	-6.25E+02	-1.20E+04	0.00E+00	0.00E+00	0.00E+0

6.6 ACCURACY AND CONVERGENCE STUDIES

The checks that we have performed so far indicate that the results are reasonable. The question about their accuracy is still unanswered. Assessing the accuracy of the results requires some comparisons and possibly a convergence study.

When checking the accuracy, we must first decide whether we are mostly interested in high accuracy of the displacements, the stresses, or both. If we are primarily interested in displacements, it is not necessary to examine the convergence of the stresses. When our goal is to calculate the stresses with "sufficient" accuracy, then we must perform a convergence study to examine the stresses at the areas of interest. In this case, examining the displacements is not necessary, since the convergence of stresses is slower than the convergence of displacements. Apparently, when we are interested in the accuracy of both displacements and stresses, achieving the desired accuracy for the stresses is sufficient.

Available experimental data would be the best means to validate the finite element model and the accuracy of the results. However, if only experimentally measured displacements are available, obtaining the same displacements with the finite element model does not assure that the stresses are also validated. In such a case it might be necessary to perform a convergence study for the stresses. Validation of the stresses with experimental data confirms the accuracy of the displacements.

It would be ideal if the analytical solution of the problem was available so that a comparison could be made with the finite element solution. In this case, however, we might not have resorted to a finite element solution, except if we wanted to check the accuracy of the finite element software. Even though it is very unlikely that an analytical solution of a complex problem would be available, checking our results with a simple analytical model is a prudent step in assessing the accuracy of the finite element analysis. A comparison using a simple model can only be quantitative, providing us with some additional insight on the system's behavior as well as an estimate of the order of magnitude of the anticipated results. For example, a simple model that simulates the short beam could be a simply supported steel beam subjected to a uniform load q. The moment of inertia about the X-axis would be given by

$$I = \frac{1x6^3}{12} = 18$$

Using this simple model, the vertical displacement at mid-span can be calculated from beam theory that ignores shear deformation, to obtain

$$\delta = \frac{5q(2L)^4}{384EI} = \frac{(5000)(24)^4}{(384)(3X10^7)(18)} = 0.008 \ in \qquad (6.6.1)$$

This value is smaller than the vertical deflection at node 5 of the finite element model, i.e., 0.0088 in. This difference is anticipated since the finite element method can account for the additional deformation due to shear.

The bending stresses of the lower fibers at mid-span are given by

$$\sigma = \frac{Mc}{I} = \frac{q(2L)^2c}{8I} = \frac{(1000)(24)^2(3)}{(8)(18)} = 11{,}484 \ psi \qquad (6.6.2)$$

The sixteen-element model gave σ_{11} = 1.148E+04 psi at node 5, see σ_{11} at node i for element 5 in Table 6.5.2.1. Comparison of the displacements and stresses between the approximate analytical solution and the finite element analysis may serve as an additional validation of our model.

The next step of our analysis is to perform a convergence study. The procedure involves developing various mesh refinements and examining the results until they converge. The process may require a number of iterations, which, however, should be relatively few, if we follow the modeling guidelines of Chapter 3.

In deciding on the mesh refinements we can be assisted by the variation of the precision in conjunction with either the von Mises stresses or the principal stresses. The procedure is illustrated starting with the sixteen-element model. Initially, we must choose the nodes to be examined in the convergence study. Nodes 5 and 3 at the lower side of the beam are selected, see Fig. 6.2.2. Node 5 lies on the axis of symmetry, while node 3 is 6 in to the left of node 5. We avoided choosing any nodes close to the concentrated loads, such as the reactions, since stresses close to concentrated loads can be artificially increased with mesh refinement.

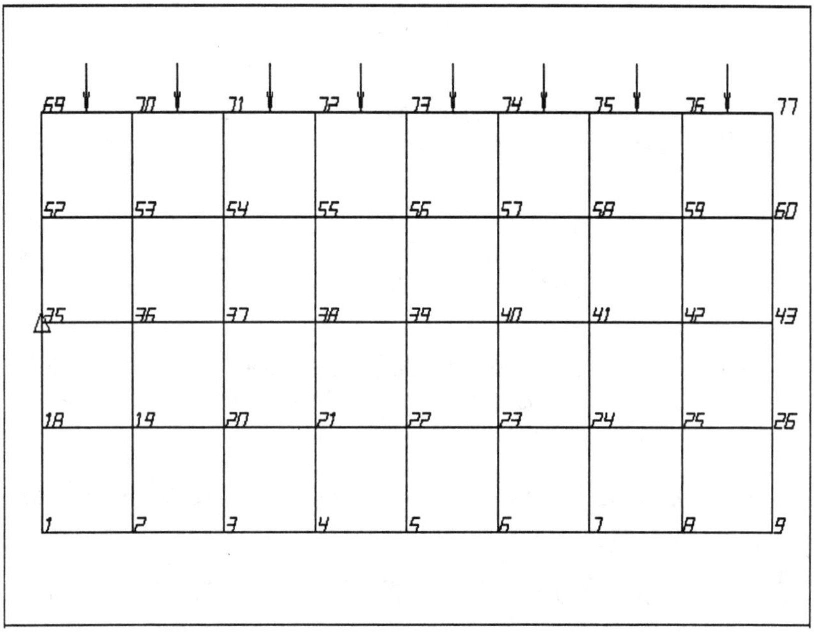

***Figure 6.6.1** One-half of sixty four-element model.*

Table 6.6.1 lists the displacements along the Z-axis, the von Mises stresses, and the precision at nodes 5 and 3. As expected, the precision at node 5 is zero, since this node lies on the axis of symmetry. This does not imply that the region around the axis of symmetry does not need any further refinement. Since the von Mises stresses are the largest along the axis of symmetry, a more refined mesh would be advised. If we accept results with a precision less that 0.088, see Fig. 6.4.6.1, further refinement would also be necessary in the vicinity of node 3.

A sixty four-element mesh is our second test model. The file name given to the sixty four-element model is "EX6264. ". For clarity, only half the model is shown in Fig. 6.6.1. The results of the analysis are also shown in Table 6.6.1. Notice that nodes 3 and 5 of the sixteen-element model correspond to nodes 5 and 9 of the sixty four-element model. Comparison of the results indicates differences in the order of 3-4%. We may conclude that the sixteen-element model would be sufficient to calculate both displacements and stresses.

Table 6.6.1

No. Elem.	Node	Displacem.	V. Mises	Precision
16	3	-6.33E-3	8730.9	0.106
16	5	-8.78E-3	11620	0.0
64	5	-6.61E-3	9085.4	0.053
64	9	-9.13E-3	12006.2	0.0

6.7 ANALYSIS OF PLATE STRUCTURE

Consider the plate shown in Fig. 6.7.1. It is simply supported along its long edges and clamped along the other edges. The 1-inch thick steel plate is characterized by a Young's modulus E = $3X10^7$ psi, a Poisson's ratio $\nu = 0.30$,

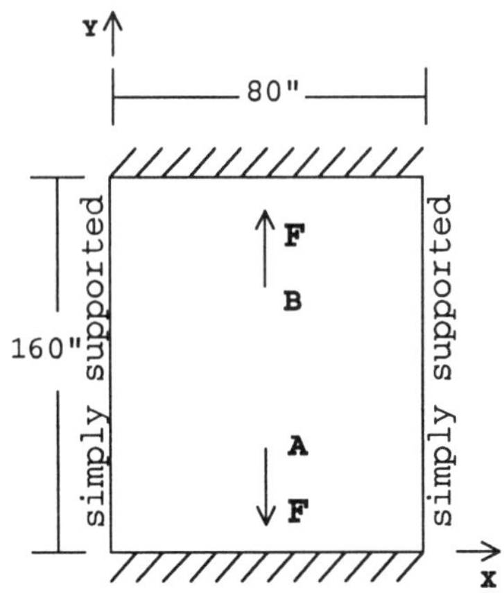

Figure 6.7.1 *Plate subjected to uniform and concentrated loads.*

and a weight density $\gamma = 0.2836$ lb/in^3. It is subjected to a uniform pressure q = 4 psi along the Z-axis and to the concentrated loads F = 10 kips applied along the axis of symmetry at the points A and B with coordinates (40,40) and

(40,120), respectively, see Fig. 6.7.1.

Since the system has two axes of symmetry, it is preferable to model only a quarter-plate with boundary conditions as specified in Section 3.9, see Fig. 6.7.2. Notice that in the quarter model, the load at point A acting along the axis of symmetry has been reduced to 5000 lbs.

The presence of both normal and in-plane loads generates bending and membrane deformations. Since the plate structure lies on only one plane and the loads are either normal or parallel to the mid-surface, the deformations and stresses that develop in the plate are decoupled. This implies that the action of the uniform pressure does not create any membrane stresses and deformations, and the concentrated loads do not affect the bending behavior of the plate, see Section 2.7. Later, we will refer to this observation to compare the results of the finite element analysis with an analytical solution.

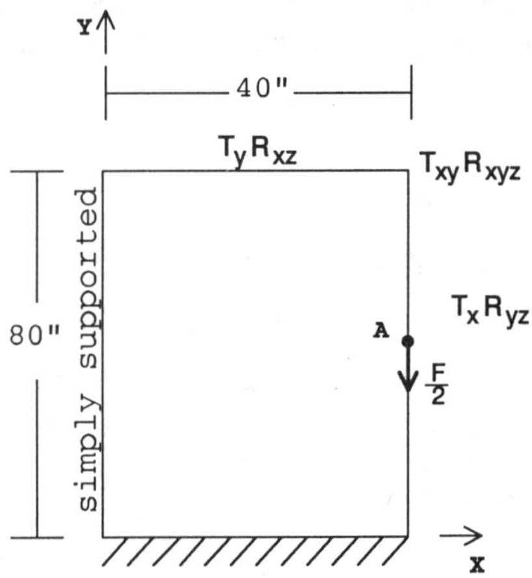

Figure 6.7.2 Quarter model of plate.

6.7.1 Analysis with Algor

Since several basic commands to develop a model with Algor have been presented in Section 6.3, we will emphasize only additional aspects of the procedure.

The quarter model created in SD2H is shown in Fig. 6.7.1.1. It consists of eighteen equal quadrilateral elements drawn in the XY plane. Since the plate consists of one material and has uniform thickness, only one value is assigned to COLOR (green No.1) and GROUP (green No.1). After we have assigned the boundary conditions and applied the concentrated load, we transfer the file to the decoder.

In DECODS we make the following selections: in ELEMENTS we select TYPE: Plate, then INFO: Veubeke; in GROUP we input the weight density, the Young's modulus, and the Poissson's ratio; in COLOR we set the thickness, the pressure, $PP_x = PP_y = 0$ and $PP_z = 1$ to specify the direction of the uniform pressure; in GLOBAL we input 1 for the pressure multiplier in the 1st load case (LC), and finally, in DECODE we select RUN. Note that by setting 0 in the B(Accel) column in GLOBAL, we are not including the effect of gravity loads. Table 6.7.1.1 lists the input data in DECODS. Execution of DECODS converts the geometric model created in SD2H to a finite element model. The ASCII file "EX67. " created by DECODS is listed in Table 6.7.1.2. As discussed in Section 6.3, before processing the "EX67. " with the linear static processor SSAP0H it is good practice to examine the model with the SVIEWH.

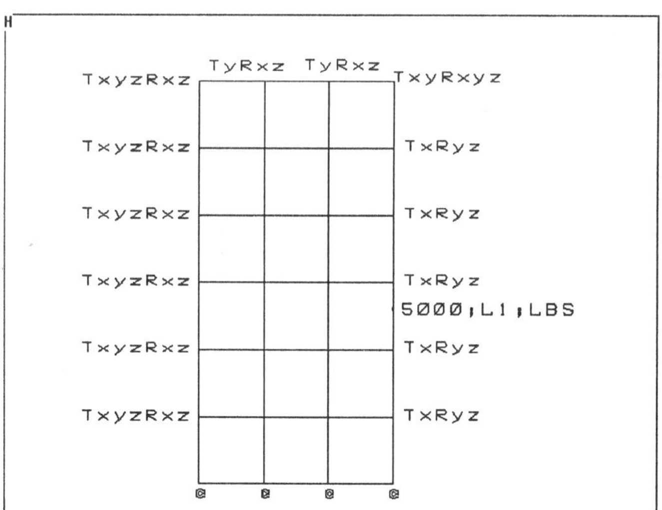

Figure 6.7.1.1 Eighteen-element mesh of quarter model.

Table 6.7.1.1

Elements	Type	Information	Group [ ]	Color [ ]
	6) Plate	o) Veubeke		

Group						
Name	Lib	Density	Young's	Poisson	Alpha	G
steel	yes	0.2836	3E7	0.3	6.5E-6	

Color						
Color	Thickness	dTref	dT/dh	Pressure	PPx	PPy
1	1			4		

Color (continued)						
PPz	Orientation	IPx	IPy	IPz		
1						

Analysis: Static: Stress Analysis			
Grav	Ax	Ay	Az
386.4	0	0	-1

Global: Load case [ ]				
Load Case	A(press)	B(Accel)	C(Disp)	D(Therm)
1	1	0	0	0

After carefully reviewing our model with SVIEWH, we process the "EX67. " with SSAP0H. The procedure to examine the binary file "EX67.sst", and the ASCII files "EX67.1" and EX67.s" generated by SSAP0H is discussed in the following sections. As elaborated in Section 6.3, the linear static processor generates several other ASCII and binary files.

Table 6.7.1.2

Prepared by DECODS 2.10-S

```
28     1     1     0  0 0     0     0     0     0  000    0    0    0    0  386.4
   1  1 1 1 1 1 1  0.000000E+00 0.000000E+00 0.000000E+00   0      0. 0 0
   2  1 1 1 1 1 1  1.333333E+01 0.000000E+00 0.000000E+00   0      0.
   3  1 1 1 1 1 1  2.666667E+01 0.000000E+00 0.000000E+00   0      0.
   4  1 1 1 1 1 1  4.000000E+01 0.000000E+00 0.000000E+00   0      0.
   5  1 1 1 1 0 1  0.000000E+00 1.333333E+01 0.000000E+00   0      0.
   6  0 0 0 0 0 0  1.333333E+01 1.333333E+01 0.000000E+00   0      0.
   7  0 0 0 0 0 0  2.666667E+01 1.333333E+01 0.000000E+00   0      0.
   8  1 0 0 0 1 1  4.000000E+01 1.333333E+01 0.000000E+00   0      0.
   9  1 1 1 1 0 1  0.000000E+00 2.666667E+01 0.000000E+00   0      0.
  10  0 0 0 0 0 0  1.333333E+01 2.666667E+01 0.000000E+00   0      0.
  11  0 0 0 0 0 0  2.666667E+01 2.666667E+01 0.000000E+00   0      0.
  12  1 0 0 0 1 1  4.000000E+01 2.666667E+01 0.000000E+00   0      0.
  13  1 1 1 1 0 1  0.000000E+00 4.000000E+01 0.000000E+00   0      0.
  14  0 0 0 0 0 0  1.333333E+01 4.000000E+01 0.000000E+00   0      0.
  15  0 0 0 0 0 0  2.666667E+01 4.000000E+01 0.000000E+00   0      0.
  16  1 0 0 0 1 1  4.000000E+01 4.000000E+01 0.000000E+00   0      0.
  17  1 1 1 1 0 1  0.000000E+00 5.333334E+01 0.000000E+00   0      0.
  18  0 0 0 0 0 0  1.333333E+01 5.333334E+01 0.000000E+00   0      0.
  19  0 0 0 0 0 0  2.666667E+01 5.333334E+01 0.000000E+00   0      0.
  20  1 0 0 0 1 1  4.000000E+01 5.333334E+01 0.000000E+00   0      0.
  21  1 1 1 1 0 1  0.000000E+00 6.666666E+01 0.000000E+00   0      0.
  22  0 0 0 0 0 0  1.333333E+01 6.666666E+01 0.000000E+00   0      0.
  23  0 0 0 0 0 0  2.666667E+01 6.666666E+01 0.000000E+00   0      0.
  24  1 0 0 0 1 1  4.000000E+01 6.666666E+01 0.000000E+00   0      0.
  25  1 1 1 1 0 1  0.000000E+00 8.000000E+01 0.000000E+00   0      0.
  26  0 1 0 1 0 1  1.333333E+01 8.000000E+01 0.000000E+00   0      0.
  27  0 1 0 1 0 1  2.666667E+01 8.000000E+01 0.000000E+00   0      0.
  28  1 1 0 1 1 1  4.000000E+01 8.000000E+01 0.000000E+00   0      0.
   6    18     1     0     0
         1          0.      0.2836   7.34E-04    6.5E-06    6.5E-06         0.
3.297E+07   9.89E+06          0. 3.297E+07          0. 1.154E+07
        1.          0.          0.          0.
        0.          0.          0.          1.
        0.          0.          0.          0.
        0.          0.          0.          0.
        0.      -386.4          0.          0.
   1     2     1     5     6     0     1 0        1.         4.         0.         0.
   2     3     2     6     7     0     1 0        1.         4.         0.         0.
   3     4     3     7     8     0     1 0        1.         4.         0.         0.
   4     6     5     9    10     0     1 0        1.         4.         0.         0.
   5     7     6    10    11     0     1 0        1.         4.         0.         0.
   6     8     7    11    12     0     1 0        1.         4.         0.         0.
   7    10     9    13    14     0     1 0        1.         4.         0.         0.
   8    11    10    14    15     0     1 0        1.         4.         0.         0.
   9    12    11    15    16     0     1 0        1.         4.         0.         0.
  10    14    13    17    18     0     1 0        1.         4.         0.         0.
  11    15    14    18    19     0     1 0        1.         4.         0.         0.
  12    16    15    19    20     0     1 0        1.         4.         0.         0.
  13    18    17    21    22     0     1 0        1.         4.         0.         0.
  14    19    18    22    23     0     1 0        1.         4.         0.         0.
  15    20    19    23    24     0     1 0        1.         4.         0.         0.
  16    22    21    25    26     0     1 0        1.         4.         0.         0.
  17    23    22    26    27     0     1 0        1.         4.         0.         0.
  18    24    23    27    28     0     1 0        1.         4.         0.         0.
  16     1          0.      -5000.          0.         0.          0.         0. 0
   0     0
        1.          0.          0.          0.
```

6.7.2 Interpretation of Graphical Results

The deformations can be examined with SVIEWH following the procedure presented in Section 6.4.1. Figure 6.7.2.1 shows the deformed and the undeformed shapes of the plate. As expected, the maximum vertical deformation develops at the center of the plate, i.e., at node 28, and is -0.485 in, see underlined value in Table 6.7.3.1. Since the vertical displacements are not affected by the in-plane forces, a comparison with analytical results for the plate subjected to only the uniform pressure is possible. Several references (e.g., Case No. 5, Table 26 in Chapter 10 of *Roark's Formulas for Stress & Strain*, Sixth Edition) provide expressions to calculate the maximum vertical displacement at the center. Such an expression gave -0.503 in for the vertical deflection at the center of the plate. The 3.58% difference between the results obtained with the eighteen-element model and the analytical solution can be reduced if a more refined mesh is used.

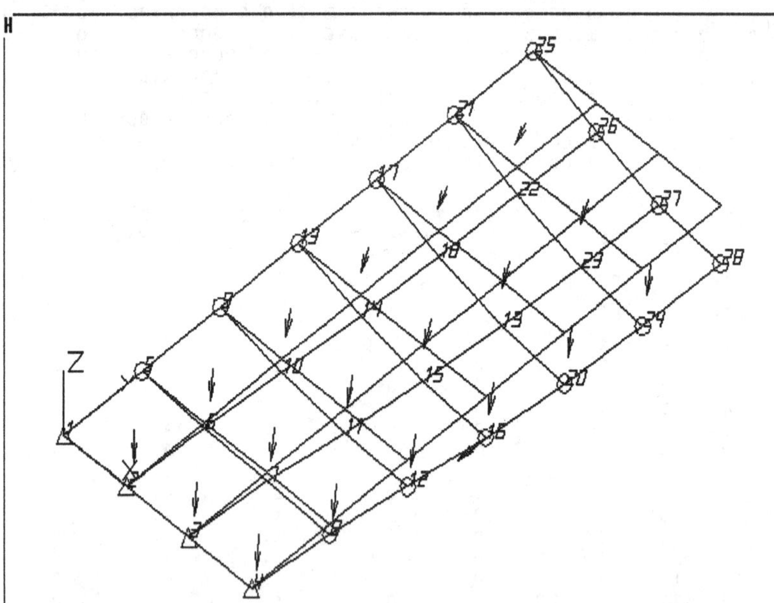

Figure 6.7.2.1 *Deformed/underformed shape of quarter plate.*

Examination of stresses in plate elements is more involved than that of other types of elements. This is attributed to the bending and the membrane behavior of plates that generate several stress components, that is, *bending, moment*, and *membrane* stresses, see Fig. 2.7.2 for the bending stresses and Fig. 2.4.2 for the in-plane stresses. The combinations of bending and membrane stresses at the top and bottom surfaces of a plate are the so-called *surface* stresses. The process of examining the stresses will be demonstrated for node 4 of Fig. 6.7.2.1.

As discussed in Section 6.4.5, Algor provides the stresses in local or element systems. Thus, the first step would be to identify the element systems through the ORIENT command. The command sequence is provided in Section 6.4.5 with the only difference being the selection of PLATE instead of ELAST in the ELEM OPT menu. Figure 6.7.2.2 shows the orientation lines together with the element and node numbers. Notice that the local 1-axis for all elements is parallel to the global X-axis, and consequently the local 2-axis is parallel to the global Y-axis. We can now proceed to view the local S_{ij} stresses.

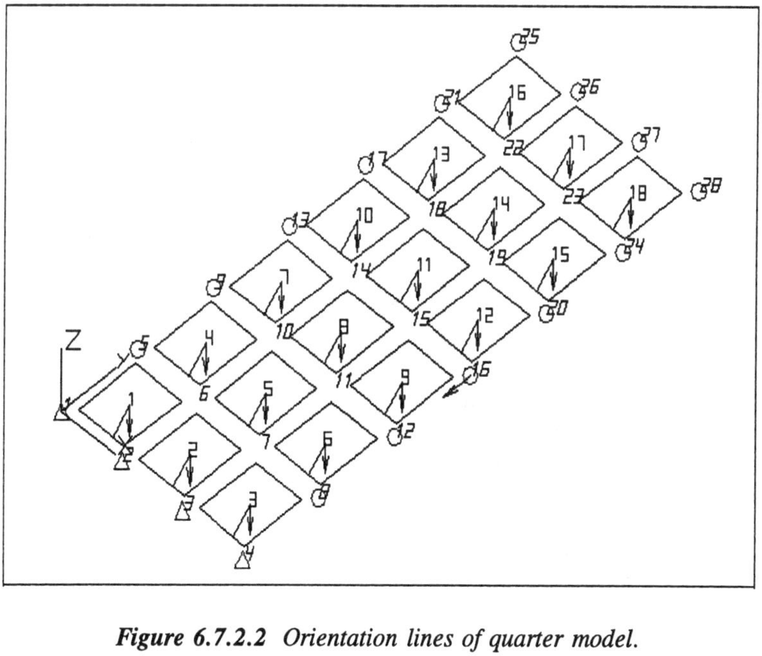

Figure 6.7.2.2 *Orientation lines of quarter model.*

Figure 6.7.2.3 depicts the S_{22} bending stresses. Starting from the main menu in SVIEWH, Fig. 6.7.2.3 has been drawn through the following sequence of commands:

(a) STRESS-DI;POST;S TENSOR;S22
(b) STRESS-DI;LGND BOX;WIND ADJ
(c) STRESS-DI;AUX POST;TYPE 6 SW;BENDING
(d) STRESS-DI;DO DITHER.

The stress at a specific node can be requested with the sequence STRESS DI;AUX POST;GET VAL.

The process can be repeated to obtain the S_{11} bending stresses with the obvious modification in the (a) sequence of commands. In order to view the membrane, the moment, or the surface stresses along the 2-axis, the sequence of commands (c) should be modified to: STRESS-DI;AUX POST;TYPE 6 SW;MEMBRANE, MOMENT, or SURFACE, respectively. Note that the stresses that are viewed with the above sequence are the ones on the "front side" of the structure.

It is also worth mentioning the use of the BACKSIDE command within the AUX POST menu. When the BACKSIDE is activated, SVIEWH shows the surface stresses on the "back side" of the model. The BACKSIDE has no effect

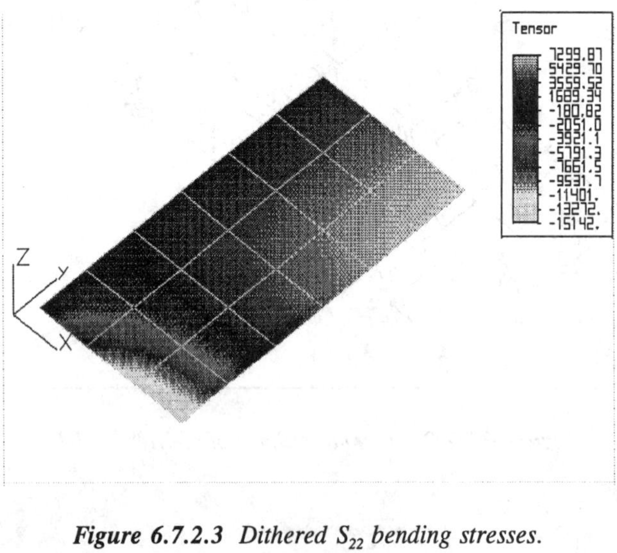

Figure 6.7.2.3 Dithered S_{22} bending stresses.

on the bending, the moment, and the membrane stresses. It can also be used to view the von Mises stresses on the plate surfaces. For example, in order to view the von Mises surface backside stresses we must change the commands (a) and (c) as follows:

(a) STRESS-DI;POST;VON MISES
(c) STRESS-DI;AUX POST;BACKSIDE;TYPE 6 SW;SURFACE.

Table 6.7.2.1 lists the S_{22} bending (b), the S_{22} membrane (m), the S_{22} front surface (fs), the S_{22} back surface (bs), the von Mises front surface (vMfs), and the von Mises back surface (vMbs) stresses at node 4.

Table 6.7.2.1

stress	b	m	fs	bs	vMfs	vMbs
psi	-15142	-97	15045	-15239	13297	13470

The negative sign of the (bs) stress indicates that the backside of the plate in the vicinity of node 4 is in compression. Analytical expressions in the literature (e.g., Case No. 5, Table 26 in Chapter 10 of *Roark's Formulas for Stress & Strain*, Sixth Edition) gave a bending stress S_{22} = -18294 psi. The difference is attributed to the rather coarse mesh of the eighteen-element model. Use of a more refined mesh based on the recommendations in Chapter 3 would improve the solution accuracy. The (fs) and (bs) stresses are related to the (b) and (m) stresses as follows:

$$15045 = 15142 - 97$$

$$-15239 = -15142 - 97$$

In design, the most commonly used stresses are the moment stresses per unit length, see Section 2.7. The moment stress per unit length along the 2-axis at node 4 is M_2 = -2524 lb-in/in. According to eqns. (2.7.3), it is related to bending stress as given by

$$\frac{(-15142)(1)^2}{6} = -2524$$

The surface von Mises stresses of the model should be compared with the yield stress σ_y to identify any regions that have yielded.

Finally, the reaction forces and precision can be examined as elaborated in Sections 6.4.4 and 6.4.6.

6.7.3 Interpretation of Numerical Results

The linear processor SSAP0H generates the ASCII files "EX67.1," "EX67.s," and "EX67.rl." Table 6.7.3.1 presents excerpts from the "EX67.1" file. It lists the control information that describes in a general sense the type of the model and analysis; the nodal data with the boundary conditions and nodal coordinates; the number of elements and their material properties; the element load multipliers and connectivity data; the equation parameters and hard-disc space information to assess the size of the model; the nodal loads and masses; the element load multipliers; the stiffness matrix parameters that can be used to detect instability and also properly model "soft" and "stiff" areas of the model; and finally, the nodal displacements and rotations.

Table 6.7.3.1

1**** Algor (c) Linear Stress Analysis - SSAP0 Rel. Ver. 11.02-3H

INPUT FILE............ex67

Prepared by DECODS 2.10-S

1**** CONTROL INFORMATION

number of node points	(NUMNP) =	28
number of element types	(NELTYP) =	1
number of load cases	(LL) =	1
number of frequencies	(NF) =	0
geometric stiffness flag	(GEOSTF) =	0
analysis type code	(NDYN) =	0
solution mode	(MODEX) =	0
equations per block	(KEQB) =	0
weight and c.g. flag	(IWTCG) =	0
bandwidth minimization flag	(MINBND) =	0
gravitational constant	(GRAV) =	3.8640E+02

bandwidth minimization specified

1**** NODAL DATA

NODE NO.	BOUNDARY CONDITION CODES DX	DY	DZ	RX	RY	RZ	NODAL POINT COORDINATES X	Y	Z	T
1	1	1	1	1	1	1	0.000E+00	0.000E+00	0.000E+00	0.000E+00
2	1	1	1	1	1	1	1.333E+01	0.000E+00	0.000E+00	0.000E+00
3	1	1	1	1	1	1	2.667E+01	0.000E+00	0.000E+00	0.000E+00
4	1	1	1	1	1	1	4.000E+01	0.000E+00	0.000E+00	0.000E+00
5	1	1	1	1	0	1	0.000E+00	1.333E+01	0.000E+00	0.000E+00
6	0	0	0	0	0	0	1.333E+01	1.333E+01	0.000E+00	0.000E+00
7	0	0	0	0	0	0	2.667E+01	1.333E+01	0.000E+00	0.000E+00
8	1	0	0	0	1	1	4.000E+01	1.333E+01	0.000E+00	0.000E+00
9	1	1	1	1	0	1	0.000E+00	2.667E+01	0.000E+00	0.000E+00
10	0	0	0	0	0	0	1.333E+01	2.667E+01	0.000E+00	0.000E+00
11	0	0	0	0	0	0	2.667E+01	2.667E+01	0.000E+00	0.000E+00
12	1	0	0	0	1	1	4.000E+01	2.667E+01	0.000E+00	0.000E+00
13	1	1	1	1	0	1	0.000E+00	4.000E+01	0.000E+00	0.000E+00
14	0	0	0	0	0	0	1.333E+01	4.000E+01	0.000E+00	0.000E+00
15	0	0	0	0	0	0	2.667E+01	4.000E+01	0.000E+00	0.000E+00
16	1	0	0	0	1	1	4.000E+01	4.000E+01	0.000E+00	0.000E+00
17	1	1	1	1	0	1	0.000E+00	5.333E+01	0.000E+00	0.000E+00
18	0	0	0	0	0	0	1.333E+01	5.333E+01	0.000E+00	0.000E+00
19	0	0	0	0	0	0	2.667E+01	5.333E+01	0.000E+00	0.000E+00
20	1	0	0	0	1	1	4.000E+01	5.333E+01	0.000E+00	0.000E+00
21	1	1	1	1	0	1	0.000E+00	6.667E+01	0.000E+00	0.000E+00
22	0	0	0	0	0	0	1.333E+01	6.667E+01	0.000E+00	0.000E+00
23	0	0	0	0	0	0	2.667E+01	6.667E+01	0.000E+00	0.000E+00
24	1	0	0	0	1	1	4.000E+01	6.667E+01	0.000E+00	0.000E+00
25	1	1	1	1	0	1	0.000E+00	8.000E+01	0.000E+00	0.000E+00
26	0	1	0	1	0	1	1.333E+01	8.000E+01	0.000E+00	0.000E+00
27	0	1	0	1	0	1	2.667E+01	8.000E+01	0.000E+00	0.000E+00
28	1	1	0	1	1	1	4.000E+01	8.000E+01	0.000E+00	0.000E+00

**** PRINT OF EQUATION NUMBERS SUPPRESSED

1**** THIN PLATE/SHELL ELEMENTS

number of elements = 18
number of materials = 1
Element formulation flag = 0
Mean temperature computation flag = 0

1**** MATERIAL PROPERTIES

material i.d. number = 1
weight density = 2.8360E-01
mass density = 7.3400E-04

Cxx/ alpha(X)	Cxy/ alpha(Y)	Cxs	Cyy	Cys	Gxy
3.297E+07	9.890E+06	0.000E+00	3.297E+07	0.000E+00	1.154E+07
6.500E-06	6.500E-06				

1**** ELEMENT LOAD MULTIPLIERS

	case a	case b	case c	case d
PRES	1.000E+00	0.000E+00	0.000E+00	0.000E+00
TEMP	0.000E+00	0.000E+00	0.000E+00	1.000E+00
X-DIR	0.000E+00	0.000E+00	0.000E+00	0.000E+00
Y-DIR	0.000E+00	0.000E+00	0.000E+00	0.000E+00
Z-DIR	0.000E+00	-3.864E+02	0.000E+00	0.000E+00

1**** NODAL LOADS (STATIC) OR MASSES (DYNAMIC)

NODE NUMBER	LOAD CASE	X-AXIS FORCE	Y-AXIS FORCE	Z-AXIS FORCE	X-AXIS MOMENT	Y-AXIS MOMENT	Z-AXIS MOMENT
16	1	0.000E+00	-5.000E+03	0.000E+00	0.000E+00	0.000E+00	0.000E+00

1**** ELEMENT LOAD MULTIPLIERS

load case	case A	case B	case C	case D
1	1.000E+00	0.000E+00	0.000E+00	0.000E+00

1**** STIFFNESS MATRIX PARAMETERS

minimum non-zero diagonal element = 7.4182E+02
maximum diagonal element = 5.4511E+07
maximum/minimum = 7.3482E+04
average diagonal element = 2.0956E+07
density of the matrix = 2.6339E+01

1**** STATIC ANALYSIS

LOAD CASE = 1

Displacements/Rotations(degrees) of nodes

NODE number	X- translation	Y- translation	Z- translation	X- rotation	Y- rotation	Z- rotation
1	0.0000E+00	0.0000E+00	0.0000E+00	0.0000E+00	0.0000E+00	0.0000E+00
2	0.0000E+00	0.0000E+00	0.0000E+00	0.0000E+00	0.0000E+00	0.0000E+00
3	0.0000E+00	0.0000E+00	0.0000E+00	0.0000E+00	0.0000E+00	0.0000E+00
4	0.0000E+00	0.0000E+00	0.0000E+00	0.0000E+00	0.0000E+00	0.0000E+00
5	0.0000E+00	0.0000E+00	0.0000E+00	0.0000E+00	1.4595E-01	0.0000E+00
6	-1.3769E-05	-1.4633E-05	-3.5589E-02	-2.5388E-01	1.1982E-01	0.0000E+00
7	-1.2039E-05	-3.3175E-05	-5.9495E-02	-4.2976E-01	6.2737E-02	0.0000E+00
8	0.0000E+00	-4.1624E-05	-6.7715E-02	-4.9100E-01	0.0000E+00	0.0000E+00
9	0.0000E+00	0.0000E+00	0.0000E+00	0.0000E+00	4.5562E-01	0.0000E+00
10	-1.0169E-05	-1.4546E-05	-1.0008E-01	-2.9154E-01	3.8236E-01	0.0000E+00
11	-2.7998E-05	-7.0546E-05	-1.6969E-01	-4.9980E-01	2.0838E-01	0.0000E+00
12	0.0000E+00	-1.1014E-04	-1.9415E-01	-5.7429E-01	0.0000E+00	0.0000E+00
13	0.0000E+00	0.0000E+00	0.0000E+00	0.0000E+00	7.3769E-01	0.0000E+00
14	-1.5939E-07	-2.0037E-05	-1.6178E-01	-2.4305E-01	6.2449E-01	0.0000E+00
15	-2.0049E-07	-4.9659E-05	-2.7591E-01	-4.1905E-01	3.4583E-01	0.0000E+00
16	0.0000E+00	-2.9270E-04	-3.1645E-01	-4.8281E-01	0.0000E+00	0.0000E+00
17	0.0000E+00	0.0000E+00	0.0000E+00	0.0000E+00	9.4992E-01	0.0000E+00
18	9.9586E-06	-1.4178E-05	-2.0859E-01	-1.6625E-01	8.0758E-01	0.0000E+00
19	2.7589E-05	-7.1145E-05	-3.5676E-01	-2.8734E-01	4.5075E-01	0.0000E+00
20	0.0000E+00	-1.1144E-04	-4.0968E-01	-3.3144E-01	0.0000E+00	0.0000E+00
21	0.0000E+00	0.0000E+00	0.0000E+00	0.0000E+00	1.0780E+00	0.0000E+00
22	1.4687E-05	-1.3349E-05	-2.3699E-01	-8.3156E-02	9.1826E-01	0.0000E+00
23	1.3056E-05	-3.4467E-05	-4.0588E-01	-1.4386E-01	5.1441E-01	0.0000E+00
24	0.0000E+00	-4.4193E-05	-4.6636E-01	-1.6602E-01	0.0000E+00	0.0000E+00
25	0.0000E+00	0.0000E+00	0.0000E+00	0.0000E+00	1.1205E+00	0.0000E+00
26	1.1045E-05	0.0000E+00	-2.4644E-01	0.0000E+00	9.5504E-01	0.0000E+00
27	1.1159E-05	0.0000E+00	-4.2224E-01	0.0000E+00	5.3560E-01	0.0000E+00
28	0.0000E+00	0.0000E+00	-4.8523E-01	0.0000E+00	0.0000E+00	0.0000E+00

Excerpts from the "EX67.s" file are presented in Table 6.7.3.2. The list includes the nodal bending and membrane stresses for elements 1 through 3, see Fig. 6.7.2.2. Remember that decoupling of bending and membrane actions allows separate calculation of bending and membrane stresses. Also, observe that the stresses underlined in Table 6.7.3.2 are also listed in columns (m) and (b) of Table 6.7.2.1.

The reaction forces are listed in the "EX67.rl" file. This file is not provided in this text; however, a discussion on its proper use is given is Section 6.5.3.

Table 6.7.3.2

1**** Algor (c) FEA Stress Processor MKNSO, Ver 11.06-3H

INPUT FILE.............ex67

1**** THIN PLATE/SHELL ELEMENTS

number of elements = 18
number of materials = 1
Element formulation flag = 0
Mean temperature computation flag = 0

1**** THIN PLATE/SHELL ELEMENT STRESSES

ELEM NO.	CASE (MODE)	NODE NO.	MEMBRANE STRESS COMPONENTS SM11	SM22	SM12	BENDING STRESS COMPONENTS SB11	SB22	SB12
1	1	1	-6.959E+00	-3.966E+01	1.229E+01	-3.938E+03	-7.444E+03	1.953E+02
1	1	2	-6.959E+00	-6.735E+00	1.229E+01	1.453E+03	-8.415E+02	5.900E+02
1	1	3	-3.794E+01	-6.735E+00	1.229E+01	2.859E+03	2.135E+03	4.425E+03
1	1	4	-3.794E+01	-3.966E+01	1.229E+01	-2.533E+03	-4.468E+03	4.030E+03
2	1	1	-1.754E+01	-7.933E+01	1.919E+01	-5.235E+03	-1.319E+04	-2.077E+02
2	1	2	-1.754E+01	-3.760E+01	1.919E+01	-1.289E+03	-8.555E+03	6.546E+02
2	1	3	-1.364E+01	-3.760E+01	1.919E+01	2.041E+03	-1.190E+03	3.311E+03
2	1	4	-1.364E+01	-7.933E+01	1.919E+01	-1.904E+03	-5.827E+03	2.449E+03
3	1	1	-2.640E+01	-9.752E+01	8.867E+00	-4.988E+03	-1.514E+04	-4.029E+02
3	1	2	-2.640E+01	-7.851E+01	8.867E+00	-3.611E+03	-1.353E+04	5.448E+02
3	1	3	6.896E-01	-7.851E+01	8.867E+00	3.821E+02	-4.321E+03	1.470E+03
3	1	4	6.896E-01	-9.752E+01	8.867E+00	-9.949E+02	-5.936E+03	5.223E+02

7

Modal Analysis: Numerical Examples

7.1 INTRODUCTION

Modal analysis is the simplest type of dynamic analysis. The results obtained through modal analysis are natural frequencies and mode shapes. A discussion on the method and an analytical procedure to calculate the natural frequencies and mode shapes of vibrating systems are presented in Chapter 5.

All types of dynamic analysis presented in Chapter 5, except time history direct integration, require calculation of natural frequencies and mode shapes. Nevertheless, as elaborated in Section 5.12, it is very useful to perform modal analysis even if we intend to solve the problem with time history direct integration. Therefore, we can say that regardless of what type of analysis is used to solve a dynamics problem, modal analysis should be the first step.

In modal analysis, the most important concern is to decide on how many modes must be determined. Once this decision is made, then the effort is placed on developing the proper model that can provide the desired number of modes with acceptable accuracy. In the following we attempt to shed some light on making these decisions.

a. Number of modes: As discussed in Section 5.4, dynamics problems are usually classified into *wave propagation* and *structural dynamics*. The most common structural dynamics problems include vibration excitation, blast and shock, wind, and earthquake loads. In vibration excitation analysis one is primarily concerned with avoiding resonance, usually at a few frequencies. In this case, modal analysis is generally used to calculate a small number of natural frequencies, e.g., vibrations of machine parts and machine foundations. As a general rule, in vibration excitation analysis lower and intermediate modes must be determined. For blast and shock loading more modes are required. In fact, for an N-degree-of-freedom system, up to 2N/3 modes may be needed. For earthquake loads, only the first few modes, usually up to the first 10 with natural frequencies less than 50 Hz are calculated.

The effectiveness of time history, frequency response, response spectrum, and random vibration analysis depends on the number of modes that must be included in the analysis. This number depends on the modal characteristics of the structure as well as the spatial distribution and the frequency content of the loads. For additional information refer also to Sections 5.7 and 5.11. The numerical examples presented in Chapters 7, 8, and 9 further elucidate considerations on the number of modes that should be obtained with modal analysis.

b. Proper model: The model used for modal analysis should have sufficient nodes to accurately represent the requested mode shapes. In general, a finite element model that provides acceptable results for a static analysis will also be satisfactory for a dynamic analysis unless the contribution of the higher modes to the response is significant. Higher frequency response models require a refined mesh to simulate the local deformation patterns.

In modal analysis, it is crucial to emphasize the significance of modeling the boundary conditions with particular accuracy and detail. Of course, it is always important to model the boundary conditions accurately whether the problem is static or dynamic. However, in statics the effect of boundary conditions decreases with distance from the boundaries, and as a result the stresses and the displacements away from the support may not be substantially affected. In modal analysis, however, changes in the boundary conditions can modify the modal properties and consequently the structural behavior greatly.

In performing modal analysis, the Algor user should also be aware of the following:

i. From the assigned material or weight densities Algor creates either lumped (diagonal) or distributed mass matrices automatically. For example, either a distributed or a lumped mass formulation can be selected in AccuPak, while in DECODS and BEDITH a lumped mass formulation is used for linear dynamic analysis.

ii. Each non-zero term of the mass matrices corresponds to a translational degree-of-freedom. Additional translational lumped masses and rotational lumped mass moments of inertia can be inputed at the nodes in a manner similar to the way concentrated forces are applied in static analysis, see Sections 7.4 and 7.5.

iii. The SSAP1H processor should be used for all cases except for "axial load stiffening." The axial load stiffening case is applicable only to beam and plate/shell element models subjected to tensile or compressive loads, and is illustrated in Section 7.7.

iv. Gap elements should not be used in a model for which we intend to perform modal analysis. Modal analysis is valid for linear systems only, and the presence of gap elements introduces non-linearities to the system. However, if gaps or cables are present in a structure, an estimate of the variation of the natural frequencies can be obtained if modal analysis is performed without gap elements, and then repeated with elastic elements equally stiff to the gap elements.

The following sections present numerical examples that elucidate important modeling and interpretation aspects of modal analysis. Specifically, Sections 7.2 through 7.4 describe the process of developing a proper model for modal analysis, explain the steps to perform the analysis with Algor, and demonstrate the significance in accounting for shear effects in frame structures. Section 7.3 illustrates the methodology to exploit symmetry in modal analysis. Section 7.4 demonstrates the effect of additional lumped masses on modal behavior and the use of simple analytical models to check finite element solutions. The modal analysis of a frame, that was introduced in Chapter 5 and extensively used in Chapters 8 and 9, is presented in Section 7.5. An example of modal analysis with rigid body modes is the topic of Section 7.6. Finally, Section 7.7 presents an example on load stiffening and demonstrates the significance of axial loads in modal analysis of slender structures.

7.2 INTRODUCTORY EXAMPLE

Consider the simply supported steel beam (W4X13) shown in Fig. 7.2.1. The properties of the beam are

I_z = moment of inertia about the Z-axis: 11.3 in^4
I_y = moment of inertia about the Y-axis: 3.86 in^4
J_x = torsional constant about the X-axis: 0.150 in^4
Z_z = section modulus for Z-axis: 5.46 in^3
Z_y = section modulus for Y-axis: 1.90 in^3
A = axial area: 3.83 in^2
S_f = shear area for flanges: 2.30 in^2
S_w = shear area for web: 1.165 in^2
E = Young's Modulus: $30x10^6$ psi
ν = Poisson's ratio: 0.3
γ = weight density: 0.2836 lb/in^3
L = beam length: 96 in.

The beam is allowed to deform only in the XY plane. We will perform a modal analysis to extract the seven lowest natural frequencies for two cases:

Case I. The effects of shear are considered in the analysis. This is realized by assigning non-zero values to the shear areas. For this example, the shear areas are given the values S_f and S_w that refer to the Y and Z-axes, respectively, and have been calculated according to Table 3.2.1.

Case II. No shear effects are included. In this case the shear areas are set equal to zero: $S_f = S_w = 0$.

Our first task is to develop the proper model that could capture the seven lowest natural frequencies accurately. Since the beam is allowed to deflect only in the XY plane, each node has three degrees-of-freedom; namely, translations along the X and Y-axes and a rotation about the Z-axis. Also, given the fact that a lumped mass formulation is used in modal analysis, only lumped masses associated with the translational degrees-of-freedom will be generated at the nodes. In addition, we recognize that the beam will experience decoupled flexural and axial vibrations. Thus, the inertia of the lumped masses along the Y-axis will be associated with the flexural response, while the inertia of the lumped masses along the X-axis will be related to the axial vibration.

In our first attempt to estimate the number of elements, we develop a model based on the assumption that out of the seven requested natural frequencies, five frequencies will be flexural and two will be axial. In order to calculate the five flexural frequencies, we need a minimum of five translational degrees-of-freedom along the Y-axis. In general, for frame structures, we must at least double the number of nodes in order to capture the corresponding mode shapes with reasonable accuracy. Thus, our first estimate is that ten nodes free to move translationally along the Y-axis are needed. Given the fact that our model must also have two additional nodes at the ends, the number of elements should be at least 11. We decide to model the beam with sixteen equal elements.

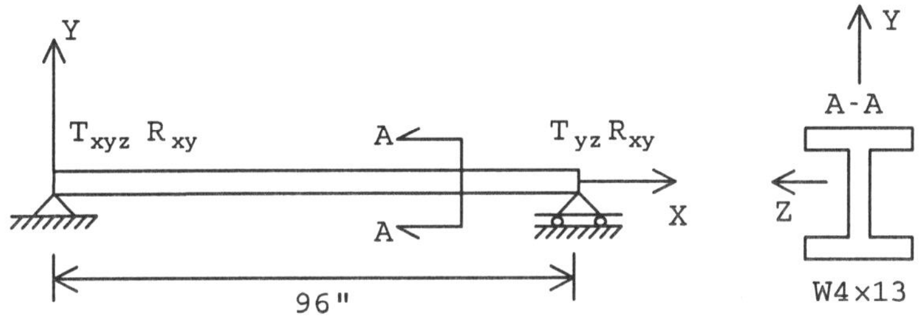

Figure 7.2.1 *Simply supported beam and cross section.*

7.2.1 Analysis with Algor

For the beam shown in Fig. 7.2.1.1 all intermediate nodes have the degrees-of-freedom T_zR_{xy}, thus, allowing free translations along the X and Y-axes and rotation about the Z-axis. The support at the origin is allowed to rotate about the Z-axis ($T_{xyz}R_{xy}$), and the other support can displace along the X and rotate about the Z-axis ($T_{yz}R_{xy}$).

In SD2H we draw the geometry and divide the model into sixteen equal-length line elements. Then, we specify the number of different material properties, cross-sectional properties, and orientation with GROUP, LAYER, and COLOR, respectively. Note that older versions of Algor used the GROUP, LAYER, and COLOR to specify orientation, material, and cross-sectional properties, respectively. Figure 7.2.1.1 shows the element and node numbers of the sixteen-element mesh. In this model we have set GROUP = 1, since the beam consists of one type of material; LAYER = 0, since the beam has a uniform cross section; and COLOR = 2, so that in BEDITH we will assign the $I_Z = I_2 = 11.3$ in^4. A detailed description on the proper use of COLOR,

LAYER and GROUP is given in the Algor Beam Design Editor Release Notes. Execution of SD2H creates the binary file "EX72.esd". This file is further processed in BEDITH.

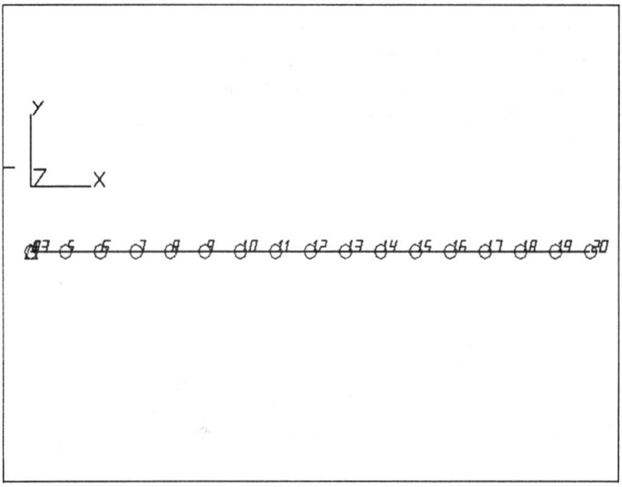

Figure 7.2.1.1 *Node numbers of sixteen-element beam model.*

In BEDITH we assign the boundary conditions to all nodes, the material, and the cross-sectional properties. It should be mentioned that BEDITH plays the role of DECODS for beam elements. In order to avoid making any mistakes in assigning the moments of inertia, it is prudent to check the orientation of the elements. For beam elements, the orientation line indicates the axis that the I_2 and S_2 are assigned. In BEDITH we also specify the type of analysis that we wish to perform. In our case, we select ANAL.TYPE; MODAL; NO FREQ and request seven natural frequencies. The OTHER in the MODAL menu provides several options that are described in the Algor Mode Shape Release Notes. The options "cutoff" and "rigid body modes" are also elaborated in Chapter 5. It should be noted that by setting in the OTHER menu the "shift = f_a" and the "cutoff = f_b", where the f_a and f_b are two selected frequencies, we can request the calculation of the natural frequencies f that satisfy the $f_a < f < f_b$.

As explained in Section 5.4, no loads are applied to the structure when we perform modal analysis. The only exception is discussed in Section 5.5 and elucidated with an example in Section 7.7.

Execution of BEDITH creates a binary file with the extension ".bed." Similarly to DECODS, that is elaborated in the analysis of the short beam in Chapter 6, BEDITH also creates an ASCII file with no extension. For our example, these files are the "EX72.bed" and the "EX72. ". The "EX72. " for case II is listed in Table 7.2.1.1. A description of the input data for the file "EX72. " is given in the Algor Linear Stress and Vibration Analysis Processor Reference Manual.

The file "EX72. " is processed with the modal analysis processor SSAP1H to calculate the requested natural frequencies and mode shapes. The most useful files generated by SSAP1H are a binary file "EX72.sst", and the ASCII files "EX72.frq" and "EX72.l". The "EX72.sst" is accessed by SVIEWH, while the use of the "EX72.frq" and "EX72.l" files is presented in Section 7.2.2. It is recommended to activate TRANS, one of the options provided by HELP, before running SSAP1H. By activating TRANS the mode shapes will also be written in the "EX72.l" file.

Table 7.2.1.1

File ex72 created by BEDIT 4.08-3H

```
23   1   1   7  0 1    0   0   0   0  000   0   0   0   0 386.39999
 1   1 1 1 1 1 1        0.          0. -1.00e+14    0          0. 0 0
 2   1 1 1 1 1 1        0. -1.00e+14        0.      0          0.
 3   1 1 1 1 1 1 -1.00e+14        0.        0.      0          0.
 4   1 1 1 1 1 0        0.          0.        0.      0          0.
 5   0 0 1 1 1 0        6.          0.        0.      0          0.
 6   0 0 1 1 1 0       12.          0.        0.      0          0.
 7   0 0 1 1 1 0       18.          0.        0.      0          0.
 8   0 0 1 1 1 0       24.          0.        0.      0          0.
 9   0 0 1 1 1 0       30.          0.        0.      0          0.
10   0 0 1 1 1 0       36.          0.        0.      0          0.
11   0 0 1 1 1 0       42.          0.        0.      0          0.
12   0 0 1 1 1 0       48.          0.        0.      0          0.
13   0 0 1 1 1 0       54.          0.        0.      0          0.
14   0 0 1 1 1 0       60.          0.        0.      0          0.
15   0 0 1 1 1 0       66.          0.        0.      0          0.
16   0 0 1 1 1 0       72.          0.        0.      0          0.
17   0 0 1 1 1 0       78.          0.        0.      0          0.
18   0 0 1 1 1 0       84.          0.        0.      0          0.
19   0 0 1 1 1 0       90.          0.        0.      0          0.
20   0 1 1 1 1 0       96.          0.        0.      0          0.
21   1 1 1 1 1 1 1.000e+14        0.        0.      0          0.
22   1 1 1 1 1 1        0. 1.000e+14        0.      0          0.
23   1 1 1 1 1 1        0.          0. 1.000e+14    0          0.
 2  16    1    0    1    0
 1 3.000e+07    0.3    0.    0.283    0.     0.     0.     0.
 1     3.83      0.    0.  0.15     11.3   3.865.4601.900   5.46  1.9 a
    0.         0.         0.         0.
    0.         0.         0.         0.
```

```
      0.          0.          0.          0.
1    4    5   23    1    1    0    0    0    0    0    0    0    0
2    5    6   23    1    1    0    0    0    0    0    0    0    0
3    6    7   23    1    1    0    0    0    0    0    0    0    0
4    7    8   23    1    1    0    0    0    0    0    0    0    0
5    8    9   23    1    1    0    0    0    0    0    0    0    0
6    9   10   23    1    1    0    0    0    0    0    0    0    0
7   10   11   23    1    1    0    0    0    0    0    0    0    0
8   11   12   23    1    1    0    0    0    0    0    0    0    0
9   12   13   23    1    1    0    0    0    0    0    0    0    0
10  13   14   23    1    1    0    0    0    0    0    0    0    0
11  14   15   23    1    1    0    0    0    0    0    0    0    0
12  15   16   23    1    1    0    0    0    0    0    0    0    0
13  16   17   23    1    1    0    0    0    0    0    0    0    0
14  17   18   23    1    1    0    0    0    0    0    0    0    0
15  18   19   23    1    1    0    0    0    0    0    0    0    0
16  19   20   23    1    1    0    0    0    0    0    0    0    0
  0    0 0.00E+00 0.00E+00 0.00E+00 0.00E+00 0.00E+00 0.00E+00
     0.    1.         0.        1.
      0      0     0         0.         0.     0     0        0.    0
```

7.2.2 Interpretation of Results

The first step in interpreting the results is to view the deformed shapes. For beam or frame analysis one may employ the options within the POST menu of BEDIT to examine the results. In this example we used the SVIEWH. Seven load cases are reported in SVIEWH. Each load case depicts one mode shape and its corresponding natural frequency. Viewing the mode shapes with the commands "displaced" and "with und" activated within the DISPLACED menu, helps us visualize the form of each mode shape. Right clicking on the Algor logo, while in the DISPLACED menu, displays the next mode shape. Left clicking shows the previous mode shape. Each mode shape and the corresponding natural frequency can also be selected within the LOAD CASE menu. It should be noted that in many instances, especially if we increase the graphic "scale" in the DISPLACED menu, we may identify inadvertent modelling errors, such as inappropriate or missing boundary conditions.

The natural frequencies for cases I and II are listed in Table 7.2.3.1. For case II, Fig. 7.2.2.1 presents the four lowest mode shapes of the beam that correspond to flexural vibrations. From the remaining three, the third and seventh are axial, while the sixth is flexural. The physical significance and use of natural frequencies and mode shapes are discussed in Section 5.4. As explained in Section 5.4, mode shapes cannot be viewed as displacements and only the overall form of a mode shape is of interest to the analyst.

The numerical results are listed in the ASCII files "EX72.frq" and "EX72.1" that can be accessed with a text editor. The "EX72.frq" is a very short file that lists only the natural circular frequencies. As in static analysis,

"EX72.1" presents the input data, such as the nodal coordinates and the element material properties. It also lists data characterizing the numerical solution of

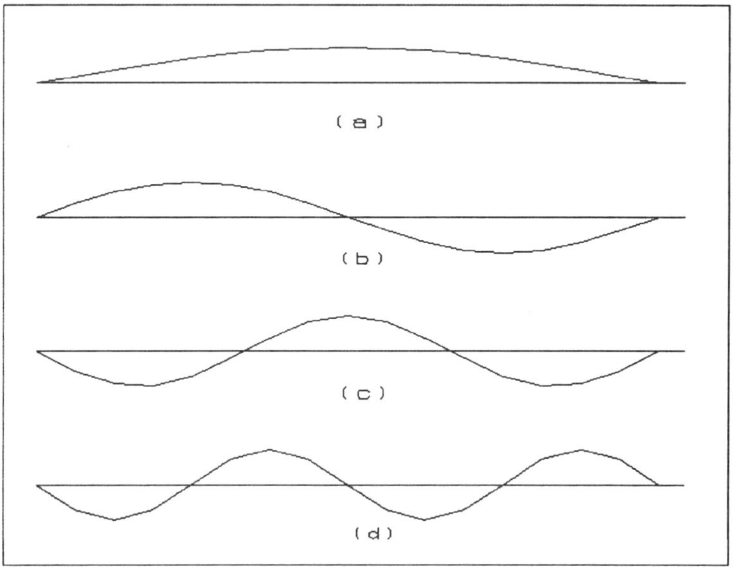

Figure 7.2.2.1 *Flexural mode shapes: (a) First; (b) Second; (c) Fourth; (d) Fifth. Note: The third mode shape is axial.*

the problem. The last part of "EX72.1" presents the mode shapes, natural circular frequencies, natural frequencies, natural periods, and tolerance for all requested modes. Tolerance is an indicator of the solution accuracy in calculating natural frequencies. Even though the mode shapes can be displayed in SVIEWH, in several instances, especially when the modes are coupled, it could be more convenient to examine them in the "EX72.1" with a text editor.

When the Sturm frequency check is activated, the "EX72.1" reports whether the solution algorithm has missed any natural frequencies. If we want to calculate the missing natural frequencies, we must repeat the modal analysis after increasing the "maximum number of iterations (NITER)" and the "number of vectors used in sub-space iteration (NAD)." These parameters can be specified in BEDITH and DECODS. A few iterations with larger values for either the NITER or the NAD usually suffice.

Table 7.2.2.1 lists excerpts from the "EX72.1" file. They include the control information, the first and the third mode shapes, and the seven lowest natural frequencies. Notice that the Y-translations and the Z-rotations of the first mode shape are non-zero, indicating that this mode is flexural. The third mode with non-zero translations along the longitudinal axis is axial.

Table 7.2.2.1

```
1**** Algor (c) Dynamic Modal Analysis  SSAP1H ver 10.02-3H

INPUT FILE............ex72
-------------------------------------------------

File ex72 created by BEDIT 4.08-3H

1**** CONTROL INFORMATION

     number of node points          (NUMNP) =   23
     number of element types        (NELTYP) =  1
     number of load cases           (LL) =      1
     number of frequencies          (NF) =      7
     geometric stiffness flag       (GEOSTF) =  0
     analysis type code             (NDYN) =    1
     solution mode                  (MODEX) =   0
     equations per block            (KEQB) =    0
     weight and c.g. flag           (IWTCG) =   0
     bandwidth minimization flag    (MINBND) =  0
     gravitational constant         (GRAV) =    3.8640E+02


1**** MODAL ANALYSIS

     mode number = 1

     eigenvectors normalized to unit mass matrix

     Displacements/Rotations(degrees) of nodes
```

NODE number	X-translation	Y-translation	Z-translation	X-rotation	Y-rotation	Z-rotation
4	0.0000E+00	0.0000E+00	0.0000E+00	0.0000E+00	0.0000E+00	5.1098E+00
5	0.0000E+00	5.3167E-01	0.0000E+00	0.0000E+00	0.0000E+00	5.0116E+00
6	0.0000E+00	1.0429E+00	0.0000E+00	0.0000E+00	0.0000E+00	4.7208E+00
7	0.0000E+00	1.5141E+00	0.0000E+00	0.0000E+00	0.0000E+00	4.2486E+00
8	0.0000E+00	1.9270E+00	0.0000E+00	0.0000E+00	0.0000E+00	3.6132E+00
9	0.0000E+00	2.2660E+00	0.0000E+00	0.0000E+00	0.0000E+00	2.8388E+00
10	0.0000E+00	2.5178E+00	0.0000E+00	0.0000E+00	0.0000E+00	1.9554E+00
11	0.0000E+00	2.6729E+00	0.0000E+00	0.0000E+00	0.0000E+00	9.9687E-01
12	0.0000E+00	2.7252E+00	0.0000E+00	0.0000E+00	0.0000E+00	-7.9234E-14
13	0.0000E+00	2.6729E+00	0.0000E+00	0.0000E+00	0.0000E+00	-9.9687E-01
14	0.0000E+00	2.5178E+00	0.0000E+00	0.0000E+00	0.0000E+00	-1.9554E+00
15	0.0000E+00	2.2660E+00	0.0000E+00	0.0000E+00	0.0000E+00	-2.8388E+00
16	0.0000E+00	1.9270E+00	0.0000E+00	0.0000E+00	0.0000E+00	-3.6132E+00
17	0.0000E+00	1.5141E+00	0.0000E+00	0.0000E+00	0.0000E+00	-4.2486E+00
18	0.0000E+00	1.0429E+00	0.0000E+00	0.0000E+00	0.0000E+00	-4.7208E+00
19	0.0000E+00	5.3167E-01	0.0000E+00	0.0000E+00	0.0000E+00	-5.0116E+00
20	0.0000E+00	0.0000E+00	0.0000E+00	0.0000E+00	0.0000E+00	-5.1098E+00

1**** MODAL ANALYSIS

mode number = 3

eigenvectors normalized to unit mass matrix

Displacements/Rotations(degrees) of nodes

NODE number	X-translation	Y-translation	Z-translation	X-rotation	Y-rotation	Z-rotation
4	0.0000E+00	0.0000E+00	0.0000E+00	0.0000E+00	0.0000E+00	0.0000E+00
5	-2.6712E-01	0.0000E+00	0.0000E+00	0.0000E+00	0.0000E+00	0.0000E+00
6	-5.3167E-01	0.0000E+00	0.0000E+00	0.0000E+00	0.0000E+00	0.0000E+00
7	-7.9110E-01	0.0000E+00	0.0000E+00	0.0000E+00	0.0000E+00	0.0000E+00
8	-1.0429E+00	0.0000E+00	0.0000E+00	0.0000E+00	0.0000E+00	0.0000E+00
9	-1.2847E+00	0.0000E+00	0.0000E+00	0.0000E+00	0.0000E+00	0.0000E+00
10	-1.5141E+00	0.0000E+00	0.0000E+00	0.0000E+00	0.0000E+00	0.0000E+00
11	-1.7289E+00	0.0000E+00	0.0000E+00	0.0000E+00	0.0000E+00	0.0000E+00
12	-1.9270E+00	0.0000E+00	0.0000E+00	0.0000E+00	0.0000E+00	0.0000E+00
13	-2.1066E+00	0.0000E+00	0.0000E+00	0.0000E+00	0.0000E+00	0.0000E+00
14	-2.2660E+00	0.0000E+00	0.0000E+00	0.0000E+00	0.0000E+00	0.0000E+00
15	-2.4035E+00	0.0000E+00	0.0000E+00	0.0000E+00	0.0000E+00	0.0000E+00
16	-2.5178E+00	0.0000E+00	0.0000E+00	0.0000E+00	0.0000E+00	0.0000E+00
17	-2.6079E+00	0.0000E+00	0.0000E+00	0.0000E+00	0.0000E+00	0.0000E+00
18	-2.6729E+00	0.0000E+00	0.0000E+00	0.0000E+00	0.0000E+00	0.0000E+00
19	-2.7121E+00	0.0000E+00	0.0000E+00	0.0000E+00	0.0000E+00	0.0000E+00
20	-2.7252E+00	0.0000E+00	0.0000E+00	0.0000E+00	0.0000E+00	0.0000E+00

1**** PRINT OF NATURAL FREQUENCIES

mode number	circular frequency (rad/sec)	frequency (Hertz)	period (sec)	tolerance
------	----------------	-----------------	---------------	----------------
1	3.7229E+02	5.9252E+01	1.6877E-02	4.1997E-16
2	1.4891E+03	2.3700E+02	4.2193E-03	2.0999E-16
3	3.3102E+03	5.2684E+02	1.8981E-03	2.1928E-13
4	3.3503E+03	5.3322E+02	1.8754E-03	3.3189E-16
5	5.9548E+03	9.4774E+02	1.0551E-03	0.0000E+00
6	9.2998E+03	1.4801E+03	6.7563E-04	1.7230E-16
7	9.8989E+03	1.5755E+03	6.3474E-04	4.5111E-07

The Sturm sequence check didn't find any missing frequency.

7.2.3 Numerical Accuracy and Comparisons

The results of modal analyses for cases I and II are presented in Table 7.2.3.1. The "F" and "A" indicate flexural and axial modes, respectively. The four lowest flexural mode shapes are shown in Fig. 7.2.2.1. The "exact" natural frequencies for the axial and flexural response of the beam are also given in Table 7.2.3.1. They have been obtained from the Bernoulli-Euler beam theory that ignores the effects of shear deformation and assumes that plane cross sections remain plane during flexure.

The exact flexural natural frequencies are evaluated from (e.g., Table 8-1, Case 5, *Formulas for Natural Frequency and Mode Shape* by Blevis)

$$f_n = \frac{n^2\pi}{2}\sqrt{\frac{EIg}{\gamma A L^4}}$$

and the exact axial natural frequencies are calculated from (e.g., Table 8-16, Case 2, *Formulas for Natural Frequency and Mode Shape* by Blevis)

$$f_n = \frac{(n-0.5)}{2L}\sqrt{\frac{Eg}{\gamma}}$$

Table 7.2.3.1

ANALYSIS TYPE	MODE (Hz)	1	2	3	4	5	6	7
I shear	f	58.5	225	477	526	786	1125	1473
	TYPE	F	F	F	A	F	F	F
II No shear	f	59.2	237	526	533	948	1480	1575
	TYPE	F	F	A	F	F	F	A
"Exact" no shear	f	58.7	235	521	528	939	1468	1564
	TYPE	F	F	A	F	F	F	A

The following conclusions on modal analysis of frame structures can be drawn from the above table:

a. Shear effects decrease the flexural natural frequencies and should be accounted for in higher modes.

b. Shear can change the sequence of the mode-type. For example, the fourth mode, which is axial in case I, becomes the third mode in case II.

c. Shear does not affect the axial modes. Observe that the magnitudes of the axial natural frequencies in cases I and II do not change.

7.3 SYMMETRIC STRUCTURES

Sections 7.3.1 and 7.3.2 present the modal analysis of a structure with one axis of symmetry. The procedure to perform modal analysis of a system with more than one axis of symmetry is discussed in Sections 7.3.3 and 7.3.4.

7.3.1 Beam Structure

The modes of the beam analyzed in Section 7.2 can be obtained with much smaller models. This can be accomplished since the structure fulfills the three requirements for symmetry presented in Section 3.9. Obviously, in this example, the axis of symmetry is parallel to the Y-axis and is passing through the mid-span. In modal analysis the only loading on the structure is the inertia, and since the mass is uniform along the beam, symmetry is also satisfied.

The structure is symmetric only in flexure. Thus, instead of the whole beam we can model only half the beam as shown in Figs. 7.3.1.1 and 7.3.1.2. The beam of Fig. 7.3.1.1, called the *symmetric model,* has the displacement and rotation boundary conditions for symmetric deformation at mid-span. Figure 7.3.1.2 shows the *antisymmetric model* with displacements and rotations for antisymmetric deformation at mid-span. The results of modal analysis without considering shear effects are shown in Tables 7.3.1.1 and 7.3.1.2 for the symmetric and antisymmetric models, respectively.

Table 7.3.1.1

MODE	1	2	3
f (Hz)	59.2	533	1480
TYPE	F	F	F

Table 7.3.1.2

MODE	1	2	3
f (Hz)	237	948	1052
TYPE	F	F	A

All modes in Table 7.3.1.1 are flexural modes. The modes 1,2, and 3 in Table 7.3.1.1 correspond to the modes 1,4,6 of case II in Table 7.2.3.1, respectively. In Table 7.3.1.2, the modes 1 and 2 correspond to the modes 2 and 5 in Table 7.2.3.1 for case II, respectively.

Since the axial boundary conditions are not symmetric, symmetry cannot be used to obtain the axial natural frequencies by modeling half the structure. In this example, the axial natural frequencies of the antisymmetric model must be divided by 2 to account for modeling half the beam. Therefore, the frequency f_3 = 1052 Hz in Table 7.3.1.2 must be divided by 2 to yield the f_3 = 526 Hz in Table 7.2.3.1.

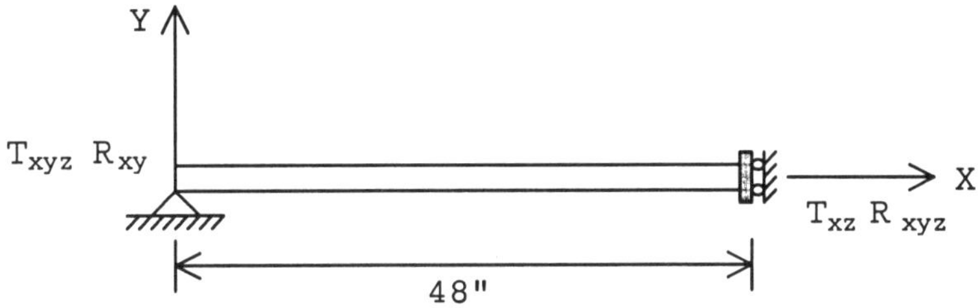

Figure 7.3.1.1 Symmetric model.

The following modeling generalizations can be deducted:

a. If a structure has one axis of symmetry, its modal properties can be extracted by using two models of half the structure as divided by the axis of symmetry. The first model must have symmetric boundary conditions at the axis of symmetry, while antisymmetric boundary conditions must be defined for the second model at the axis of symmetry.

b. Consideration must be given to modes corresponding to deformations that are decoupled; for example, axial from flexural in beams, and membrane from flexural in plates.

Since symmetry can significantly reduce the modeling effort in dynamics as well as in statics problems, it should be exploited. An example of a system with more than one axes of symmetry is presented in Section 7.3.3.

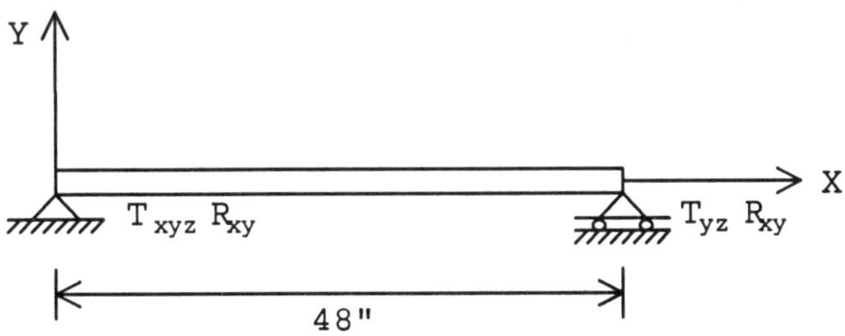

Figure 7.3.1.2 Antisymmetric model.

7.3.2 Analysis with Algor

In SD2H each model in Figs. 7.3.1.1 and 7.3.1.2 is divided into eight equal-length elements. The degrees-of-freedom defined in BEDITH for each intermediate node of the symmetric and antisymmetric models are T_zR_{xy} since the beam is allowed to deform only in the XY plane. The only difference between the symmetric and the antisymmetric model is the boundary conditions at the axis of symmetry.

The procedure to develop the two beam models and to view the results is described in Section 7.2.1. Be reminded that, as a modal analysis problem, a non-zero weight or mass density must be defined for the beam elements; no loads are applied to the models; and, as discussed in Section 5.4, the mode shapes viewed in SVIEWH are not the real displacements of the structure. The files for the symmetric and antisymmetric models are named "EX73S. " and "EX73A. ", respectively.

7.3.3 Plate Structure

Consider the plate shown in Fig. 7.3.3.1. The 4-inch thick plate is characterized by a Poisson's ratio $\nu = 0.15$, a modulus of elasticity $E = 3x10^6$ psi, and a weight density $\gamma = 0.0868$ lb/in^3. It is clamped along the top and bottom sides, and is simply supported along the other sides. We will perform a modal analysis to calculate the six lowest natural frequencies.

First, the analysis is performed with the seventy two-element model of the whole plate shown in Fig. 7.3.3.2. This is followed by four modal analyses, each performed with a quarter model having boundary conditions as defined in Figs. 3.9.7(b) through (e). For example, the quarter model corresponding to Fig. 3.9.7(b) is shown in Fig. 7.3.3.3.

Table 7.3.3.1 lists the six lowest natural frequencies of the plate obtained from analyzing the whole structure and the quarter models. The natural frequencies of the whole plate-model are listed in column (FREQ). The columns (B) through (E) correspond to the models of Figs. 3.9.7(b) - (e). In Table 7.3.3.1 the "X" indicates the natural frequencies obtained from the modal analysis of the corresponding model. The natural frequencies calculated from the analytical solution of the problem are also listed in the last column (F) (e.g., Table 11.4, Case 19, *Formulas for Natural Frequency and Mode Shape* by Blevis). The results clearly indicate that one must perform a modal analysis for each one of the four quarter models in order to obtain all the natural frequencies of a structure with two axes of symmetry. It should be noted, however, that the modeling, computational, and postprocessing advantages of using symmetry outweigh the effort to create the quarter models and perform four separate modal analyses.

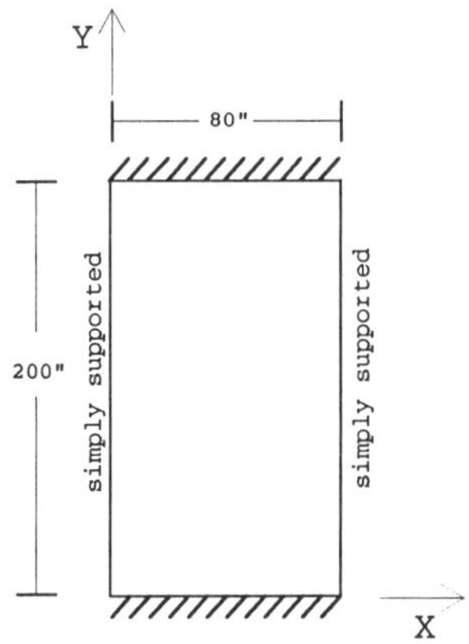

Figure 7.3.3.1 *Plate dimensions and boundary conditions.*

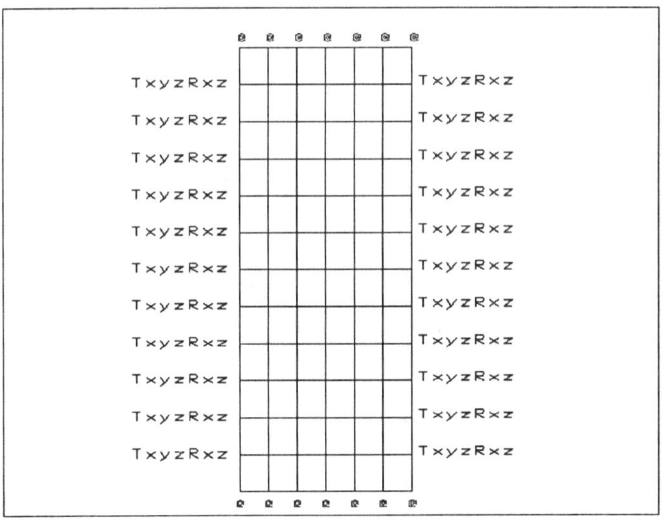

Figure 7.3.3.2 *Seventy two-element plate model.*

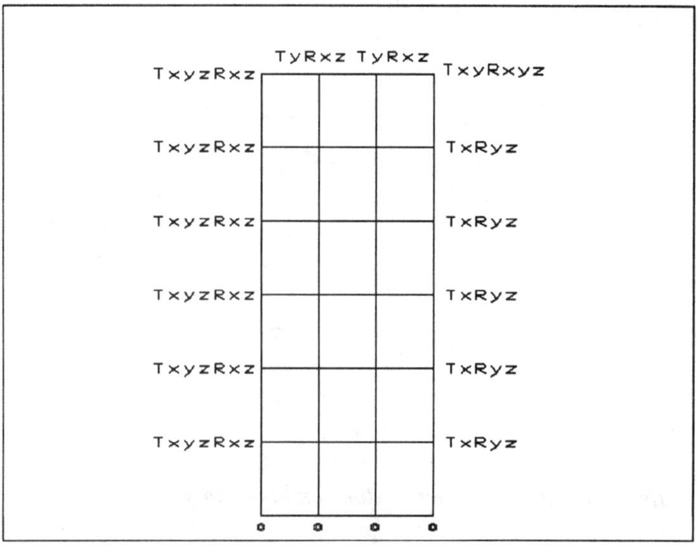

Figure 7.3.3.3 Quarter model and nodal degrees-of-freedom.

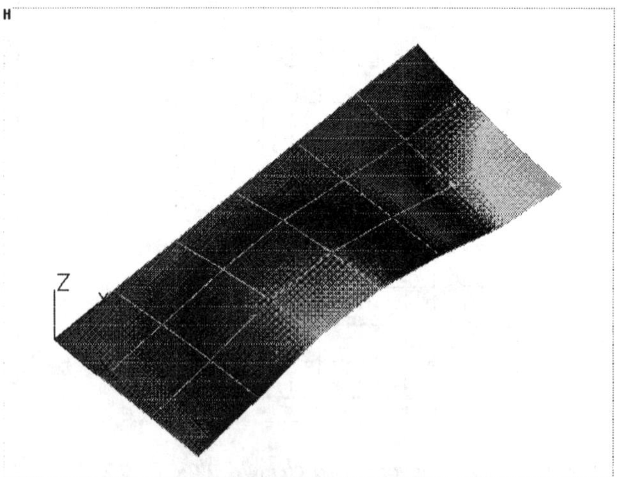

Figure 7.3.3.4 Dithered second mode shape of quarter model.

Table 7.3.3.1

MODE	FREQ.	B	C	D	E	F
1	41.02	X				40.72
2	62.71		X			61.63
3	96.05	X				93.89
4	139.51				X	136.79
5	140.17		X			138.91
6	160.64			X		157.78

7.3.4 Analysis with Algor

The four quarter models of the plate have been developed so that their boundary conditions along the axes of symmetry include all possible combinations of symmetry and antisymmetry.

The material properties as well as the type of analysis are defined in DECODS. Specifically, the thickness is defined in ELEMENTS;COLOR, and E, ν, and γ are specified in ELEMENTS;GROUP. In ANALYSIS;MODAL six natural frequencies are requested for the whole structure, while only two natural frequencies are requested for each quarter model. The use of several parameters within the MODAL menu, such as the cutoff and the Sturm frequency check has been elaborated in Chapter 5. No values are assigned in GLOBAL; LOAD CASE. Table 7.3.4.1 lists the file "EX733Q. " created by DECODS that corresponds to the model shown in Fig. 7.3.3.3.

Modal analysis is performed with SSAP1H. It is recommended to activate TRANS, one of the options provided by HELP, before running SSAP1H so that the mode shapes will be written in the "EX733Q.1" file. As stated in Sections 7.2.1 and 7.2.2, activating TRANS could be especially helpful in interpreting the mode shapes of two- and three-dimensional structures. The second mode shape of the model shown in Fig. 7.3.3.3 corresponds to $f_2 = 96.05$ Hz and is shown in Fig. 7.3.3.4.

Once SSAP1H has calculated the natural frequencies and mode shapes, they can be examined in SVIEWH following the procedure presented in Section

7.2.2. It should be noted that the file name given to the seventy two-element model is "EX733. ".

Table 7.3.4.1

Prepared by DECODS 2.10-S

```
28  1    0  2 0 1  0   0   0   0  000   0   0   0   0    386.4
 1  1 1 1 1 1 1  0.000000E+00 0.000000E+00 0.000000E+00  0      0. 0 0
 2  1 1 1 1 1 1  1.333333E+01 0.000000E+00 0.000000E+00  0      0.
 3  1 1 1 1 1 1  2.666667E+01 0.000000E+00 0.000000E+00  0      0.
 4  1 1 1 1 1 1  4.000000E+01 0.000000E+00 0.000000E+00  0      0.
 5  1 1 1 1 0 1  0.000000E+00 1.666667E+01 0.000000E+00  0      0.
 6  0 0 0 0 0 0  1.333333E+01 1.666667E+01 0.000000E+00  0      0.
 7  0 0 0 0 0 0  2.666667E+01 1.666667E+01 0.000000E+00  0      0.
 8  1 0 0 0 1 1  4.000000E+01 1.666667E+01 0.000000E+00  0      0.
 9  1 1 1 1 0 1  0.000000E+00 3.333334E+01 0.000000E+00  0      0.
10  0 0 0 0 0 0  1.333333E+01 3.333334E+01 0.000000E+00  0      0.
11  0 0 0 0 0 0  2.666667E+01 3.333334E+01 0.000000E+00  0      0.
12  1 0 0 0 1 1  4.000000E+01 3.333334E+01 0.000000E+00  0      0.
13  1 1 1 1 0 1  0.000000E+00 5.000000E+01 0.000000E+00  0      0.
14  0 0 0 0 0 0  1.333333E+01 5.000000E+01 0.000000E+00  0      0.
15  0 0 0 0 0 0  2.666667E+01 5.000000E+01 0.000000E+00  0      0.
16  1 0 0 0 1 1  4.000000E+01 5.000000E+01 0.000000E+00  0      0.
17  1 1 1 1 0 1  0.000000E+00 6.666667E+01 0.000000E+00  0      0.
18  0 0 0 0 0 0  1.333333E+01 6.666667E+01 0.000000E+00  0      0.
19  0 0 0 0 0 0  2.666667E+01 6.666667E+01 0.000000E+00  0      0.
20  1 0 0 0 1 1  4.000000E+01 6.666667E+01 0.000000E+00  0      0.
21  1 1 1 1 0 1  0.000000E+00 8.333333E+01 0.000000E+00  0      0.
22  0 0 0 0 0 0  1.333333E+01 8.333333E+01 0.000000E+00  0      0.
23  0 0 0 0 0 0  2.666667E+01 8.333333E+01 0.000000E+00  0      0.
24  1 0 0 0 1 1  4.000000E+01 8.333333E+01 0.000000E+00  0      0.
25  1 1 1 1 0 1  0.000000E+00 1.000000E+02 0.000000E+00  0      0.
26  0 1 0 1 0 1  1.333333E+01 1.000000E+02 0.000000E+00  0      0.
27  0 1 0 1 0 1  2.666667E+01 1.000000E+02 0.000000E+00  0      0.
28  1 1 0 1 1 1  4.000000E+01 1.000000E+02 0.000000E+00  0      0.
 6  18  1   0    0
    1      0.    0.0868 2.246E-04       0.        0.        0.
 3.069E+06 4.604E+05         0. 3.069E+06         0. 1.304E+06
    1.         0.         0.         0.
    0.         0.         0.         1.
    0.         0.         0.         0.
    0.         0.         0.         0.
    0.     -386.4         0.         0.
 1   1   2   6   5   0   1 0        4.        0.        0.        0.
 2   2   3   7   6   0   1 0        4.        0.        0.        0.
 3   3   4   8   7   0   1 0        4.        0.        0.        0.
 4   5   6  10   9   0   1 0        4.        0.        0.        0.
 5   6   7  11  10   0   1 0        4.        0.        0.        0.
 6   7   8  12  11   0   1 0        4.        0.        0.        0.
 7   9  10  14  13   0   1 0        4.        0.        0.        0.
 8  10  11  15  14   0   1 0        4.        0.        0.        0.
 9  11  12  16  15   0   1 0        4.        0.        0.        0.
10  13  14  18  17   0   1 0        4.        0.        0.        0.
11  14  15  19  18   0   1 0        4.        0.        0.        0.
12  15  16  20  19   0   1 0        4.        0.        0.        0.
13  17  18  22  21   0   1 0        4.        0.        0.        0.
14  18  19  23  22   0   1 0        4.        0.        0.        0.
15  19  20  24  23   0   1 0        4.        0.        0.        0.
16  21  22  26  25   0   1 0        4.        0.        0.        0.
```

```
17  22  23  27  26   0   1 0        4.        0.        0.        0.
18  23  24  28  27   0   1 0        4.        0.        0.        0.
 0   0
   0.      0.      0.      0.
  0   0   0      0.      0.   0   0      0.   0
```

7.4 CONCENTRATED MASS AND STIFFNESS

Consider the beam analyzed in Section 7.2 with the following two modifications:

a. A weight with a mass M = 100 lb-sec^2/in is placed at mid-span.
b. An elastic spring with stiffness k_s = 20000 lb/in is also placed at mid-span.

Figure 7.4.1(a) shows the beam with the attached spring and concentrated mass. We will calculate the fundamental frequency of the system.

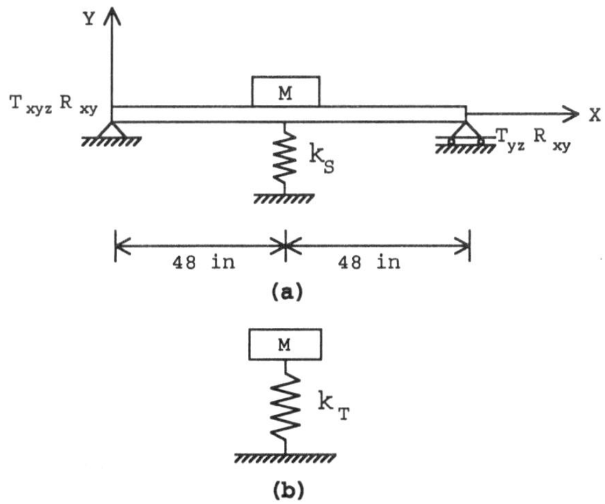

Figure 7.4.1 *(a) Beam with concentrated mass and spring; (b) Equivalent lumped mass SDOF system.*

Before proceeding to the analysis, the following observations can be made:

i. The total mass of the beam is: m = (A)(L)(γ/g) = 0.27 lb-sec^2/in. Thus, m being considerably smaller than the concentrated mass M, it is not expected to affect the flexural response of the beam.
ii. The stiffness of the beam at mid-span and the spring supporting M can be approximated by:

$$k_T = 48\frac{EI}{L^3} + k_s = 38{,}392 \ \ lb/in$$

The two observations imply that the fundamental frequency of the system in Fig. 7.4.1(a) can be estimated from the SDOF system in Fig. 7.4.1(b). By combining eqns. (4.3.3) and (4.3.11), the fundamental frequency of the SDOF system can be calculated from

$$f = \frac{1}{2\pi}\sqrt{\frac{k_T}{M}} = 3.12 \quad Hz$$

If we perform a modal analysis using the same number of elements as presented in Section 7.2.1, we will obtain a fundamental frequency

$$f = 3.116 \quad Hz$$

The results are almost identical! We also anticipate that, because of the presence of the large lumped mass at mid-span, the higher natural frequencies will be much greater than the fundamental frequency. In conclusion, we can state that, before performing a FEM analysis, it is good practice to estimate the solution with a simple model.

7.4.1 Analysis with Algor

The beam model with the concentrated mass and spring at mid-span can be readily developed along the lines described in Section 7.2.1. Here, we outline only the main differences that refer to the elastic stiffness and the concentrated mass.

The elastic spring is defined in BEDITH following the sequence of commands: ADD/MOD; BDRY ELE. In the BDRY ELE menu we can assign the elastic spring properties by selecting VALUE and setting KD = 1 and TRACE = 20000. In the BDRY ELE menu, we also set the direction of the elastic element along the Y-axis through the sequence of commands: ENTER VEC;Y;NEGATE and LENGTH = 4. The magnitude of the length was chosen for illustrative purposes.

In order to specify the translational mass in BEDITH, we follow the sequence: ADD/MOD;LUMPD MAS;KEY VAL;MASS. Also in the LUMPD MAS menu we select VALUE to define F_y = 100, i.e, the translational mass along the Y-axis. In ANAL. TYPE;MODAL;NO FREQ, we request only one frequency, and then we run the analysis with the sequence TRANSFER;RUN SAP1 from the main menu. The file that is generated by BEDITH and executed with SSAP1H is the "EX74. ". The "EX74." is listed in Table 7.4.1.1

Table 7.4.1.1

File ex74 created by BEDIT 4.08-3H

```
24     2    1    1 0 1   0    0    0    0   000    0    0    0     0  386.39999
 1    1 1 1 1 1 1          0.          0. -1.00e+14     0          0. 0 0
 2    1 1 1 1 1 1          0. -1.00e+14          0.     0          0.
 3    1 1 1 1 1 1 -1.00e+14          0.          0.     0          0.
 4    1 1 1 1 1 0          0.          0.          0.     0          0.
 5    0 0 1 1 1 0          6.          0.          0.     0          0.
 6    0 0 1 1 1 0         12.          0.          0.     0          0.
 7    0 0 1 1 1 0         18.          0.          0.     0          0.
 8    0 0 1 1 1 0         24.          0.          0.     0          0.
 9    0 0 1 1 1 0         30.          0.          0.     0          0.
10    0 0 1 1 1 0         36.          0.          0.     0          0.
11    0 0 1 1 1 0         42.          0.          0.     0          0.
12    0 0 1 1 1 0         48.          0.          0.     0          0.
13    0 0 1 1 1 0         54.          0.          0.     0          0.
14    0 0 1 1 1 0         60.          0.          0.     0          0.
15    0 0 1 1 1 0         66.          0.          0.     0          0.
16    0 0 1 1 1 0         72.          0.          0.     0          0.
17    0 0 1 1 1 0         78.          0.          0.     0          0.
18    0 0 1 1 1 0         84.          0.          0.     0          0.
19    0 0 1 1 1 0         90.          0.          0.     0          0.
20    0 1 1 1 1 0         96.          0.          0.     0          0.
21    1 1 1 1 1 1 1.000e+14          0.          0.     0          0.
22    1 1 1 1 1 1          0. 1.000e+14          0.     0          0.
23    1 1 1 1 1 1          0.          0. 1.000e+14     0          0.
24    1 1 1 1 1 1         48.         -4.          0.     0          0.
2   16    1    0    1    0
1 3.000e+07    0.3      0.      0.283     0.     0.      0.      0.
1     3.83     0.       0.      0.15     11.3   3.865.4601.900  5.46   1.9 a
      0.       0.       0.        0.
      0.       0.       0.        0.
      0.       0.       0.        0.
 1     4     5    23     1     1     0     0     0     0     0     0     0     0
 2     5     6    23     1     1     0     0     0     0     0     0     0     0
 3     6     7    23     1     1     0     0     0     0     0     0     0     0
 4     7     8    23     1     1     0     0     0     0     0     0     0     0
 5     8     9    23     1     1     0     0     0     0     0     0     0     0
 6     9    10    23     1     1     0     0     0     0     0     0     0     0
 7    10    11    23     1     1     0     0     0     0     0     0     0     0
 8    11    12    23     1     1     0     0     0     0     0     0     0     0
 9    12    13    23     1     1     0     0     0     0     0     0     0     0
10    13    14    23     1     1     0     0     0     0     0     0     0     0
11    14    15    23     1     1     0     0     0     0     0     0     0     0
12    15    16    23     1     1     0     0     0     0     0     0     0     0
13    16    17    23     1     1     0     0     0     0     0     0     0     0
14    17    18    23     1     1     0     0     0     0     0     0     0     0
```

```
15   18   19   23    1    1    0    0    0    0    0    0    0    0
16   19   20   23    1    1    0    0    0    0    0    0    0    0
 7    1
      0.          0.         0.        1.
12   24 0 0 0      1    0  0          0.         0.    20000.
12    0          0.        100.       0.         0.       0.       0. 0
 0    0  0.00E+00  0.00E+00  0.00E+00  0.00E+00  0.00E+00  0.00E+00
      0.          1.         0.        1.
 0    0    0          0.        0.     0    0         0.   0
```

7.5 FRAME WITH CONCENTRATED MASSES

This section presents the finite element modal analysis of Example 5.14.1. The frame shown in Fig. 5.14.1 can be modeled as a cantilever beam with two lumped masses placed at 17 ft and 29 ft, respectively, from the support of the beam, see Fig. 7.5.1. The lumped masses are m_1 = 116.46 lb-sec^2/in and m_2 = 38.82 lb-sec^2/in.

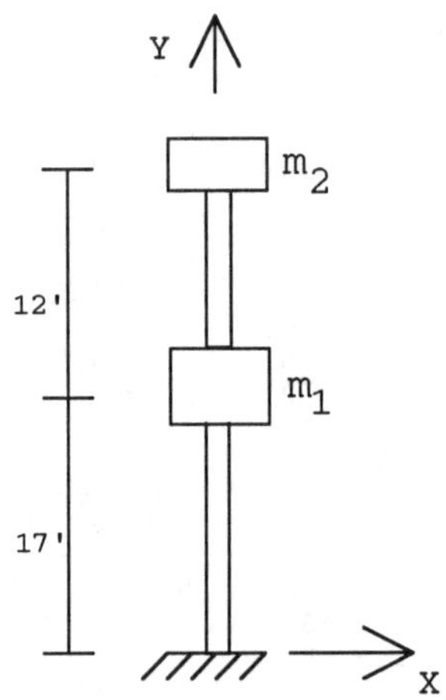

Figure 7.5.1 *"Stick" model of frame.*

Since the beam model shown in Fig. 7.5.1 represents the two steel columns, its properties are given by

$$I_{z1} = 2 \text{ X } 272 = 544 \text{ in}^4 \qquad I_{z2} = 2 \text{ X } 144 = 288 \text{ in}^4$$

$$A_1 = 2 \text{ X } 14.4 = 28.80 \text{ in}^2 \quad A_2 = 2 \text{ X } 7.61 = 15.22 \text{ in}^2$$

where, I_{z1} and I_{z2} are the moments of inertia about the Z-axis of the first 17 ft and the ensuing 12 ft of the beam, respectively. The A_1 and A_2 are the corresponding cross-sectional areas. The beam is assigned a weight density: $\gamma = 0.283$ lb/in^3.

The response of the system is dominated by the two translational inertia forces corresponding to the lumped masses m_1 and m_2 along the X-axis. The two natural frequencies obtained through the analysis presented in the following section are: $f_1 = 1.876$ Hz and $f_2 = 5.586$ Hz. Notice that f_1 and f_2 are almost identical to the natural frequencies obtained from the analytical solution presented in step 2 of Section 5.14.1. The small difference is primarily attributed to the mass of the columns that was not considered in the analytical solution.

7.5.1 Analysis with Algor

Since modeling of beam structures has been discussed in previous sections, we outline here only the aspects that require some additional attention. Each combined column of the "stick model" is modeled with four equal beam elements, see Fig. 7.5.1.1. In SD2H, the lower part of the cantilever is assigned a different layer from the top in order to define the corresponding cross-sectional properties in BEDITH. We assign color No. 2 to all lines so that we can set $I_{zi} = I_{2i}$ $(i = 1,2)$ in BEDITH. Table 7.5.1.1 lists the COLOR, GROUP, and LAYER assigned in SD2H. It should be noted that older versions of Algor associated COLOR, GROUP, and LAYER to cross-sectional properties, orientation, and material properties, respectively.

Table 7.5.1.1

Element No.	COLOR	GROUP	LAYER
1-4	2	1	0
5-8	2	1	1

Since the structure is allowed to deform only in XY plane, in BEDITH only the I_{z1}, I_{z2}, A_1, and A_2 must be defined, the rest of the cross-sectional properties can be assigned the default values. It should be emphasized that calculation of stresses in modal analysis is meaningless; thus, the section moduli can be assigned the default values. The node at the origin is fixed, while nodes 9 and 13 at y = 17 ft and y = 29 ft, respectively, are allowed to translate along the X and Y-axes. The intermediate nodes can translate along the X and Y-axes and rotate about the Z-axis. The degrees-of freedom specified in BEDITH are shown in Fig. 7.5.1.1. The translational masses m_1 and m_2 along the X and Y-axis are placed at nodes 9 and 13, respectively, as described in Section 7.4.1. Specifically, we set $F_x = F_y = 116.46$ at node 9 and $F_x = F_y = 38.82$ at node 13. Also, in BEDITH we request the calculation of two natural frequencies. Table 7.5.1.2 lists the "EX75. " file generated by BEDITH and executed with SSAP1H. The two mode shapes displayed in SVIEWH are shown in Fig. 7.5.1.2. The procedure to view the mode shapes is described in Section 7.2.2. Finally, Table 7.5.1.3 presents the natural frequencies and mode shapes listed in the "EX75.1" file.

As a validation of our analysis, recall that the $\{\phi\}_1$ and $\{\phi\}_2$ of the two-degree-of-freedom model of Example 5.14.1 are given by eqn. (5.14.1f), and compare them with their corresponding values underlined in Table 7.5.1.3. They are practically identical.

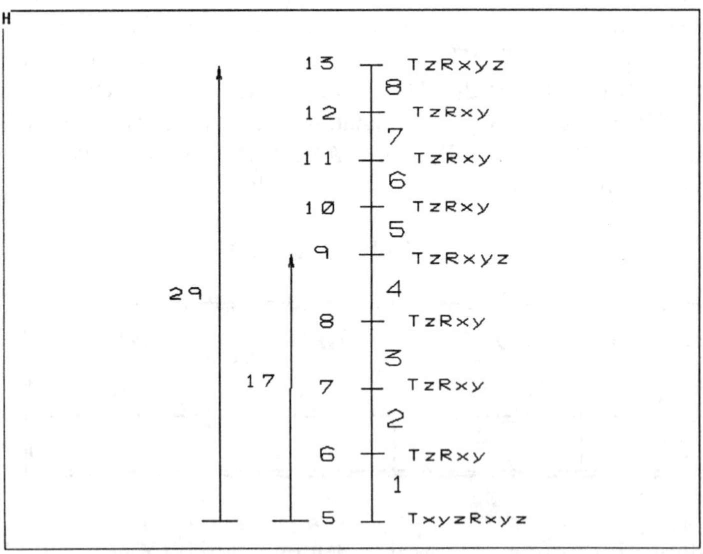

Figure 7.5.1.1 Eight-element beam model.

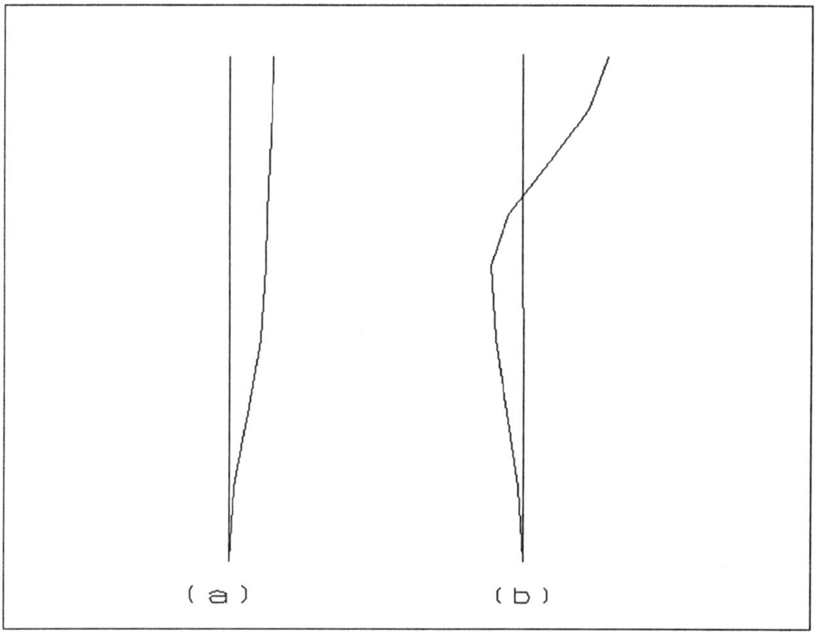

Figure 7.5.1.2 Mode shapes: (a) Fundamental; (b) Second.

Table 7.5.1.2

File EX75 created by BEDIT 4.08-3H

```
15   1   1   2  0 1   0   0   0   0  000   0   0   0   0 386.39999
 1  1 1 1 1 1 1          0.        0. -1.00e+14     0          0. 0 0
 2  1 1 1 1 1 1          0. -1.00e+14        0.     0          0.
 3  1 1 1 1 1 1 -1.00e+14        0.        0.     0          0.
 4  1 1 1 1 1 1          0.        0.        0.     0          0.
 5  1 1 1 1 1 1 1.000e+14        0.        0.     0          0.
 6  0 0 1 1 1 0          0.       51.        0.     0          0.
 7  0 0 1 1 1 0          0.      102.        0.     0          0.
 8  0 0 1 1 1 0          0.      153.        0.     0          0.
 9  0 0 1 1 1 1          0.      204.        0.     0          0.
10  0 0 1 1 1 0          0.      240.        0.     0          0.
11  0 0 1 1 1 0          0.      276.        0.     0          0.
12  0 0 1 1 1 0          0.      312.        0.     0          0.
13  0 0 1 1 1 1          0.      348.        0.     0          0.
14  1 1 1 1 1 1          0. 1.000e+14        0.     0          0.
15  1 1 1 1 1 1          0.        0. 1.000e+14     0          0.
 2  8   2   0    1    0
```

```
1 3.000e+07        0.3         0.     0.283      0.     0.      0.      0.
1  28.799999       0.          0.     0.14       544.   0.0833331.0001.000
1.    1. a
2  15.22           0.          0.     0.14       288.   0.0833331.0001.000
1.    1. a
    0.         0.         0.         0.
    0.         0.         0.         0.
    0.         0.         0.         0.
  1    4    6   15    1    1    0    0    0    0    0    0    0    0
  2    6    7   15    1    1    0    0    0    0    0    0    0    0
  3    7    8   15    1    1    0    0    0    0    0    0    0    0
  4    8    9   15    1    1    0    0    0    0    0    0    0    0
  5    9   10   15    1    2    0    0    0    0    0    0    0    0
  6   10   11   15    1    2    0    0    0    0    0    0    0    0
  7   11   12   15    1    2    0    0    0    0    0    0    0    0
  8   12   13   15    1    2    0    0    0    0    0    0    0    0
  9    0 116.45999 116.45999      0.        0.         0.        0. 0
 13    0      38.82      38.82      0.        0.         0.        0. 0
  0     0  0.00E+00  0.00E+00  0.00E+00  0.00E+00  0.00E+00  0.00E+00
      0.         1.         0.         1.
 0    0    0         0.        0.    0    0        0.    0
```

Table 7.5.1.3

1**** Algor (c) Dynamic Modal Analysis SSAP1H ver 10.02-3H

INPUT FILE............ex75

File EX75 created by BEDIT 4.08-3H

1**** MODAL ANALYSIS

mode number = 1

eigenvectors normalized to unit mass matrix

Displacements/Rotations(degrees) of nodes

NODE number	X- translation	Y- translation	Z- translation	X- rotation	Y- rotation	Z- rotation
6	1.1833E-02	9.4283E-12	0.0000E+00	0.0000E+00	0.0000E+00	-2.3923E-02
7	3.7848E-02	1.3163E-11	0.0000E+00	0.0000E+00	0.0000E+00	-3.1869E-02
8	6.3836E-02	9.3229E-12	0.0000E+00	0.0000E+00	0.0000E+00	-2.3871E-02
9	7.5634E-02	-1.5918E-13	0.0000E+00	0.0000E+00	0.0000E+00	0.0000E+00
10	7.7869E-02	2.1779E-12	0.0000E+00	0.0000E+00	0.0000E+00	-6.3964E-03
11	8.2773E-02	3.1393E-12	0.0000E+00	0.0000E+00	0.0000E+00	-8.5025E-03
12	8.7663E-02	2.2449E-12	0.0000E+00	0.0000E+00	0.0000E+00	-6.3567E-03
13	8.9880E-02	-3.9152E-14	0.0000E+00	0.0000E+00	0.0000E+00	0.0000E+00

1**** MODAL ANALYSIS

mode number = 2

eigenvectors normalized to unit mass matrix

Displacements/Rotations(degrees) of nodes

NODE number	X- translation	Y- translation	Z- translation	X- rotation	Y- rotation	Z- rotation
6	-8.2212E-03	1.0323E-10	0.0000E+00	0.0000E+00	0.0000E+00	1.6591E-02
7	-2.6203E-02	1.4412E-10	0.0000E+00	0.0000E+00	0.0000E+00	2.1947E-02
8	-4.4020E-02	1.0207E-10	0.0000E+00	0.0000E+00	0.0000E+00	1.6275E-02
9	-5.2030E-02	-1.7417E-12	0.0000E+00	0.0000E+00	0.0000E+00	0.0000E+00
10	-2.3326E-02	2.3846E-11	0.0000E+00	0.0000E+00	0.0000E+00	-8.2227E-02
11	3.9805E-02	3.4372E-11	0.0000E+00	0.0000E+00	0.0000E+00	-1.0957E-01
12	1.0287E-01	2.4579E-11	0.0000E+00	0.0000E+00	0.0000E+00	-8.2059E-02
13	1.3150E-01	-4.2730E-13	0.0000E+00	0.0000E+00	0.0000E+00	0.0000E+00

1**** PRINT OF NATURAL FREQUENCIES

mode number	circular frequency (rad/sec)	frequency (Hertz)	period (sec)	tolerance
1	1.1790E+01	1.8765E+00	5.3292E-01	5.2074E-07
2	3.5100E+01	5.5863E+00	1.7901E-01	4.5194E-06

The Sturm sequence check didn't find any missing frequency.

7.6 SYSTEM WITH RIGID BODY MODES

Consider an 18-in circular steel crankshaft with E = $3x10^7$ psi and weight density $\gamma = 0.283$ lb/in^3. It connects two cylinders with polar mass moments of inertia $I_{G1} = 10$ lb-in-sec^2 and $I_{G2} = 6$ lb-in-sec^2, see Fig. 7.6.1. The 12-in long section of the shaft has a $d_1=1$ in diameter with a cross-sectional area $A_1 = 0.785$ in^2, moment of inertia $I_1 = 0.049$ in^4, and torsional constant $J_1 = 0.098$ in^4. The 6-in long section has a $d_2=3/4$ in diameter with $A_2 = 0.442$ in^2, $I_2 = 0.016$ in^4, and $J_2 = 0.031$ in^4. The shaft can rotate freely about the X-axis, and all other possible motions are restrained.

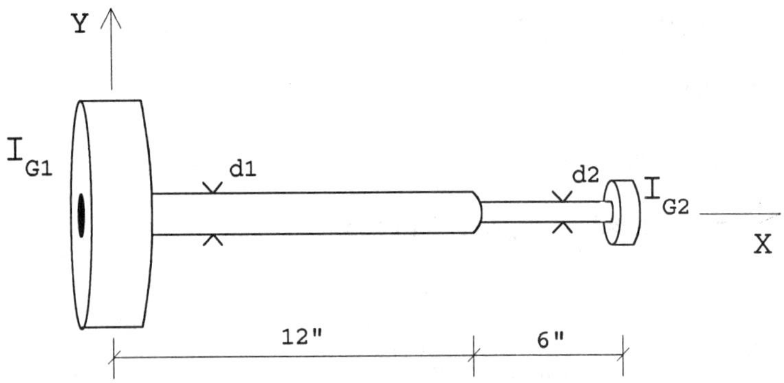

Figure 7.6.1 *Motor generator system.*

Since there is no torsional restraint, the shaft can rotate as a rigid body about the X-axis and also deform torsionally. The polar moment of inertia of the shaft is only a small fraction of the polar moment of inertia of the cylinders, thus, having a minimal effect on the modal properties and the dynamic response of the system. The two dominant natural frequencies that are requested in modal analysis are the rigid-body mode and the lowest non-zero torsional mode. It should be noted that unsupported models could exhibit up to six rigid body modes with zero natural frequencies.

7.6.1 Analysis with Algor

In order to distinguish between the material properties of the 12-in and the 6-in diameter sections of the shaft, two different layers are assigned to the model in SD2H. The 12-in long section is assigned layer No. 0, and the 6-in section layer No.1. Since the cross-section of the shaft is circular, there is no need to modify the default value for the group and color. The two sections are divided into four equal elements each.

The degrees-of-freedom of all nodes defined in BEDITH are $T_{xyz}R_{yz}$, thus, allowing only the torsional rotation about the X-axis. Since the default is steel, the material properties are not modified in the ADD/MOD;PROPERTY menu. Once we enter the MODIFY;SECTIONAL menu, the beams that were assigned the layer No. 0 in SD2H are automatically selected. By selecting VALUE we can enter the corresponding properties. The properties of the remaining part of the shaft are entered after selecting NEXT. The torsional inertias about the X-axis M_x at the end-nodes 4 and 12 as well as the KEY VAL are assigned within the ADD/MOD;LUMPD MAS menu, as presented in Section 7.4.1. The eight-element model of the shaft is shown in Fig. 7.6.1.1.

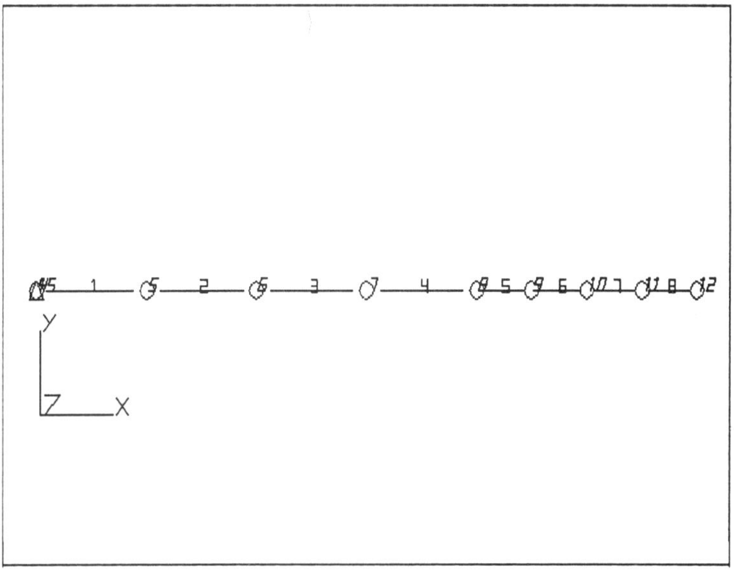

Figure 7.6.1.1 *Eight-element model of motor generator.*

To request two natural frequencies including the rigid body mode, we follow the sequence of commands ANAL. TYPE;MODAL. We set the NO FREQ = -2 and the RGD MODE = 1. The minus sign of the NO FREQ indicates that rigid body modes are requested, and the value assigned to RGD MODE indicates the number of rigid body modes anticipated. Finally we run the file "EX76. " through the sequence TRANSFER;RUN SAP1 to obtain the two lowest natural frequencies f_1 and f_2. Table 7.6.1.1 lists the file "EX76. " developed with BEDITH and executed with SSAP1H. The natural frequencies obtained with SSAP1H are: $f_1 = 1.59X10^{-11}$ Hz, which is practically zero and corresponds to the rigid body mode, and $f_2 = 15.70$ Hz, which is the lowest non-zero natural torsional frequency. Examining the mode shapes in SVIEWH may be difficult since the modes are torsional. In this case, it is easier to identify the type of the mode shapes by checking their amplitudes in the "EX76.1" file.

Table 7.6.1.1

File EX76 created by BEDIT 4.08-3H

```
15   1   1  -2 0 1   0   0   0   0  000    0    0    0   0  386.39999
 1  1 1 1 1 1 1          0.          0. -1.00e+14    0         0. 0 0
 2  1 1 1 1 1 1          0. -1.00e+14         0.     0         0.
 3  1 1 1 1 1 1 -1.00e+14          0.         0.     0         0.
 4  1 1 1 0 1 1          0.          0.         0.     0         0.
 5  1 1 1 0 1 1          3.          0.         0.     0         0.
 6  1 1 1 0 1 1          6.          0.         0.     0         0.
 7  1 1 1 0 1 1          9.          0.         0.     0         0.
 8  1 1 1 0 1 1         12.          0.         0.     0         0.
 9  1 1 1 0 1 1        13.5          0.         0.     0         0.
10  1 1 1 0 1 1         15.          0.         0.     0         0.
11  1 1 1 0 1 1        16.5          0.         0.     0         0.
12  1 1 1 0 1 1         18.          0.         0.     0         0.
13  1 1 1 1 1 1 1.000e+14          0.         0.     0         0.
14  1 1 1 1 1 1          0. 1.000e+14         0.     0         0.
15  1 1 1 1 1 1          0.          0. 1.000e+14    0         0.
 2    8    2     0     1     0
1 3.000e+07        0.3          0.       0.283     0.       0.       0.       0.
1    0.785      0.         0.  0.098    0.049     0.0491.0001.000  1.    1. a
2    0.442      0.         0.  0.031    0.016     0.0161.0001.000  1.    1. a
     0.         0.         0.        0.
     0.         0.         0.        0.
     0.         0.         0.        0.
 1    4    5   14    1    1   0   0    0   0    0    0    0  0
 2    5    6   14    1    1   0   0    0   0    0    0    0  0
 3    6    7   14    1    1   0   0    0   0    0    0    0  0
 4    7    8   14    1    1   0   0    0   0    0    0    0  0
 5    8    9   14    1    2   0   0    0   0    0    0    0  0
 6    9   10   14    1    2   0   0    0   0    0    0    0  0
 7   10   11   14    1    2   0   0    0   0    0    0    0  0
 8   11   12   14    1    2   0   0    0   0    0    0    0  0
 4    0        0.        0.        0.       10.       0.       0. 0
12    0        0.        0.        0.        6.       0.       0. 0
 0    0  0.00E+00  0.00E+00  0.00E+00  0.00E+00  0.00E+00  0.00E+00
      0.         1.        0.        1.
 0    0    0          0.        0.    0    1          0.   0
```

7.7 LOAD STIFFENED STRUCTURE

Consider the simply supported beam with the properties given in Section 7.2. The beam is subjected to a compressive axial load P, as shown in Fig. 7.7.1. We request the seven lowest natural frequencies for each of the loads listed in the first column of Table 7.7.1.

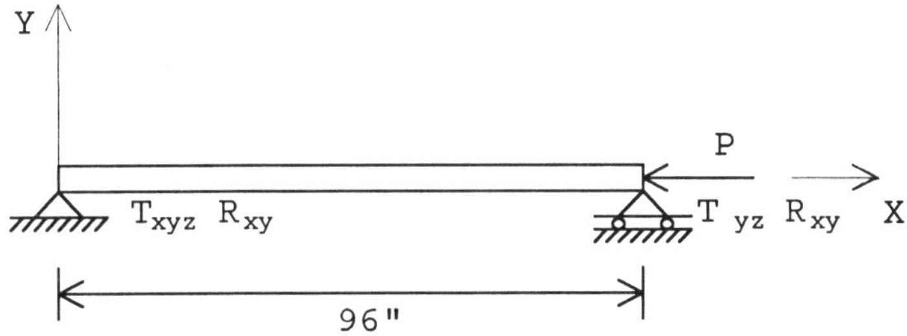

Figure 7.7.1 *Axially loaded beam.*

As discussed in Section 5.5, the presence of a compressive or a tensile axial load can significantly affect the modal characteristics of a slender structure, such as the beam in our example. The axial load must always be less than the critical buckling load to perform a valid linear analysis. For this reason, it is always prudent that a buckling analysis precedes either a static or a dynamic analysis of a system with slender members subjected to compressive loads. Thus, our first step is to calculate the critical load P_{cr} of the beam. The P_{cr} can be obtained from either a buckling analysis using Algor or an analytical solution, e.g., Chajes. For this simple system, it is given by

$$P_{cr} = \frac{\pi^2 EI}{L^2}$$

where for E = 30×10^6 psi, I = 11.3 in^4, and L = 96 in, the above equation yields

$$P_{cr} = 363{,}042 \; lbs$$

The results of modal analysis for each one of the compressive and the tensile loads P are given in Table 7.7.1. The minus sign indicates the compressive loads P. The "A" denotes axial modes, all the other modes are flexural. The effects of shear have not been included in this study.

Table 7.7.1

LOAD P (lb)	MODES (Frequency Hz)						
	1	2	3	4	5	6	7
0	59.25	237.0	526.8 A	533.2	947.7	1480	1575 A
- 10000	58.43	236.2	526.4 A	532.4	946.9	1479	1575 A
-180000	42.1	221.8	518.3	526.8 A	932.9	1465	1575 A
- 360000	5.42	205.5	502.9	526.8 A	917.9	1450	1575 A
-460000	RIGID BODY MODES						
10000	60.1	237.8	526.8 A	534.0	948.6	1480	1575 A
180000	72.5	251.3	526.8 A	547.7	962.3	1494	1575 A
360000	83.6	264.8	526.8 A	561.8	976.6	1509	1575 A
460000	89.2	271.9	526.8 A	569.5	984.5	1517	1575 A
TYPE	F	F	A	F	F	F	A

The following conclusions can be drawn:

a. The magnitude of an axial load leaves the natural frequencies of the axial modes unaffected. This is attributed to the fact that the axial deformations are decoupled from the flexural. Notice the change in sequence of the axial and the flexural modes for increasing compressive loads.

b. The effect of a compressive load on the natural frequencies is significant only when its magnitude is close to the buckling load. Observe the dramatic decrease of the fundamental frequency for increasing magnitude of the axial load. Compressive loads can greatly affect the lower modes; however, their influence on the higher modes is relatively small.

c. Modal analysis is not possible when the axial load is greater than P_{cr}. In such case, the beam fails, and the failure is detected as a rigid body mode. As shown in Table 7.7.1, no natural frequencies are reported for P = - 460,000 lb, a load that is greater than P_{cr}.

d. Tensile loads increase the fundamental frequency of slender structures. Observe, however, that the effect of tensile loads on the fundamental frequency is substantial, while it is relatively small on high natural frequencies.

7.7.1 Analysis with Algor

The FEM modeling of the beam is described in Section 7.2.1, and the sixteen-element model is shown in Fig. 7.2.1.1. There are a few modeling differences, which are discussed for a representative compressive load, i.e, P = -180,000 lb.

In BEDITH an axial compressive load is applied at the right support after following the sequence: ADD/MOD;LOAD;FORCE;VALUE. In VALUE we set FY = -180000. For illustrative purposes, the length of the force vector may be adjusted.

The modal analysis is not performed with SSAP1H. Instead, either the processor SSAP8H (Modal analysis with load stiffening (Beams only)) or the SSAP8SH (Modal Analysis with load stiffening (Plates and Beams)) is used. The files generated by running "EX77. " can be viewed and interpreted as elaborated in Section 7.2.2. Table 7.7.1.1 lists the "EX77. " generated by BEDITH and executed with SSAP8H for the compressive load of -180,000 lb. Finally, it should be mentioned that the critical buckling load calculated with the closed form expression in Section 7.7 can be verified through a buckling analysis using the SSAP6H processor, see the Algor Buckling Analysis Release Notes.

Table 7.7.1.1

File ex72 created by BEDIT 4.08-3H

```
23    1    1    7 0 1   0    0    0    0  000    0    0    0    0 386.39999
 1   1 1 1 1 1 1          0.          0. -1.00e+14    0              0. 0 0
 2   1 1 1 1 1 1          0. -1.00e+14          0.    0              0.
 3   1 1 1 1 1 1 -1.00e+14          0.          0.    0              0.
 4   1 1 1 1 1 0          0.          0.          0.    0              0.
 5   0 0 1 1 1 0          6.          0.          0.    0              0.
 6   0 0 1 1 1 0         12.          0.          0.    0              0.
 7   0 0 1 1 1 0         18.          0.          0.    0              0.
 8   0 0 1 1 1 0         24.          0.          0.    0              0.
 9   0 0 1 1 1 0         30.          0.          0.    0              0.
10   0 0 1 1 1 0         36.          0.          0.    0              0.
11   0 0 1 1 1 0         42.          0.          0.    0              0.
12   0 0 1 1 1 0         48.          0.          0.    0              0.
13   0 0 1 1 1 0         54.          0.          0.    0              0.
14   0 0 1 1 1 0         60.          0.          0.    0              0.
15   0 0 1 1 1 0         66.          0.          0.    0              0.
16   0 0 1 1 1 0         72.          0.          0.    0              0.
17   0 0 1 1 1 0         78.          0.          0.    0              0.
18   0 0 1 1 1 0         84.          0.          0.    0              0.
19   0 0 1 1 1 0         90.          0.          0.    0              0.
20   0 1 1 1 1 0         96.          0.          0.    0              0.
21   1 1 1 1 1 1 1.000e+14          0.          0.    0              0.
22   1 1 1 1 1 1          0. 1.000e+14          0.    0              0.
23   1 1 1 1 1 1          0.          0. 1.000e+14    0              0.
2   16    1    0    1    0
1 3.000e+07    0.3 0.    0.283     0.      0.       0.      0.
1 3.83    0.    0.  0.15  11.3    3.865.4601.900    5.46    1.9 a
    0.         0.         0.         0.
    0.         0.         0.         0.
    0.         0.         0.         0.
 1    4    5   23    1    1    0    0    0    0      0      0      0   0
 2    5    6   23    1    1    0    0    0    0      0      0      0   0
 3    6    7   23    1    1    0    0    0    0      0      0      0   0
 4    7    8   23    1    1    0    0    0    0      0      0      0   0
 5    8    9   23    1    1    0    0    0    0      0      0      0   0
 6    9   10   23    1    1    0    0    0    0      0      0      0   0
 7   10   11   23    1    1    0    0    0    0      0      0      0   0
 8   11   12   23    1    1    0    0    0    0      0      0      0   0
 9   12   13   23    1    1    0    0    0    0      0      0      0   0
10   13   14   23    1    1    0    0    0    0      0      0      0   0
11   14   15   23    1    1    0    0    0    0      0      0      0   0
12   15   16   23    1    1    0    0    0    0      0      0      0   0
13   16   17   23    1    1    0    0    0    0      0      0      0   0
14   17   18   23    1    1    0    0    0    0      0      0      0   0
15   18   19   23    1    1    0    0    0    0      0      0      0   0
16   19   20   23    1    1    0    0    0    0      0      0      0   0
20    1   -180000.         0.          0.         0.         0.        0.
0    0  0.00E+00  0.00E+00  0.00E+00  0.00E+00  0.00E+00  0.00E+00
    0.         1.         0.         1.
0    0    0         0.        0.    0     0        0.    0
```

8

Time History Analysis: Numerical Examples

8.1 INTRODUCTION

This chapter presents numerical examples on time history modal superposition and time history direct integration. Before studying the examples, it is suggested to review Sections 5.10 and 5.11 that refer to the time history modal superposition and Section 5.12 that presents the time history direct integration. For a comparison of the two methods refer to Section 5.13.

8.2 TIME HISTORY MODAL SUPERPOSITION: INTRODUCTORY EXAMPLE

This example demonstrates the procedure of Section 5.11 as well as the significance of damping in forced vibration.

Consider the simply supported steel beam with properties specified in Section 7.2. The beam is subjected to a step load applied at mid-span as shown in Fig. 8.2.1. Obtain the beam response for a damping ratio $\xi = 0.01$. Repeat the analysis for $\xi = 0.05$, and also for $\xi = 0.20$. Note that the beam is allowed to deform only in the XY plane.

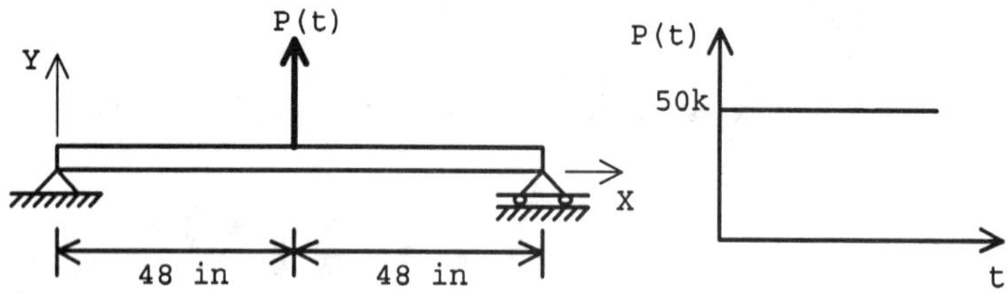

Figure 8.2.1 *Beam subjected to a step load at mid-span.*

We must first perform a modal analysis. The results of modal analysis for the W4X13 beam are given in Table 7.2.3.1. The natural frequencies and mode shapes can now be used to perform a time history modal superposition for the three different damping ratios. In all cases the response has been determined for the time period of 0.150 sec, which corresponds to 150 time steps with a time step Dt = 0.001 sec. Shear effects have not been considered.

In selecting the appropriate Dt, three considerations were taken into account:

a. The structure fulfills all requirements of symmetry, see Sections 7.3 and 3.9; thus, for the load acting on the axis of symmetry, only the odd modes contribute to the response, see Fig. 7.2.2.1.

b. Since the third flexural natural frequency (f_3 = 533 Hz) is much greater than the fundamental (f_1 = 59.2 Hz) and the load is not harmonic with a natural frequency close to any high natural frequency of the system, we can anticipate that the total response is primarily attributed to the first mode. It is reasonable then to consider the fundamental frequency as the critical frequency ($f_1 = f_c$). Consequently, the selection of Dt can be based only on f_1.

c. As recommended in Section 5.11, Dt should be chosen such that Dt $< 0.1T_1$. The selection of Dt = 0.001 sec is in agreement with this recommendation.

In order to demonstrate the importance of using a time step that is small enough to determine the response with an insignificant numerical error, we have calculated the response at mid-span for ξ = 0.05, Dt = 0.001 sec and Dt = 0.005 sec. The results are shown in Fig. 8.2.2. Observe that the responses differ in both amplitude and time variation. The response obtained for Dt = 0.005 has decreased in amplitude and increased in period. These changes are attributed to the selection of a Dt > (0.1 / 59.2 sec), see the third recommendation in Section 5.11, and recall that $f_c = f_1$ = 59.2 Hz.

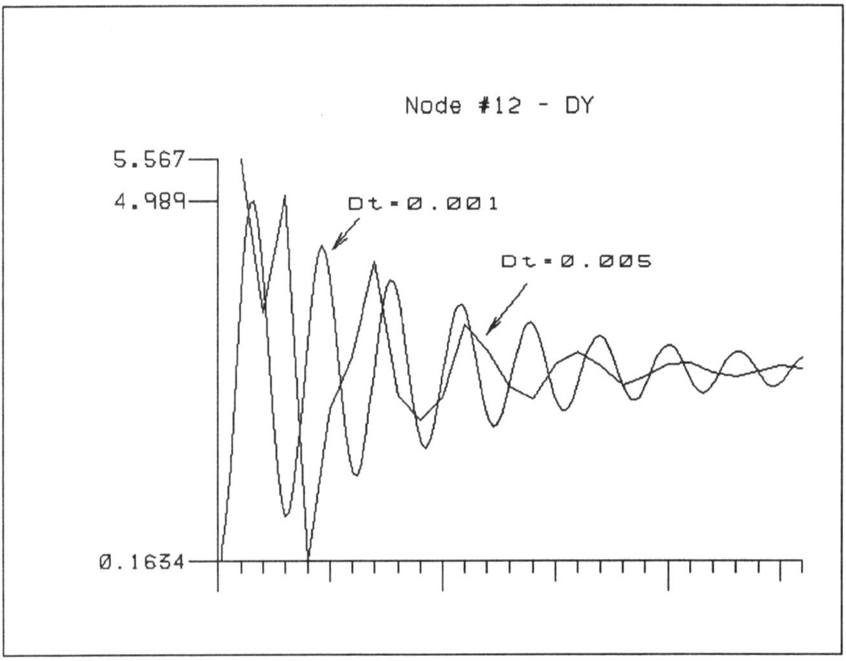

Figure 8.2.2 *Response at mid-span for: (a) Dt = 0.001 sec; (b) Dt = 0.005 sec.*

In order to demonstrate the effects of damping, representative results of three modal superposition analyses are shown in Fig. 8.2.3. This figure presents the response at mid-span for each damping ratio ξ. As expected, the transient response of the beam decreases for increasing damping. It is worth mentioning that all three responses oscillate about the static deflection amplitude y_s at mid-span, where

$$y_s = \frac{P_o L^3}{48EI} = 2.72 \ inches$$

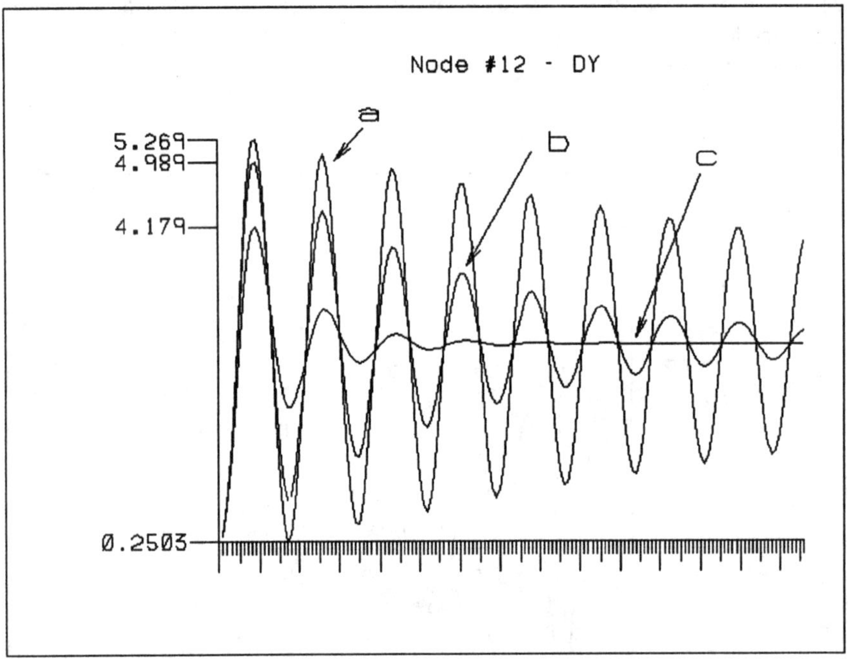

Figure 8.2.3 Response at mid-span for: (a) ξ = 0.01; (b) ξ = 0.05; (c) ξ = 0.2.

8.2.1 Analysis with Algor

Since the beam is allowed to deform in the XY plane, the degrees-of-freedom for all nodes between the supports specified in BEDITH are T_zR_{xy}. The degrees-of-freedom at the supports are shown in Fig. 7.2.1. The material and the cross-sectional properties of the beam are also defined in BEDITH as presented is Section 7.2.1. Note that no load is applied to the structure in BEDITH. As explained in the following, the loads are defined in the preprocessor TIMELOAD. For two- and three-dimensional structures, the dynamic loads are also defined in TIMELOAD not in SD2H.

A three-step procedure must be followed to perform time history modal superposition:

a. Modal analysis.
b. Preparation of the loading data with TIMELOAD.
c. Execution of the time history modal superposition processor SSAP2H.

Since the first step, i.e., modal analysis, has been presented in Sections 7.2.1 through 7.2.3, we will elaborate on the second and third steps of the procedure.

The TIMELOAD creates a file with the loading data through the question and answer procedure presented in the Algor Classic Modeling Tools Release Notes. For convenience, the procedure is given in an abbreviated form for our example. Bold characters are used for the answers.

	Prompt		Answer
a.	Enter processor input file name: (i.e., analysis file with no extension).		**EX72**
b.	Enter dynamic analysis file name (to be created): (i.e., name of output file with no extension).		**EX82**
c.	Enter dynamic analysis code (NDYN):		**-2**
	Time history restart	-2	
	Direct integration	4	
	Transient heat transfer	11	
d.	Enter desired function type:		
	Ramp or constant	1	
	Piece-wise linear	2	
	Sinusoidal	3	
	Function number:		**1**
e.	Solution time step (Delta t)		**0.001**
f.	Initial function value:		**50,000**
g.	Final function value:		**<CR>**
h.	Time for final function value:		**0.150**
i.	Time to end the analysis:		**<CR>**
j.	Results are only... time steps for output:		**<CR>**
k.	Damping factor: (fraction of critical)		**0.05**

l. Enter nodal load data
First node number: **12**
m. Last node number: **<CR>**
n. Enter DOF: **2**
o. Enter scale factor: **1**
p. Enter DOF to continue: **<CR>**
q. First node number: **<CR>**
r. Specify displacement output type
Tabular 1
Printer plot 2
select Maxima only (default): 3 **<CR>**:
s. Enter displacement output requests...
First node number: **<CR>**
t. No nodes have been entered.
Enter nodal data again (Y/N): **N**

The file "EX82. " created by TIMELOAD is listed in Table 8.2.1.1. Once TIMELOAD is executed, we proceed to run the SSAP2H processor. The first file requested by SSAP2H is the one created by TIMELOAD. The name of this file is the response to question (b), i.e., "EX82. ". The second file requested by SSAP2H is the modal analysis file, i.e., "EX72. ". Before we execute the SSAP2H with the RUN command, it is prudent to activate NOS. By activating NOS we avoid creating the very lengthy file "EX82.s" that contains the stress time history. It is more efficient to activate NOS in the final analysis, after we have evaluated and are satisfied with the results. Also by activating TRANS, we can generate the time history of the displacements at the nodes specified in TIMELOAD. The displacement time history is listed in the "EX82.l" file as elaborated in Section 8.2.3. It should be mentioned that execution of SSAP2H creates several binary and ASCII files. Use of these files is discussed in the following section.

Table 8.2.1.1

```
Copyright (c) Algor, Inc.  All rights reserved.
File ex72 created by BEDIT 4.08-3H
23   1   1   7 0-2   0   0   0   0  000   0   0   0   0   386.4
 1   0   1 147   1 1.000E-03 5.000E-02
12   2   1   0 1.000E+00
 0
0.00E+00
21.00000000 Constant function : F(t) = 50000.
  050000..1500050000.
 3   1
 0
 3   1
 0
```

It should be noted that each of the responses in Figs 8.2.2 and 8.2.3 has been obtained with SDNODE and merged in SD2H as instructed in the Algor Processor Reference Manual.

8.2.2 Interpretation of Graphical Results

The first step in interpreting the results is to examine the displacements in SVIEWH. There are 146 load cases that can be seen in SVIEWH. Each load case corresponds to a time step. Note that even though we were expecting 150 time steps according to the "Solution time step" and "Time to end the analysis" values that were specified in TIMELOAD, 146 steps were evaluated by the numerical algorithm. This difference can be accommodated by specifying a larger number for the "time to end the analysis." In the DISPLACED menu, by activating the DISPLACED and WITH UND and using the options in the LOAD CASE menu, we can view the deformed and the undeformed structure for each time step. The procedure is identical to the one presented in Section 7.2.2 where the mode shapes were examined. In order to view the sequential deformation of the system, we can use an SVIEWH macro. First, we view a few load cases in the DISPLACED menu to adjust the scale. Then, from the main menu we follow the sequence of commands: LOAD CASE;FIRST;ALT L;N;ALT O and specify a "repetition count" 146 followed by <CR>.

In static and dynamic analysis, evaluating forces and stresses involves examination of the: **(a)** von Mises, Tresca, and principal stresses; **(b)** reaction forces; **(c)** stress components; and **(d)** precision. Sections 6.4 through 6.6 presented the procedure using a plane stress static analysis example. In dynamic analysis, however, the effort is much greater, since several load cases must be examined.

If the structural material is ductile, we can examine through either the von Mises or the Tresca stresses whether a part of the structure has yielded. If the material is brittle, we must examine the principal stresses. Note that we must check all load cases or at least the ones for which we expect high stresses in order to assure that the structure remains elastic during the whole time period. For a discussion on the von Mises, the Tresca, and the principal stresses refer to Sections 2.9 and 6.4.3. Also, for the definition of the "worst stresses" that play the role of von Mises and Tresca stresses in beams and frames, refer to the Algor Processor Reference Manual or access the "help" option (F_1) in SVIEWH.

There is no need to examine the reaction forces for more than one load case. That is, verifying equilibrium between reactions and external loads for only one load case can serve as a check that the loading and boundary conditions have been applied correctly. When we sense that the reactions are not as anticipated, we should examine either the boundary conditions or the applied loads with the aid of the options within the INQUIRE menu in SVIEWH.

Finally, precision should be evaluated in regions of the model and for the load cases that either the von Mises or the principal stresses are high. It is important that our model is capable of capturing the stresses accurately for those load cases and regions. If precision is high (which implies low accuracy), it may be necessary to repeat the analysis using a smaller time step and/or a finer mesh. When the precision is low and the stresses are high in areas of interest, it may still be necessary to refine the mesh in those areas and repeat the analysis.

8.2.3 Numerical Results

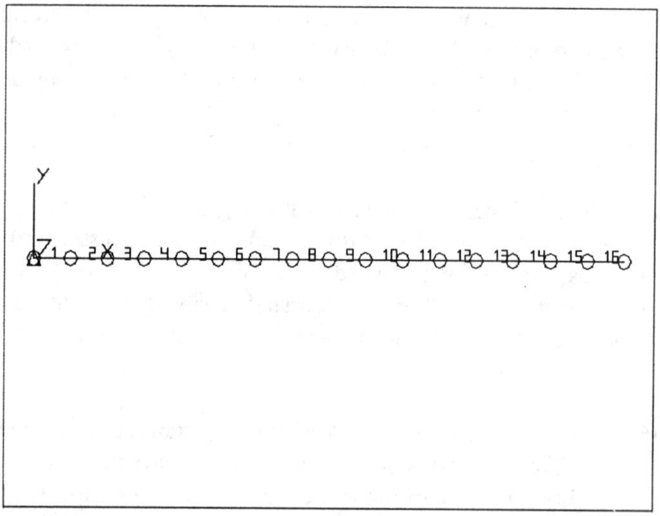

Figure 8.2.3.1 Element numbers of beam model.

Execution of SSAP2H generates the ASCII files with extensions ".l" and ".s". Table 8.2.3.1 presents a part of the "EX82.l" file. It lists the information data inputed in TIMELOAD, the natural frequencies from modal analysis, the maximum absolute values of the displacements, and the time that the maxima occurred. Depending on the answers to questions (r) and (s) in Section 8.2.1, one can generate the time history of the displacements at selected nodes by specifying the nodes in TIMELOAD and activating TRANS before running SSAP2H.

Table 8.2.3.2 lists a part of the "EX82.s" file. This file is rather lengthy and contains information on the stiffness and inertia properties of the structure as well as the element forces and stresses for each node and for every time step. Table 8.2.3.2 lists the element forces and stresses for elements 1 and 8 for the first four time steps. The element numbers of the sixteen-element beam model are shown in Fig. 8.2.3.1.

Table 8.2.3.1

```
1**** Algor (c) SSAP2 Modal superposition time history Ver 10.04-3H

INPUT FILE.............EX82
MODAL ANALYSIS FILE....EX72
------------------------------------------------
File ex72 created by BEDIT 4.08-3H

1**** CONTROL INFORMATION

     number of node points          (NUMNP) =   23
     number of element types        (NELTYP) =  1
     number of load cases           (LL) =      1
     number of frequencies          (NF) =      7
     geometric stiffness flag       (GEOSTF) =  0
     analysis type code             (NDYN) =   -2
     solution mode                  (MODEX) =   0
     equations per block            (KEQB) =    0
     weight and c.g. flag           (IWTCG) =   0
     bandwidth minimization flag    (MINBND) =  0
     gravitational constant         (GRAV) =    3.8640E+02

 **** TIME HISTORY CONTROL INFORMATION

     number of time functions       (NFN) =     1
     ground motion indicator        (NGM) =     0
     number of arrival times        (NAT) =     1
     number of time steps           (NT) =      147
     output print interval          (NOT) =     1
     time step                      (DT) =      1.0000E-03
     damping factor                 (DAMP) =    5.0000E-02
```

1**** DYNAMIC NODAL FORCES/MOMENTS

NODE NUMBER	NODAL D-O-F INDEX	TIME FUNCTION NO.	ARRIVAL TIME NO.	FUNCTION SCALE FACTOR
12	2	1	1	1.000E+00
0	0	0	1	0.000E+00

1**** ARRIVAL TIME DEFINITIONS

NUMBER	ARRIVAL TIME
1	0.0000E+00

1**** TIME FUNCTION DEFINITIONS

time function number = 1
number of abscissae = 2
function scale factor = 1.0000E+00
FUNCTION DESCRIPTION: Constant function : F(t) = 50000.

	TIME	AMPLITUDE
1	0.0000E+00	5.0000E+04
2	1.5000E-01	5.0000E+04

Maximum absolute values of displacement from output time steps

	X transl.	Y transl.	Z transl.	X rot(deg)	Y rot(deg)	Z rot(deg)
Node no.	20	12	0	0	0	4
MAXIMA	4.241E-14	4.989E+00	0.000E+00	0.000E+00	0.000E+00	9.133E+00
TIME	1.000E-03	9.000E-03	0.000E+00	0.000E+00	0.000E+00	9.000E-03

Table 8.2.3.2

1**** Algor (c) FEA Stress Processor MKNSO 15-DEC-93, Ver 11.06-3H
INPUT FILE............ex82

1**** BEAM ELEMENT FORCES AND MOMENTS

ELEM. NO.	CASE (MODE)	AXIAL FORCE R1	SHEAR FORCE R2	SHEAR FORCE R3	TORSION MOMENT M1	BENDING MOMENT M2	BENDING MOMENT M3
1	1	0.000E+00	0.000E+00	5.609E+04	0.000E+00	0.000E+00	0.000E+00
		0.000E+00	0.000E+00	-5.609E+04	0.000E+00	-3.365E+05	0.000E+00
1	2	0.000E+00	0.000E+00	-4.465E+04	0.000E+00	0.000E+00	0.000E+00
		0.000E+00	0.000E+00	4.465E+04	0.000E+00	2.679E+05	0.000E+00
1	3	0.000E+00	0.000E+00	4.164E+04	0.000E+00	0.000E+00	0.000E+00
		0.000E+00	0.000E+00	-4.164E+04	0.000E+00	-2.498E+05	0.000E+00

```
1    4  0.000E+00 0.000E+00  1.632E+04 0.000E+00  0.000E+00 0.000E+00
        0.000E+00 0.000E+00 -1.632E+04 0.000E+00 -9.792E+04 0.000E+00
8    1  0.000E+00 0.000E+00  5.051E+04 0.000E+00  6.125E+05 0.000E+00
        0.000E+00 0.000E+00 -5.051E+04 0.000E+00 -9.156E+05 0.000E+00
8    2  0.000E+00 0.000E+00 -1.251E+04 0.000E+00  2.602E+05 0.000E+00
        0.000E+00 0.000E+00  1.251E+04 0.000E+00 -1.852E+05 0.000E+00
8    3  0.000E+00 0.000E+00  1.545E+04 0.000E+00  6.121E+05 0.000E+00
        0.000E+00 0.000E+00 -1.545E+04 0.000E+00 -7.048E+05 0.000E+00
8    4  0.000E+00 0.000E+00  5.847E+03 0.000E+00  9.122E+05 0.000E+00
```

8.3 USE OF SYMMETRY IN TIME HISTORY MODAL SUPERPOSITION

The two examples in Sections 8.3.1 and 8.3.2 demonstrate the proper use of symmetry in time history modal superposition. The methodology is also applicable to time history direct integration analysis.

In the first example, the response of a symmetric structure subjected to a dynamic load acting at the axis of symmetry is calculated. The methodology can be extended to structures with more than one axis of symmetry with a loading acting at the center of symmetry. The second example considers the more general case of a symmetric structure subjected to dynamic loads which do not act along any axis of symmetry. As illustrated in these examples, taking advantage of symmetry allows the use of smaller models. This results in substantial savings in modeling, processing, and examination of the results.

8.3.1 Loads on Axis of Symmetry

The beam analyzed in Section 8.2.1 fulfills the requirements of a symmetric structure presented in Section 3.9. It is then possible to determine its response to the step load P(t) by modeling half the beam subjected to half the load 0.5P(t) at the axis of symmetry and with the boundary conditions shown in Fig. 8.3.1.1.

The natural frequencies of the symmetric model are listed in Table 7.3.1.1. If we perform the analysis with the time step Dt = 0.001 sec, the responses at mid-span for $\xi = 0.01$, 0.05 and 0.2 will be identical to the ones shown in Fig. 8.2.3, even though the half-beam model accounts for only the odd modes of the whole structure. This is attributed to the fact that for a two-dimensional symmetric structure subjected to a load on the axis of symmetry, the even modes do not contribute to the response, see Fig. 7.2.2.1. It should be noted that the procedure demonstrated for a beam structure can be extended to two- and three-dimensional structures subjected to loads acting at the center of symmetry.

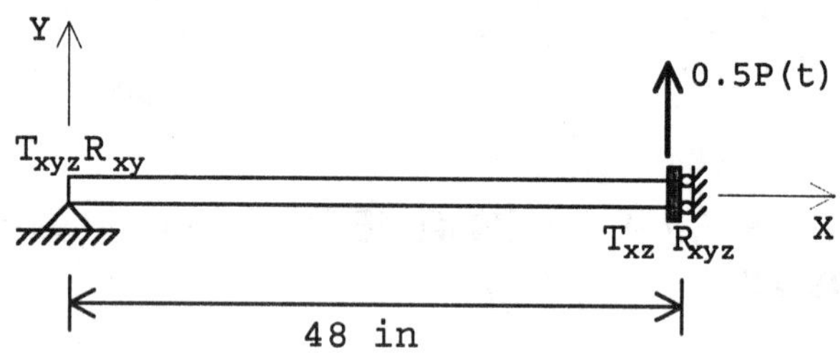

Figure 8.3.1.1 Symmetric model.

8.3.2 Use of Symmetry for Non-symmetrically Loaded Structures

This section demonstrates the use of symmetry to determine the response of symmetric structures subjected to non-symmetrically applied loads. In order to apply the following procedure, the structure must fulfill all the requirements of symmetry specified in Section 3.9 including a symmetric distribution of the mass. Only the load is not symmetrically applied.

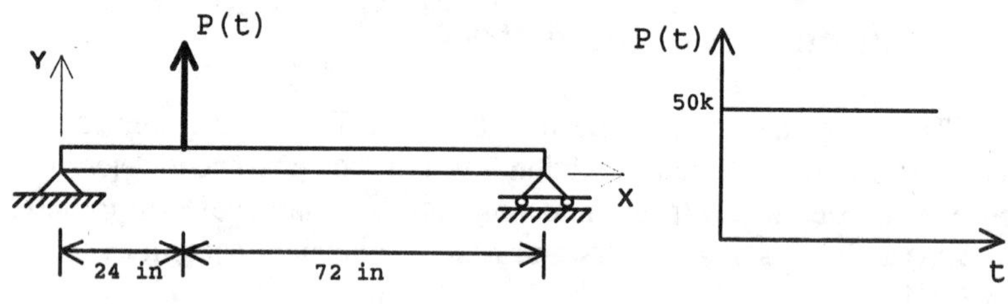

Figure 8.3.2.1 Non-symmetrically loaded beam.

Consider the W4X13 beam subjected to the step load acting at a distance of 24 in to the right of the left support. Determine the response at the point of load application for a damping ratio $\xi = 0.05$. The properties of the beam are given in Section 7.2, and its geometry and load are shown in Fig. 8.3.2.1. The natural frequencies of the beam are listed in Table 7.2.3.1. In this example we have not considered the effects of shear.

The response at the point of load application for the total duration of 0.08 sec with a time step Dt = 0.0002 sec is indicated with (a) in Fig. 8.3.2.2. The analysis employed the sixteen-element model also used for the modal analysis of the beam in Section 7.2. Referring to Fig. 7.2.1.1, the point of load application is node 8. In order to demonstrate the use of symmetry, two additional time history modal superposition analyses are performed. In the first one, the response at node 8 is obtained with the "symmetric" model shown in Fig. 8.3.2.3(a). Notice that the right-end node of this model is allowed to displace along the Y-axis only. In the second analysis, the response at node 8 of the "antisymmetric" model is determined. The right-end node of the antisymmetric model in Fig. 8.3.2.3(b) is allowed to displace along the X-axis and rotate about the Z-axis. Eight equal-length beam elements are used for both the symmetric and the antisymmetric models. In Fig. 8.3.2.2, the responses at node 8 obtained with the symmetric and the antisymmetric models are denoted with (b) and (c), respectively.

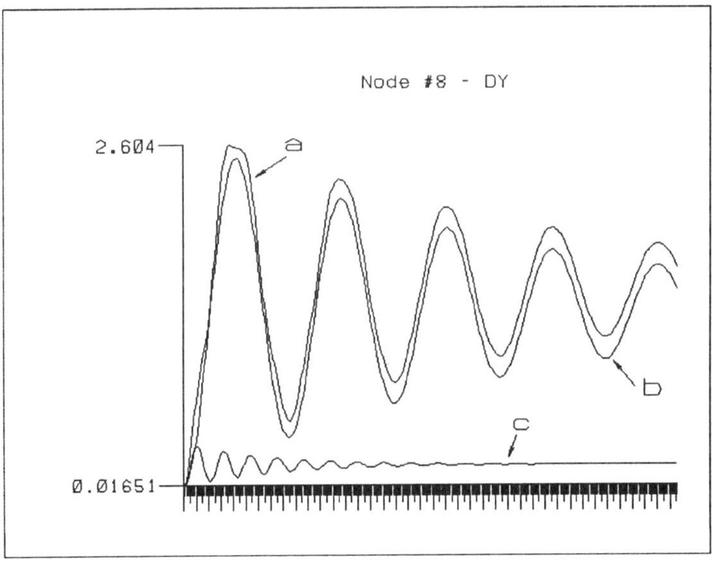

Figure 8.3.2.2 *Vertical response at mid-span from the: (a) Sixteen-element model; (b) Eight-element symmetric model (c) Eight-element antisymmetric model.*

Contrary to the Dt = 0.001 sec used in Section 8.2.1 for the sixteen-element model, a smaller Dt = 0.0002 sec is used to run the analysis that provided the responses (a), (b), and (c) in Fig. 8.3.2.2. The smaller Dt was required only for the antisymmetric model with a fundamental frequency f = 237 Hz, see Table 7.3.1.2, so that Dt < (0.1 / 237 sec) according to the recommendation No. 3 in Section 5.11. Even though the smaller Dt was required only for the antisymmetric model, it was also used for the symmetric and the sixteen-element model of the whole beam for uniformity in plotting Fig. 8.3.2.2.

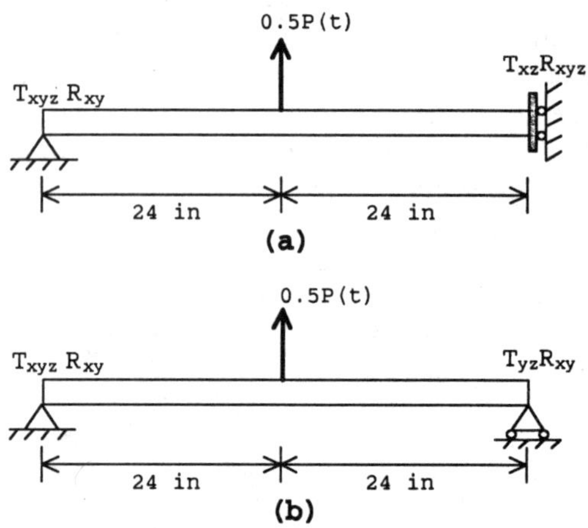

Figure 8.3.2.3 (a) Eight-element symmetric model; (b) Eight-element antisymmetric model.

In Fig. 8.3.2.2 superposition of the responses (b) and (c) yields the response (a). We can conclude that the response of the structure can be determined by combining the responses of the symmetric and antisymmetric models. The procedure can be applied not only for one concentrated load but also for multiple point and distributed loads. It is also applicable to two- and three-dimensional symmetric structures.

8.4 TIME HISTORY DIRECT INTEGRATION

Time history direct integration does not involve use of natural frequencies and mode shapes. Thus, in principle, it is not required to perform modal analysis before proceeding to time history direct integration. However, as elaborated in Section 5.12, knowledge of the system's natural frequencies can be effectively used to perform time history direct integration.

If our analysis is correct, we will obtain identical results when we use the same model and the same time step to perform either time history direct integration or modal superposition analysis. A discussion on the differences and similarities of the two methods is presented in Section 5.13. When a structure is symmetric, the procedures presented in Section 8.3 can also be applied to time history direct integration.

8.4.1 Numerical Example

This example serves two purposes: (a) to introduce the necessary steps to perform time history direct integration; and (b) to calculate the Rayleigh damping from modal damping ratios ξ.

Consider the beam used in Section 8.2 with the step load acting at mid-span as shown in Fig. 8.2.1. Calculate the response at mid-span with time history direct integration. Use the sixteen-element model of Figs. 7.2.1.1 and 8.2.3.1, and assume that the damping ratio for all modes is $\xi = 0.05$. Ignore the effects of shear.

First, we must calculate the coefficients α and β to determine the Rayleigh damping matrix [C]. According to Section 5.12, the constants α and β are obtained from eqns. (5.12.4) which for our case take the form

$$\alpha = \frac{2\,\omega_1\,\omega_2}{\omega_2^{\,2}-\omega_1^{\,2}}(\omega_2-\omega_1)\xi$$

$$\beta = \frac{2}{\omega_2^{\,2}-\omega_1^{\,2}}(\omega_2-\omega_1)\xi \tag{8.4.1.1}$$

by setting

$$\xi = 0.05$$

$$\omega_1 = (2\pi)(59.2) = 371.96 \ rad/\text{sec} \tag{8.4.1.2}$$

$$\omega_2 = (2\pi)(237) = 1489.11 \ rad/\text{sec}$$

the system of eqns. (8.4.1.1) yields

$$\alpha = 29.76$$
$$\beta = 5.4E{-}05 \qquad (8.4.1.3)$$

Determination of α and β requires the ω_1 and ω_2 which are obtained through modal analysis. For this example, the circular natural frequencies are determined from the f_1 and f_2 listed in Table 7.2.3.1. Thus, even though in principle time history direct integration does not utilize natural frequencies and mode shapes, its application requires the modal characteristics of the system in order to calculate the parameters α and β.

Second, we must select a time step Dt. Selection of the appropriate Dt is not straightforward as in time history modal superposition where a modal analysis has preceded and guidelines have been established, see Section 5.11. When the natural frequencies are unknown, it takes several trials to chose an appropriate time step. In this example, we will take advantage of prior knowledge of the natural frequencies and select a Dt = 0.001 sec, see Section 8.2 where the beam was analyzed using time history modal superposition. The response at mid-span is practically identical to the one shown in Fig. 8.2.3(b). It should not be surprising that the responses obtained with time history direct integration and modal superposition are identical.

We may conclude that even this simple example demonstrates the important advantages of modal superposition over direct integration for dynamic analysis of linear elastic systems. It also emphasizes the need to know the natural frequencies in order to determine the time step and the Rayleigh damping.

8.4.2 Analysis with Algor

Through Rayleigh damping in time history direct integration, Algor allows more freedom to account for damping than in modal superposition where only one damping ratio can be assigned to the system. As indicated by eqns. (5.12.4), Rayleigh damping permits the use of two damping ratios.

In this section we outline the procedure to obtain the response of the beam presented in Section 8.4.1. The development of the beam model has been described in Sections 7.2 and 7.2.1. It should be noted that the analysis type specified in BEDITH can be either static or modal. The file created in BEDITH for the default static analysis is given the name "EX84. " and is listed in Table 8.4.2.1. It should be noted that instead of the "EX84.", the "EX72. " file developed in Section 7.2 can also be used. As in time history modal superposition, no load should be applied to the structure in BEDITH when we intend to perform time history direct integration. The loads are defined in TIMELOAD. For two- and three-dimensional structures the dynamic loads are also defined in TIMELOAD, not in SD2H.

Once the model is created and its properties are specified in either DECODS for two- and three-dimensional systems or BEDITH for frame structures, one can proceed to define the loads and the output requests in TIMELOAD. Note that in either BEDITH or DECODS the problem can be treated as either static or modal.

In TIMELOAD we are requested to define the α and β coefficients, the time step Dt, the duration of the load, and the duration of the analysis. The procedure to define all these parameters requires a modal analysis which can be avoided when the natural frequencies are already known or can be estimated with acceptable accuracy.

TIMELOAD is presented in the Algor Classic Modeling Tools. For convenience, the sequence of the question and answer procedure prompted by TIMELOAD is also given in an abbreviated form for the example of Section 8.4.1:

a. Enter processor input file name: (i.e., analysis file with no extension).		**EX84**
b. Enter dynamic analysis file name (to be created): (i.e., name of output file with no extension).		**EX84DI**
c. Enter desired dynamic analysis code (NDYN):		**4**
Time history restart	-2	
Direct integration	4	
Transient heat transfer	11	
d. Enter desired function type:		
Ramp or constant	1	
Piece-wise linear	2	
Sinusoidal	3	
Function number:		**1**
e. Solution time step (Delta t):		**0.001**
f. Initial function value:		**50,000**

g. Final function value: **<CR>**
h. Time for final function value: **0.150**
i. Time to end the analysis: **<CR>**
j. Results are only ....
time steps for output: **<CR>**
k. Damping matrix used is...
Alpha: **29.76**
Beta: **5.4E-5**
l. Enter nodal load data
First node number: **12**
m .Last node number: **<CR>**
n. Enter DOF: **2**
o. Enter scale factor: **1**
p. Enter DOF to continue: **<CR>**
q. First node number: **<CR>**
r. Specify displacement output type
Tabular 1
Printer plot 2
Maxima only (default) 3 : **<CR>**
s. Enter displacement output requests
First node number: **<CR>**
t. No nodes have been entered.
Enter nodal data again (Y/N): **N**

The "time for final function value" must be specified at least two time steps longer than desired for output. The "EX84DI. " created by TIMELOAD is listed in Table 8.4.2.1.

After the TIMELOAD is executed, we proceed to run the time history direct integration, i.e., the SSAP4H processor. The SSAP4H requests the file that has been created by TIMELOAD, i.e., the response to question (b). Before we execute SSAP4H with RUN, it is suggested to activate NOS. Activating NOS will not create the lengthy "EX84DI.s" file that lists the stress time history. It is more efficient to request NOS and obtain a printed output after examining the results and been satisfied with the analysis. By activating TRANS, we can generate the time history of the nodal displacements specified in TIMELOAD. The displacement time history is listed in the "EX84.1" file. Execution of SSAP4H creates several binary and ASCII files. Use of these files is discussed in the following section.

Table 8.4.2.1

File ex84 created by BEDIT 4.08-3H

```
23    1    1    0 0 0    0    0    0    0  000    0    0    0    0 386.39999
 1  1 1 1 1 1 1         0.          0. -1.00e+14   0         0. 0 0
 2  1 1 1 1 1 1         0. -1.00e+14         0.   0         0.
 3  1 1 1 1 1 1 -1.00e+14         0.         0.   0         0.
 4  1 1 1 1 1 0         0.         0.         0.   0         0.
 5  0 0 1 1 1 0         6.         0.         0.   0         0.
 6  0 0 1 1 1 0        12.         0.         0.   0         0.
 7  0 0 1 1 1 0        18.         0.         0.   0         0.
 8  0 0 1 1 1 0        24.         0.         0.   0         0.
 9  0 0 1 1 1 0        30.         0.         0.   0         0.
10  0 0 1 1 1 0        36.         0.         0.   0         0.
11  0 0 1 1 1 0        42.         0.         0.   0         0.
12  0 0 1 1 1 0        48.         0.         0.   0         0.
13  0 0 1 1 1 0        54.         0.         0.   0         0.
14  0 0 1 1 1 0        60.         0.         0.   0         0.
15  0 0 1 1 1 0        66.         0.         0.   0         0.
16  0 0 1 1 1 0        72.         0.         0.   0         0.
17  0 0 1 1 1 0        78.         0.         0.   0         0.
18  0 0 1 1 1 0        84.         0.         0.   0         0.
19  0 0 1 1 1 0        90.         0.         0.   0         0.
20  0 1 1 1 1 0        96.         0.         0.   0         0.
21  1 1 1 1 1 1 1.000e+14         0.         0.   0         0.
22  1 1 1 1 1 1         0. 1.000e+14         0.   0         0.
23  1 1 1 1 1 1         0.         0. 1.000e+14   0         0.
 2  16    1    0    1    0
 1 3.000e+07       0.3         0.     0.283     0.       0.       0.       0.
 1     3.83      0.       0.    0.15     11.3    3.865.4601.900   5.46
1.9 a
      0.        0.        0.        0.
      0.        0.        0.        0.
      0.        0.        0.        0.
 1   4    5   23    1    1    0    0    0    0      0      0      0   0
 2   5    6   23    1    1    0    0    0    0      0      0      0   0
 3   6    7   23    1    1    0    0    0    0      0      0      0   0
 4   7    8   23    1    1    0    0    0    0      0      0      0   0
 5   8    9   23    1    1    0    0    0    0      0      0      0   0
 6   9   10   23    1    1    0    0    0    0      0      0      0   0
 7  10   11   23    1    1    0    0    0    0      0      0      0   0
 8  11   12   23    1    1    0    0    0    0      0      0      0   0
 9  12   13   23    1    1    0    0    0    0      0      0      0   0
10  13   14   23    1    1    0    0    0    0      0      0      0   0
11  14   15   23    1    1    0    0    0    0      0      0      0   0
12  15   16   23    1    1    0    0    0    0      0      0      0   0
13  16   17   23    1    1    0    0    0    0      0      0      0   0
14  17   18   23    1    1    0    0    0    0      0      0      0   0
15  18   19   23    1    1    0    0    0    0      0      0      0   0
16  19   20   23    1    1    0    0    0    0      0      0      0   0
 0    0 0.00E+00 0.00E+00 0.00E+00 0.00E+00 0.00E+00 0.00E+00
  0.        1.        0.        1.
```

Table 8.4.2.2

File ex84 created by BEDIT 4.08-3H

```
23    1    1    0   0 4     0     0     0     0  000      0    0    0    0  386.4
 1   1 1 1 1 1 1           0.          0. -1.00e+14   0          0. 0 0
 2   1 1 1 1 1 1           0.  -1.00e+14         0.   0          0.
 3   1 1 1 1 1 1  -1.00e+14           0.         0.   0          0.
 4   1 1 1 1 1 0           0.          0.         0.   0          0.
 5   0 0 1 1 1 0           6.          0.         0.   0          0.
 6   0 0 1 1 1 0          12.          0.         0.   0          0.
 7   0 0 1 1 1 0          18.          0.         0.   0          0.
 8   0 0 1 1 1 0          24.          0.         0.   0          0.
 9   0 0 1 1 1 0          30.          0.         0.   0          0.
10   0 0 1 1 1 0          36.          0.         0.   0          0.
11   0 0 1 1 1 0          42.          0.         0.   0          0.
12   0 0 1 1 1 0          48.          0.         0.   0          0.
13   0 0 1 1 1 0          54.          0.         0.   0          0.
14   0 0 1 1 1 0          60.          0.         0.   0          0.
15   0 0 1 1 1 0          66.          0.         0.   0          0.
16   0 0 1 1 1 0          72.          0.         0.   0          0.
17   0 0 1 1 1 0          78.          0.         0.   0          0.
18   0 0 1 1 1 0          84.          0.         0.   0          0.
19   0 0 1 1 1 0          90.          0.         0.   0          0.
20   0 1 1 1 1 0          96.          0.         0.   0          0.
21   1 1 1 1 1 1  1.000e+14           0.         0.   0          0.
22   1 1 1 1 1 1           0.  1.000e+14         0.   0          0.
23   1 1 1 1 1 1           0.          0. 1.000e+14   0          0.
 2  16    1    0    1    0
 1 3.000e+07     0.3       0.    0.283    0.     0.     0.     0.
 1   3.83     0.      0.     0.15    11.3   3.865.4601.900   5.46     1.9 a
        0.         0.         0.         0.
        0.         0.         0.         0.
        0.         0.         0.         0.
 1    4    5   23    1    1    0    0    0    0      0      0      0   0
 2    5    6   23    1    1    0    0    0    0      0      0      0   0
 3    6    7   23    1    1    0    0    0    0      0      0      0   0
 4    7    8   23    1    1    0    0    0    0      0      0      0   0
 5    8    9   23    1    1    0    0    0    0      0      0      0   0
 6    9   10   23    1    1    0    0    0    0      0      0      0   0
 7   10   11   23    1    1    0    0    0    0      0      0      0   0
 8   11   12   23    1    1    0    0    0    0      0      0      0   0
 9   12   13   23    1    1    0    0    0    0      0      0      0   0
10   13   14   23    1    1    0    0    0    0      0      0      0   0
11   14   15   23    1    1    0    0    0    0      0      0      0   0
12   15   16   23    1    1    0    0    0    0      0      0      0   0
13   16   17   23    1    1    0    0    0    0      0      0      0   0
14   17   18   23    1    1    0    0    0    0      0      0      0   0
15   18   19   23    1    1    0    0    0    0      0      0      0   0
16   19   20   23    1    1    0    0    0    0      0      0      0   0
 0    0 0.00E+00 0.00E+00 0.00E+00 0.00E+00 0.00E+00 0.00E+00
 .000E+00 .100E+01 .000E+00 .100E+01
 1    0    1 147    1 1.000E-03 2.976E+01 5.400E-05
 12   2    1    0 1.000E+00
 0
0.00E+00
21.00000000 Constant function : F(t) = 50000.
 050000..1500050000.
 3    1
 0
 3    1
 0
```

8.4.3 Numerical Results

The interpretation of the graphical results is identical to the procedure presented in Section 8.2.2. In this section we list excerpts from the "EX84DI.1" and the "EX84DI.s" ASCII files generated by SSAP4H. Since they present a few differences with the corresponding files created by the processor of time history modal superposition SSAP2H, Tables 8.4.3.1 and 8.4.3.2 list only selected parts of the "EX84DI.1" and the "EX84DI.s" files, respectively. Notice that the last part of Table 8.4.3.1 lists the maximum displacement and the time that it occurred. In Table 8.4.3.2 we list only the forces for elements 1 and 8 for the first four time steps. The element numbers are shown in Figure 8.2.3.1.

Table 8.4.3.1

```
1**** Algor (c) Time History Direct Integration  SSP4H  Ver 10.04-3H
INPUT FILE.............ex84di
------------------------------------------------

File ex84 created by BEDIT 4.08-3H  15-OCT-93
1**** CONTROL INFORMATION

    number of node points         (NUMNP) =    23
    number of element types       (NELTYP) =   1
    number of load cases          (LL) =       1
    number of frequencies         (NF) =       0
    geometric stiffness flag      (GEOSTF) =   0
    analysis type code            (NDYN) =     4
    solution mode                 (MODEX) =    0
    equations per block           (KEQB) =     0
    weight and c.g. flag          (IWTCG) =    0
    bandwidth minimization flag   (MINBND) =   0
    gravitational constant        (GRAV) =     3.8640E+02

    bandwidth minimization specified

1**** STEP-BY-STEP SOLUTION CONTROL INFORMATION

    number of time varying functions  =       1
    ground motion indicator =                 0
    number of arrival times =                 1
    number of solution time steps =           147
    output (print) interval =                 1
```

solution time increment = 1.0000E-03
damping coefficient (alpha) = 2.9760E+01
damping coefficient (beta) = 5.4000E-05

1D Y N A M I C L O A D I N P U T

NODE NUMBER	DEGREE OF FREEDOM	FUNCTION REFERENCE	ARRIVAL TIME NUMBER	FUNCTION MULTIPLIER
12	2	1	1	.1000E+01

A R R I V A L T I M E V A L U E S

INPUT ORDER	ARRIVAL TIME VALUE
1	.0000E+00

1T I M E F U N C T I O N D A T A
TIME FUNCTION NUMBER = (1)
NUMBER OF POINTS = (2)
SCALE FACTOR = (.1000E+01)
DESCRIPTION = (Constant function : F(t) = 50000.)

INPUT ORDER	TIME VALUE	FUNCTION VALUE
1	.0000E+00	.5000E+05
2	.1500E+00	.5000E+05

The maximum displacement component of 4.885E+00
occurred at time 1.000E-02, time step 10,
for the Y-translation of node 12.

Table 8.4.3.2

1**** Algor (c) FEA Stress Processor MKNSO Ver 11.06-3H

INPUT FILE.............ex84di

1**** BEAM ELEMENT FORCES AND MOMENTS

ELEM. NO.	CASE (MODE)	AXIAL FORCE R1	SHEAR FORCE R2	SHEAR FORCE R3	TORSION MOMENT M1	BENDING MOMENT M2	BENDING MOMENT M3
1	1	0.000E+00	0.000E+00	-1.089E+03	0.000E+00	0.000E+00	0.000E+00
		0.000E+00	0.000E+00	1.089E+03	0.000E+00	6.532E+03	0.000E+00
1	2	0.000E+00	0.000E+00	-4.620E+03	0.000E+00	0.000E+00	0.000E+00

		0.000E+00	0.000E+00	4.620E+03	0.000E+00	2.772E+04	0.000E+00
1	3	0.000E+00	0.000E+00	3.908E+02	0.000E+00	0.000E+00	0.000E+00
		0.000E+00	0.000E+00	-3.908E+02	0.000E+00	-2.345E+03	0.000E+00
1	4	0.000E+00	0.000E+00	1.351E+04	0.000E+00	0.000E+00	0.000E+00
		0.000E+00	0.000E+00	-1.351E+04	0.000E+00	-8.106E+04	0.000E+00
8	1	0.000E+00	0.000E+00	6.943E+03	0.000E+00	3.882E+04	0.000E+00
		0.000E+00	0.000E+00	-6.943E+03	0.000E+00	-8.048E+04	0.000E+00
8	2	0.000E+00	0.000E+00	2.980E+04	0.000E+00	2.211E+05	0.000E+00
		0.000E+00	0.000E+00	-2.980E+04	0.000E+00	-4.000E+05	0.000E+00
8	3	0.000E+00	0.000E+00	1.970E+04	0.000E+00	4.011E+05	0.000E+00
		0.000E+00	0.000E+00	-1.970E+04	0.000E+00	-5.193E+05	0.000E+00
8	4	0.000E+00	0.000E+00	2.437E+04	0.000E+00	6.192E+05	0.000E+00

9

Frequency Response, Response Spectrum, Random Vibration Analysis: Numerical Examples

9.1 FREQUENCY RESPONSE ANALYSIS: NUMERICAL EXAMPLES

When a lightly damped system is subjected to harmonic loads, the response must be obtained for a long time duration in order to reach steady-state, see Fig. 4.6.3. In this case, it could be computationally prohibitive to reach steady-state with time history, and we should resort to frequency response analysis. As a general rule, when the externally applied loads or the ground excitation are harmonic, the preferred method of analysis is the frequency response. Use of frequency response analysis is also appropriate when the duration of the transient part of the response is either very short or of no interest, for example in most machine vibration problems.

As discussed in Section 5.14 and demonstrated in Example 5.14.1, frequency response analysis provides the steady-state response when the structure is subjected to a single harmonic load. When the excitation includes more than one frequency, then frequency response analysis provides the steady-state response amplitudes by combining the individual harmonic responses. Any maxima that occur during the transient part cannot be captured with frequency response analysis. Thus, strictly speaking, the results would not be conservative when the maxima occur during the transient response. If we want to obtain both the transient and the steady-state parts of the response, then we should employ either time history modal superposition or direct integration.

9.1.1 System Subjected to Harmonic Loads

This section presents the finite element solution of Example 5.14.1 using beam elements to model the frame. The frame of Fig. 5.14.1 can be modeled as a cantilever beam with two lumped masses placed as shown in Fig. 9.1.1.1. The harmonic load acting at the top is $F_2(t) = 15000 \sin(2\pi*2.5t)$ lb.

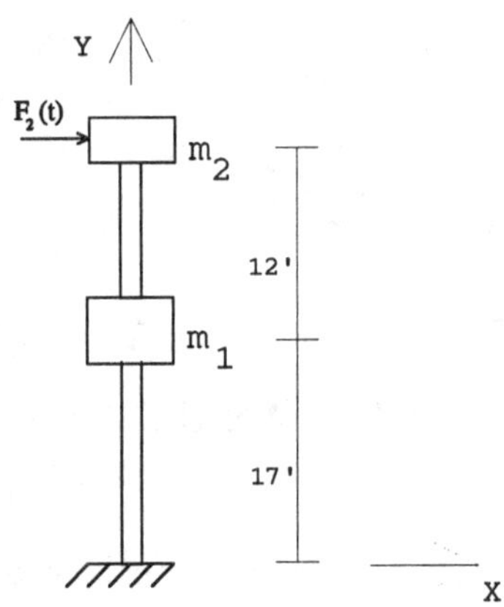

Figure 9.1.1.1 *"Stick" model of frame subjected to a harmonic load.*

The first step in obtaining the response is to perform modal analysis. The response is dominated by two translational lumped masses m_1 and m_2 along the X-axis. Thus, we request two natural frequencies and obtain: $f_1 = 1.876$ Hz and $f_2 = 5.586$ Hz. The f_1 and f_2 have been determined in Example 5.14.1 and with a finite element analysis using beam elements in Section 7.5.

We assign zero damping ($\xi = 0$) to the system and perform a frequency response analysis that includes only the two lowest modes. The analysis provides the nodal steady-state response amplitudes, see Table 9.1.4.1 for nodes 9 and 13.

In order to demonstrate the effect of damping, we assign 10% of critical damping to each mode and repeat the analysis. We include only the first two modes and obtain the solution given by eqn. (9.1.4.6). The results for zero and 10% critical damping are further discussed in Section 9.1.4.

9.1.2 Analysis with Algor

In frequency response analysis the excitation can be harmonic loads, ground motions as well as combinations of harmonic loads and ground motions. Depending on the number of loads and excitation frequencies, the response output is provided in one of the following two forms:

a. If the excitation is caused by a single or a set of harmonic loads with one specific frequency, then both the amplitude and phase angle at each node are calculated, see Example 5.14.1. In this case, frequency response analysis can be considered as an extension to multi-degree-of-freedom systems of the damped harmonic excitation presented for a SDOF system in Section 4.6.

b. In any other case, the phase angles are not retained and the amplitudes at each node are estimated using the square root of the sum of the squares (SRSS) of the individual responses. The user must be reminded that the SRSS provides acceptable results for the default value of the cluster factor = 0.1 when the natural frequency of each mode is at least 10 percent greater than the frequency of the next higher mode. Otherwise the cluster factor should be given a value of less than 0.1 to accommodate closely spaced natural frequencies. The displacements obtained with SRSS are not necessarily equal to the maximum displacements that the system will experience. They are, however, a statistically acceptable approximation of the structure's steady-state displacement amplitudes. Also, it should be remembered that the stresses obtained with SRSS would only be accurate when computed from the individual modal stresses. Such SRSS stresses can be examined in SVIEWH.

In frequency response analysis, Algor allows greater freedom than in either time history modal superposition or time history direct integration. Specifically, we can define as many damping ratios ξ as we consider necessary at frequencies of our choice. Note that, as a general rule, the damping ratios are defined at the natural frequencies of the structure.

A three-step procedure must be followed to perform frequency response analysis. It includes:

1. Modal analysis. This is discussed in Sections 7.5 and 7.5.1. The model used for both the modal and the frequency response analysis is shown in Fig. 7.5.1.1.

2. Specification of damping and loading data with the preprocessor PRESS5. Like the other preprocessors of dynamic analysis, PRESS5 generates the input data file through a question and answer procedure. For a discussion on the input data refer to the Algor Frequency Response Analysis Release Notes. An abbreviated form of the procedure is given for the example of Section 9.1.1. Bold characters are used for the answers. We have omitted repeating the <CR> response given to the question:

continue to next section   0
modify data   1 >> 0   **<CR>**

which is asked several times in order to confirm the input data. A $\xi = 0.1$ is assigned to both modes.

Copyright (c) Algor, Inc.   All rights reserved.

a. Enter modal analysis file name: (i.e., modal analysis file with no extension.)	**EX75**
b. Enter SSAP5 file name (to be created): (i.e., output file with no extension.)	**EX91**
c. No. of applied nodes (NAND):	**1**
d. Enter 1 to select same frequency loads (IFREQ) Enter 0 otherwise:	**1**
e. Cluster factor for frequencies (CLUSTF):	**<CR>**
f. Enter Node, Direction, Type, Index, Scaling and Phase:	**13,1,2,1,1,0**
g. Enter NAF, INFQ:	**1,0**
h. Enter 1 frequency (Hz) for [# 1 of 1 sets]: Notice that 2.50 = 15.7/(2π) Hz., where 15.7 rad/sec is the circular frequency of the applied load.	**2.5**
i. Enter NFDX - Number of freq. points .... Enter NFDX:	**2**
j. Enter Freq. (Hz) and ratio for [#1 of 2 sets]:	**1.876,0.1**
k. Enter Freq. (Hz) and ratio for [#2 of 2 sets]:	**5.586,0.1**
l. NFAX- Number of frequency points defining amplitude Enter NFAX	**1**
m. Enter freq (Hz), accel. and force for [#1 of 1 sets]:	**2.5,0,15000**

The "EX91. " file generated by PRESS5 is listed in Table 9.1.2.1.

3. Execute the frequency response analysis processor SSAP5H. The first file requested by SSAP5H is the file created in response to question (b). If the user desires to obtain a list of the displacement output in the "EX91.1" file, the TRANS should be activated before running SSAP5H. The second file requested by SSAP5H is the modal analysis file given in response to question (a). The SSAP5H generates several binary and ASCII files which are examined in the following two sections.

Table 9.1.2.1


```
File EX75 created by BEDIT 4.08-3H
15   1   1   2 0-5   0   0   0   0 000   0   0   0   0   386.40
** Applied nodes, direction, type, index, Scaling fac, Phase
1    1    0    0    0 0.0000E+00  0.0000E+00  0.0000E+00  0.0000E+00
1   13    1    2    1 1.0000E+00  0.0000E+00
** Frequency index definition:
1    1    0
1   2.5000E+00
** Frequencies (Hz) vs critical damping ratios
2
1   1.8760E+00  1.0000E-01
2   5.5860E+00  1.0000E-01
** Frequencies (Hz) vs accelerations (in G) and Forces
1
1   2.5000E+00  0.0000E+00  1.5000E+04
```

9.1.3 Graphical Results

When only one harmonic load acts on the system, the number of load cases that can be examined in SVIEWH is equal to the natural modes evaluated in modal analysis increased by one. Thus, in our case there exist three load cases in SVIEWH.

First we examine the displacements with the DISPLACED, WITH UND, and LOAD CASE, as discussed in Section 8.2.2. The first load case depicts the displacement component that is in-phase with the load. The second load case shows the out-of-phase part of the response. Note that under the effect of a single harmonic load, we obtain an out-of-phase response only when there is damping in the system. For zero damping, the third load case is obtained through the superposition described in step 4 of Example 5.14.1. In any other case, however, the last load case is the SRSS of the in- and out-of phase responses. Obviously as an SRSS, the nodal displacements in the last load case are positive.

Figure 9.1.3.1 shows the third load case, i.e., the amplitudes of the steady-state response. Even though the last load case is the most significant, examination of the other load cases can reveal significant information regarding the effect of the in- and out-of-phase responses to the system's steady-state response.

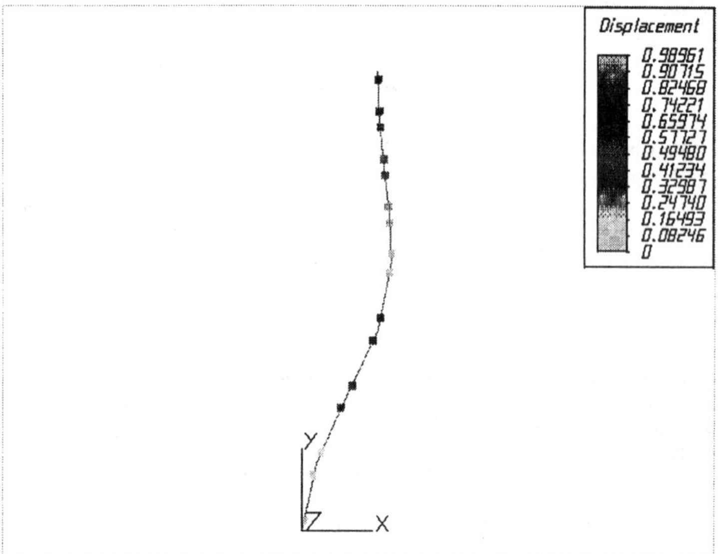

Figure 9.1.3.1 *Steady-state response amplitudes of frame.*

Next we evaluate the stress results. Similar to the displacements, the stresses of the first two load cases are the stress components that are in- and out-of phase with the applied load, respectively, and the third load case is the SRSS of the first two. The stresses examined in SVIEWH in the third load case are the stress amplitudes at steady-state. When the structure is subjected to harmonic loads with different frequencies, the last load case, being the SRSS of the corresponding in-phase and out-of-phase responses, is an estimate of the stress amplitudes that develop at steady state.

It is important to note that the last load case should be strictly used only for quantitative assessment of the von Mises and Tresca for two- and three-dimensional structures or of the "worst stresses" for frame structures to determine whether any part of the system has yielded. The critical stresses obtained as a combination of those already combined through SRSS stresses are not accurate enough to assess linearity in the system. They can only be used for a qualitative assessment that can be validated with either time history or direct integration analysis.

9.1.4 Numerical Results

A part of "EX91.1" generated by activating TRANS and running SSAP5H is presented in Table 9.1.4.2. The first part of "EX91.1" lists the values of the parameters assigned in PRESS5. This is followed by the modal participation factors for the natural frequencies obtained from modal analysis. It should be noted that the modal participation factors are used only when the structure is subjected to loads that can be expressed in terms of the same function of time, e.g., a ground excitation, see step 3 in Example 5.10.1.

The last part of "EX91.1" in Table 9.1.4.2 provides the "maximum displacement components" and the "phase angles" for all load cases. PRESS5H allows determination of the response for all possible resonance conditions, that is, when the frequency of the applied load is equal to either one of the natural frequencies of the system, i.e., 1.876 Hz or 5.586 Hz. The displacement amplitudes and phase angles for the 2.50 Hz load are given in Table 9.1.4.1 for nodes 9 and 13 and for the damping ratios $\xi = 0$ and $\xi = 0.1$. The results for the latter case are also underlined in Table 9.1.4.2.

Table 9.1.4.1

Dam. Rat ξ	NODE	A1*	A2*	A3*	A4*	A5*
0.0	9	-1.051	-5.4E-16	1.051	180.0	π
	13	-0.861	-6.4E-16	0.861	180.0	π
0.1	9	-0.949	-0.279	0.990	163.6	2.855
	13	-0.746	-0.375	0.835	153.3	2.675

*A1 = in-phase response
*A2 = out-of-phase response
*A3 = resultant response
*A4 = phase angle of total response in degrees
*A5 = phase angle of total response in radians

Using columns A3 and A5 in Table 9.1.4.1 for zero damping, the steady-state horizontal displacements of nodes 9 and 13, indicated as $y_{s9}(t)$ and $y_{s13}(t)$, respectively, are given by

$$y_{s9}(t) = 1.051 \ \sin(15.7t + \pi) = -1.051 \sin 15.7t$$
$$y_{s13}(t) = 0.861 \ \sin(15.7t + \pi) = -0.861 \sin 15.7t \tag{9.1.4.1}$$

Notice the small difference between eqns. (9.1.4.1) and eqns. (5.14.1o).

Using columns A1 and A2 for the damping ratio $\xi = 0.1$, the responses at nodes 9 and 13 are given by

$$y_{s9} = -0.949 \ \sin 15.7t - 0.279 \ \cos 15.7t$$
$$y_{s13} = -0.746 \ \sin 15.7t - 0.375 \ \cos 15.7t \tag{9.1.4.2}$$

We can transform an equation of the form

$$y(t) = A\cos\omega t + B\sin\omega t \tag{9.1.4.3}$$

to an equivalent expression through the simple trigonometric transformation:

$$y(t) = C\sin(\omega t + \theta) \tag{9.1.4.4}$$

where the amplitude C and the angle θ are given by

$$C = \sqrt{A^2 + B^2}$$
$$\tan\theta = \frac{A}{B} \tag{9.1.4.5}$$

In view of eqns. (9.1.4.3) through (9.1.4.5), we can express eqns. (9.1.4.2) in the form

$$y_{s9}(t) = 0.990 \ \sin(15.7t + 2.855)$$
$$y_{s13}(t) = 0.835 \ \sin(15.7t + 2.675) \tag{9.1.4.6}$$

Notice that eqns. (9.1.4.6) can be readily obtained from the columns A3 and A5 of Table 9.1.4.1.

By comparing eqns. (9.1.4.1) with eqns. (9.1.4.6), we observe that the presence of damping reduces the response amplitudes and introduces phase angles between the response and the applied loads.

Table 9.1.4.2

```
1**** Algor (c) Frequency Response Processor  SSAP5 Ver 10.04-3H
INPUT FILE.............ex91
MODAL ANALYSIS FILE....ex75
------------------------------------------------
File EX75 created by BEDIT 4.08-3H
1**** CONTROL INFORMATION
     number of node points          (NUMNP) =      15
     number of element types        (NELTYP) =     1
     number of load cases           (LL) =         1
     number of frequencies          (NF) =         2
     geometric stiffness flag       (GEOSTF) =     0
     analysis type code             (NDYN) =      -5
     solution mode                  (MODEX) =      0
     equations per block            (KEQB) =       0
     weight and c.g. flag           (IWTCG) =      0
     bandwidth minimization flag    (MINBND) =     0
     gravitational constant         (GRAV) =       3.8640E+02

 **** No of applied nodes (NAND) =                 1
      (Stop if NAND <= zero)

 **** Single freq. option selected (IFREQ) =       1

     Applied nodes, directions, types, freq. indices:
     Node # = 0      : Ground Excitation
     Direc. = 1,2,3  : X-, Y-, Z- direction
     Ityp = 1        : Acceleration
            2        : Force
     Iidx   = Index for applied exciting frequency
     Sclamp = Scaling factor for amplitude for this node
     Phsshf = Phase shift for this node
```

No.	Node #	Direc.	Ityp	Iidx	Sclamp	Phsshf	Remarks
1	13	1	2	1	1.0000E+00	0.0000E+00	Valid data

**** Additional exciting frequencies:
No. of frequency indices (NIDX) = 1
No. of additional frequencies (NAF) = 1
Frequency inclusion index (INFQ) = 0
INFQ = 0 : Exclude all natural frequencies
= 1 : Include all natural frequencies

Input of additional exciting frequencies:

No.	Frequencies (Hz)
1	2.5000E+00

Total requested exciting frequencies:

No.	Frequencies (Hz)
1	2.5000E+00

**** Damping ratios for all natural frequencies:
No. of input damping ratios (NFDX) = 2

Input frequency vs. damping ratio:

No.	Freq (Hz)	damping ratio
1	1.8760E+00	1.0000E-01
2	5.5860E+00	1.0000E-01

Final damping ratio for each natural frequency:

No.	Freq (Hz)	damping ratio
1	1.8765E+00	1.0000E-01
2	5.5863E+00	1.0000E-01

**** Acceleration and force for all exciting frequencies:
No. of input amplitudes (NFAX) = 1

Input freq. vs. acceleration and force:

No.	Freq(Hz)	Accel.(in G)	Force
1	2.5000E+00	0.0000E+00	1.5000E+04

Final acceleration and force for each exciting frequency:

1 Freq.(Hz): 2.5000E+00
Accel.(G): 0.0000E+00
Force : 1.5000E+04

**** Modal participation factors for ground acceleration:

No.	Nat.Freq Hz	M.P.fact_X	M.P.fact_Y	M.P.fact_Z
1	1.8765E+00	1.2593E+01	1.7183E-11	0.0000E+00
2	5.5863E+00	-1.0032E+00	1.8832E-10	0.0000E+00

**** Response (phase and amplitude) of each mode at each applied frequency:

START OF LOAD 1

Applied frequency case # 1 (Applied frequency= 2.500E+00 Hz)

Mode No.	Phase Angle (Deg.)	Amplitude
1	1.6103E+02	1.1834E+01
2	6.3859E+00	1.9896E+00

**** Final displacement response for each forcing frequency:
****** In Phase Response

Forcing frequency case # 1
Maximum displacement components:

Node	Dx	Dy	Dz	Rx	Ry	Rz
6	-1.4868E-01	9.8587E-11	0.0000E+00	0.0000E+00	0.0000E+00	3.0053E-01
7	-4.7538E-01	1.3764E-10	0.0000E+00	0.0000E+00	0.0000E+00	4.0005E-01
8	-8.0145E-01	9.7487E-11	0.0000E+00	0.0000E+00	0.0000E+00	2.9933E-01
9	-9.4933E-01	-1.6622E-12	0.0000E+00	0.0000E+00	0.0000E+00	0.0000E+00
10	-9.1759E-01	2.2775E-11	0.0000E+00	0.0000E+00	0.0000E+00	-9.0997E-02
11	-8.4765E-01	3.2829E-11	0.0000E+00	0.0000E+00	0.0000E+00	-1.2149E-01
12	-7.7767E-01	2.3476E-11	0.0000E+00	0.0000E+00	0.0000E+00	-9.1107E-02
13	-7.4588E-01	-4.0670E-13	0.0000E+00	0.0000E+00	0.0000E+00	0.0000E+00
Max.:	-7.4588E-01	1.3764E-10	0.0000E+00	0.0000E+00	0.0000E+00	4.0005E-01
Node:	13	7	0	0	0	7

Phase angles:

Node	PDx	PDy	PDz	PRx	PRy	PRz
6	1.6362E+02	3.0950E+01	0.0000E+00	0.0000E+00	0.0000E+00	3.4361E+02
7	1.6361E+02	3.0950E+01	0.0000E+00	0.0000E+00	0.0000E+00	3.4360E+02
8	1.6360E+02	3.0950E+01	0.0000E+00	0.0000E+00	0.0000E+00	3.4357E+02
9	1.6359E+02	2.1098E+02	0.0000E+00	0.0000E+00	0.0000E+00	0.0000E+00
10	1.6221E+02	3.0948E+01	0.0000E+00	0.0000E+00	0.0000E+00	2.0519E+02
11	1.5889E+02	3.0949E+01	0.0000E+00	0.0000E+00	0.0000E+00	2.0512E+02
12	1.5516E+02	3.0948E+01	0.0000E+00	0.0000E+00	0.0000E+00	2.0507E+02
13	1.5331E+02	2.1109E+02	0.0000E+00	0.0000E+00	0.0000E+00	0.0000E+00
Max.:	1.6362E+02	2.1109E+02	0.0000E+00	0.0000E+00	0.0000E+00	3.4361E+02
Node:	6	13	0	0	0	6

****** Out of phase response, 90 degrees

Forcing frequency case # 1

Maximum displacement components:

Node	Dx	Dy	Dz	Rx	Ry	Rz
6	-4.3711E-02	-5.9121E-11	0.0000E+00	0.0000E+00	0.0000E+00	8.8378E-02
7	-1.3983E-01	-8.2539E-11	0.0000E+00	0.0000E+00	0.0000E+00	1.1777E-01
8	-2.3588E-01	-5.8460E-11	0.0000E+00	0.0000E+00	0.0000E+00	8.8249E-02
9	-2.7951E-01	9.9790E-13	0.0000E+00	0.0000E+00	0.0000E+00	0.0000E+00
10	-2.9446E-01	-1.3657E-11	0.0000E+00	0.0000E+00	0.0000E+00	4.2808E-02
11	-3.2730E-01	-1.9686E-11	0.0000E+00	0.0000E+00	0.0000E+00	5.6962E-02
12	-3.6007E-01	-1.4077E-11	0.0000E+00	0.0000E+00	0.0000E+00	4.2618E-02
13	-3.7494E-01	2.4521E-13	0.0000E+00	0.0000E+00	0.0000E+00	0.0000E+00
Max.:	-3.7494E-01	2.4521E-13	0.0000E+00	0.0000E+00	0.0000E+00	1.1777E-01
Node:	13	13	0	0	0	7

Phase angles:

Node	PDx	PDy	PDz	PRx	PRy	PRz
6	1.6362E+02	3.0950E+01	0.0000E+00	0.0000E+00	0.0000E+00	3.4361E+02
7	1.6361E+02	3.0950E+01	0.0000E+00	0.0000E+00	0.0000E+00	3.4360E+02
8	1.6360E+02	3.0950E+01	0.0000E+00	0.0000E+00	0.0000E+00	3.4357E+02
9	1.6359E+02	2.1098E+02	0.0000E+00	0.0000E+00	0.0000E+00	0.0000E+00
10	1.6221E+02	3.0948E+01	0.0000E+00	0.0000E+00	0.0000E+00	2.0519E+02
11	1.5889E+02	3.0949E+01	0.0000E+00	0.0000E+00	0.0000E+00	2.0512E+02
12	1.5516E+02	3.0948E+01	0.0000E+00	0.0000E+00	0.0000E+00	2.0507E+02
13	1.5331E+02	2.1109E+02	0.0000E+00	0.0000E+00	0.0000E+00	0.0000E+00
Max.:	1.6362E+02	2.1109E+02	0.0000E+00	0.0000E+00	0.0000E+00	3.4361E+02
Node:	6	13	0	0	0	6

****** Resultant Response, variable phase

Forcing frequency case # 1

Maximum displacement components:

Node	Dx	Dy	Dz	Rx	Ry	Rz
6	1.5497E-01	1.1496E-10	0.0000E+00	0.0000E+00	0.0000E+00	3.1326E-01
7	4.9552E-01	1.6049E-10	0.0000E+00	0.0000E+00	0.0000E+00	4.1702E-01
8	8.3544E-01	1.1367E-10	0.0000E+00	0.0000E+00	0.0000E+00	3.1207E-01
9	9.8962E-01	1.9388E-12	0.0000E+00	0.0000E+00	0.0000E+00	0.0000E+00
10	9.6368E-01	2.6556E-11	0.0000E+00	0.0000E+00	0.0000E+00	1.0056E-01
11	9.0864E-01	3.8278E-11	0.0000E+00	0.0000E+00	0.0000E+00	1.3418E-01
12	8.5698E-01	2.7373E-11	0.0000E+00	0.0000E+00	0.0000E+00	1.0058E-01
13	8.3481E-01	4.7490E-13	0.0000E+00	0.0000E+00	0.0000E+00	0.0000E+00
Max.:	9.8962E-01	1.6049E-10	0.0000E+00	0.0000E+00	0.0000E+00	4.1702E-01
Node:	9	7	0	0	0	7

Phase angles:

Node	PDx	PDy	PDz	PRx	PRy	PRz
6	1.6362E+02	3.0950E+01	0.0000E+00	0.0000E+00	0.0000E+00	3.4361E+02
7	1.6361E+02	3.0950E+01	0.0000E+00	0.0000E+00	0.0000E+00	3.4360E+02
8	1.6360E+02	3.0950E+01	0.0000E+00	0.0000E+00	0.0000E+00	3.4357E+02
9	1.6359E+02	2.1098E+02	0.0000E+00	0.0000E+00	0.0000E+00	0.0000E+00
10	1.6221E+02	3.0948E+01	0.0000E+00	0.0000E+00	0.0000E+00	2.0519E+02
11	1.5889E+02	3.0949E+01	0.0000E+00	0.0000E+00	0.0000E+00	2.0512E+02
12	1.5516E+02	3.0948E+01	0.0000E+00	0.0000E+00	0.0000E+00	2.0507E+02
13	1.5331E+02	2.1109E+02	0.0000E+00	0.0000E+00	0.0000E+00	0.0000E+00
Max.:	1.6362E+02	2.1109E+02	0.0000E+00	0.0000E+00	0.0000E+00	3.4361E+02
Node:	6	13	0	0	0	6

9.2 RESPONSE SPECTRUM ANALYSIS: NUMERICAL EXAMPLES

As elaborated in Section 5.15, when the excitation is defined in the form of spectra, the response is obtained with response spectrum analysis. Response spectra provide a rational method to describe earthquakes and blast generated ground motions. The method requires prior calculation of the natural frequencies and mode shapes of the structure. In the case of earthquakes, it is usually sufficient to include only the modes with natural frequencies less than 50 Hz. This limit is primarily imposed by the cut-off frequency of earthquake accelerograms. In most cases, however, by inspecting the modal participation factors, it is possible to include fewer modes.

It should also be remembered that damping in the system is indirectly incorporated in the analysis. Specifically, the spectra are selected from the family of curves with the anticipated damping in the structure. The definition of the response spectrum as well as representative spectra are presented in Section 4.9.

9.2.1 System Subjected to Response Spectra of a Harmonic Ground Excitation

Assume that the frame analyzed in Section 9.1.1 is subjected to a harmonic ground excitation with an acceleration $\ddot{y}_g(t) = (0.4g)\sin(2\pi*2.5t)$. Assume that damping is 5% of critical ($\xi = 0.05$) and calculate the maximum relative displacements at y = 17 ft and y = 29 ft.

Figure 9.2.1.1 shows the "stick" model of the frame subjected to the ground harmonic excitation. We adopt the same eight-element model used to determine the steady-state response of the frame in Section 9.1.1. The development of the eight-element model is presented in Sections 7.5 and 7.5.1.

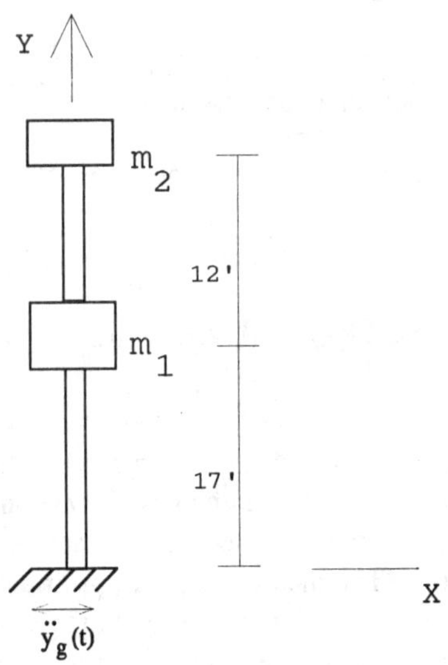

Figure 9.2.1.1 *"Stick" model of frame subjected to a harmonic ground excitation.*

This problem can be solved in several different ways. For example, if we wish to obtain both the transient and the steady-state response of the frame, we can apply either a time history modal superposition or a time history direct integration analysis. Either method can provide the response of the system at selected time intervals. If we are interested in the steady-state response only, then we can employ a frequency response analysis. The maximum relative displacements can be obtained with response spectrum analysis. As elaborated in Chapter 5, response spectrum analysis determines each modal response and the total response by combining the individual modal responses.

In order to perform a response spectrum analysis for a harmonic ground excitation, we must first develop the corresponding spectra. This can be accomplished through the procedures of Section 4.9. Note that the closed form expression of the spectral displacement is given by eqn. (4.9.3). The next step involves modal analysis. The results of modal analysis are presented in Section 7.5. Finally, the response can be obtained by combining the results of modal analysis with the response spectra as demonstrated in steps 3 and 4 of Example 5.15.1.

The maximum relative displacements at nodes 9 and 13 are presented in Table 9.2.1.1. The second and third columns list the modal relative displacements, and the "Relative Response" column displays the maximum relative displacements. The "Relative Response" has been obtained with the SRSS combination method applied to the modal relative displacements, see Section 5.16. For example, the maximum relative displacement along the X-axis at node 9 is determined as follows:

$$1.350 = \sqrt{1.350^2 + (-0.008)^2} \tag{9.2.1.1}$$

The displacements in eqn. (9.2.1.1) are also underlined in Table 9.2.4.1. Observe that the relative responses in Table 9.2.1.1 are almost equal to the relative displacements due to the first mode. In this example, the contribution of the second mode is insignificant. This is anticipated from the large value of the modal participation factor, i.e., 12.59 for the first mode along the X-direction in Table 9.2.4.1.

Table 9.2.1.1

NODE No.	MODE 1	MODE 2	RELATIVE RESP.
9	1.350	- 0.008	1.350
13	1.605	0.021	1.605

9.2.2 Analysis with Algor

In response spectrum analysis, the excitation is expressed as either spectra of a harmonic ground motion or design response spectra. In either case, the input can be defined as displacements, accelerations, and multiples of the acceleration of gravity g.

In summary, for a harmonic ground excitation Algor utilizes the mode shapes and nodal masses to calculate the modal participation factors. The response spectrum generated by the preprocessor RESONATE is then used by the processor to calculate the modal responses by multiplying the participation factors with the corresponding response spectrum values. Finally, the maximum relative responses are calculated through a modal combination method.

A three-step procedure is followed to obtain the response of a structure subjected to the spectra of a harmonic ground excitation:

1. Modal analysis: The analysis of the eight-element model and the results are given in Sections 7.5 and 7.5.1. The element and node numbers of the model are shown in Fig. 7.5.1.1.

2. Generation of spectra for the harmonic ground motion with the preprocessor RESONATE: A description of RESONATE can be found in the Algor Response Spectrum Analysis Release Notes. For convenience, we present the question and answer procedure as prompted by RESONATE for the frame subjected to the harmonic ground excitation:

a. Enter modal analysis file name: (i.e., modal analysis file with no extension.)	**EX75**
b. Enter response spectrum file name (to be created): (i.e., output file with no extension.)	**EX92**
c. Enter frequency of oscillator (Hertz):	**2.5**
d. Enter damping ratio (ξ=0.01=1 percent):	**0.05**
e. Enter type of spectrum input Displacement (0) Acceleration (1) G's (2):	**2**
f. Enter base acceleration amplitude (G's):	**0.40**
g. Global X direction active Yes (1) No (0):	**1**
h. Enter X direction scale factor:	**1**

i. Global Y direction active
Yes (1) No (0): **0**

j. Global Z direction active
Yes (1) No (0): **0**

k. Enter mode combination method to be used
Original SAP-IV method (1)
NRC reg. guide 1.92 method (2)
Modified NRC Reg. Guide 1.92 method (3)
Method of mode combination: **1**

l. Included individual modal loads in output file
Yes (1) No (0): **1**

Briefly stated, RESONATE executes the procedure graphically depicted in Fig. 4.9.1 for a harmonic ground excitation. It creates the file "EX92. " which contains the data that define the spectra for the harmonic excitation. The first part of this file, is listed in Table 9.2.2.1. As explained in the Algor Processor Reference Manual, the first line is a "comment line," and the second line is "the master control line." All the remaining lines have been created by RESONATE and are described in the Algor Response Spectrum Analysis Release Notes. Starting from the sixth line, pairs of natural periods of single-degree-of-freedom systems (SDOF) and their corresponding spectral displacements S_d are listed. For example, the underlined S_d in the last line of Table 9.2.2.1 corresponds to a SDOF system with a natural period T = 0.1272 sec, and is evaluated from eqn. (4.9.3) as follows:

$$S_d = \frac{1}{\sqrt{(1-0.318^2)^2 + [(2)(0.05)(0.318)]^2}} = 1.112 \qquad (9.2.2.1)$$

where r has been calculated from eqn. (4.5.3), that is

$$r = \frac{T}{\overline{T}} = \frac{0.1272}{0.40} = 0.318 \qquad (9.2.2.2)$$

in which $\overline{T} = 1/2.5$ sec is obtained from eqn. (4.3.11) for the frequency of the harmonic ground excitation $\overline{f} = 2.5$ Hz.

3. Calculation of the response with the response spectrum analysis processor SSAP3H: The SSAP3H prompts you to enter first the output file specified in response to question (b). If it is desired to obtain a detailed list of the displacements in the "EX92.1", TRANS should be activated before running SSAP3H. Next, SSAP3H requests the modal analysis file, which is the response to question (a).

Table 9.2.2.1

File EX75 created by BEDIT 4.08-3H

```
15   1   1   2  0-3   0   0   0    0  000    0    0    0    0    386.4
1.000      .000      .000   2   1    0   -1
SIMPLE OSCILLATOR RESPONSE SPECTRUM
90 4.000E-01
4.000E-03 1.000E+00
8.696E-02 1.049E+00
8.871E-02 1.051E+00
9.054E-02 1.054E+00
9.244E-02 1.056E+00
9.443E-02 1.059E+00
9.650E-02 1.061E+00
9.867E-02 1.064E+00
1.009E-01 1.068E+00
1.033E-01 1.071E+00
1.058E-01 1.075E+00
1.084E-01 1.079E+00
1.112E-01 1.083E+00
1.140E-01 1.088E+00
1.171E-01 1.093E+00
1.203E-01 1.099E+00
1.237E-01 1.105E+00
1.272E-01 1.112E+00
```

9.2.3 Graphical Results

There are numerous similarities in the format of the outputs of frequency response and response spectrum analysis. In fact, examination of the graphical results with SVIEWH is identical. The number of load cases generated by SSAP3H that can be examined with SVIEWH is equal to the number of modes obtained through modal analysis increased by one. In the analysis of this frame in which two modes have been determined, three load cases have been created. For each load case, the displacements can be viewed with the aid of the commands DISPLACED;CALC SCALE;DISPLACED, and the options in the

LOAD CASE menu. All load cases except the last one are the relative modal displacements, and express the contribution of each mode to the response. The last load case provides the requested maximum relative displacements or "total combined nodal deflections" of the system and is obtained by combining all the previous load cases. The displacements in the last load case are obtained through the modal combination selected in response to question (k). Note that the SRSS is the so-called "original SSAP-IV" method. Figure 9.2.3.1 shows the displacements for the last load case. The total combined deflections at nodes 9 and 13 (1.350 in and 1.605 in, respectively) are listed in the last column of Table 9.2.1.1, and are also underlined in Table 9.2.4.1.

For each of the three load cases, the stresses and the element nodal forces can be examined in SVIEWH through the STRESS-DI;POST;BEAM-TRUS commands and the selection of a stress or force of interest within the BEAM-TRUS menu. Before we examine any stress components, we should always identify the orientation of the elements with the ORIENT command. The sequence of using the ORIENT in two- and three-dimensional problems is described in Section 6.4.5. For convenience, we repeat the process here for a frame structure: Starting from the MAIN MENU, the following sequence is applied: OPTIONS;EL OPT;BEAM;ORIENT. After selecting a color, the F10;REDRAW sequence is followed to view the orientation lines that indicate the direction of the element 2-axis. Knowing that the element 1-axis is along the length of the element, and that the local 3-axis is normal to the 1- and 2-axes with a direction defined by the right-hand rule, we can completely identify the orientation of each element and the associated stresses, see Fig. 3.2.2. It is suggested to view the structure in the isometric view in order to easily identify the orientation lines. We can now proceed to view the element stresses, internal forces, and moments for each load case. For example, we can obtain a dithered plot of the shear forces in the local 3-axis, which for our problem is parallel to the global X-axis, by following the sequence: STRESS-DI;POST;BEAM-TRUS;R3 SHF, and then return to the MAIN MENU to perform a STRESS DI;DO DITHER. Shear forces of individual elements can be requested in the AUX POST menu through GET VAL and then "clicking" on the element of interest. Shear stresses can also be dithered by making the proper selection within the BEAM-TRUS menu. As for the displacements, the first two load cases are the modal forces or stresses and the last case is their modal combination. For example, the modal shear forces along the local 3-axis for node 9 of element 5 in the first two load cases and the SRSS in the third load case are related as follows:

$$9008 = \sqrt{8952^2 + 1003^2} \tag{9.2.3.1}$$

The shear forces in eqn. (9.2.3.1) are also underlined in Table 9.2.4.2.

It should be pointed out that the "worst stresses" in SVIEWH are an unreliable and often overconservative criterion to assess linearity in the frame, see Section 9.1.3. If a qualitative assessment of "worst stresses" for frames or the von Mises stresses for two- and three-dimensional elements indicates yield, we may need to resort to either time history modal superposition or direct integration for validation.

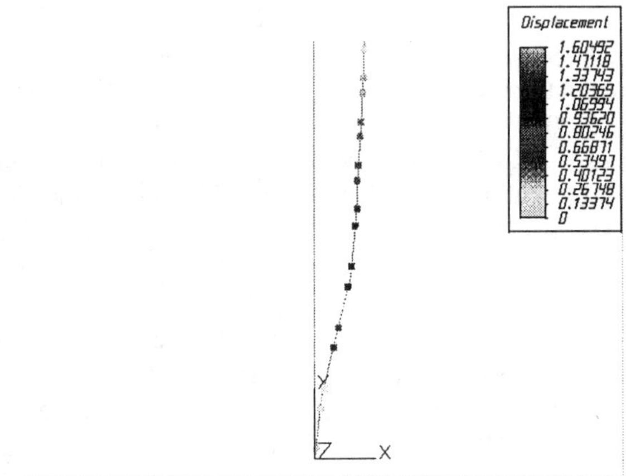

Figure 9.2.3.1 *Total combined nodal deflections.*

9.2.4 Numerical Results

Execution of SSAP3H generates the ASCII "EX92.1" and "EX92.s" files. Parts of the "EX92.1" are listed in Table 9.2.4.1, and excerpts from the "EX92.s" are given in Table 9.2.4.2.

Besides the information pertinent to the input data and the natural frequencies, the "EX92.1" contains a list defining the three modal combination methods. For a discussion on the combination methods the user is referred to Section 5.16. It should be noted that Algor uses the term "Original SAPIV

Procedure" to indicate the SRSS. It also provides two NRC guide procedures. Currently Algor does not include the CQC method.

In Table 9.2.4.1, notice that the modal participation factors along the Y- and Z-axes are practically zero, and that the maximum participation factor corresponds to the first mode. This is the main reason why the relative response is essentially due to the first mode. In general, the greater the modal participation factor the greater the contribution of the corresponding mode to the system's response. The "EX92.1" also includes a plot of the spectra generated from the harmonic ground excitation. Due to space limitations, however, this graph is not included in Table 9.2.4.1. The last part of "EX92.1" presents the total combined nodal deflections, that is, the deflections for load case 3 in SVIEWH.

The file "EX92.s" includes information about the number of elements and their type, the element load multipliers, the element forces, moments and stresses, and the resultant stress combinations. Table 9.2.4.2 lists the element forces and moments of element 5 for the three load cases. The forces listed in the last load case are the SRSS combined forces. These are the forces that should be used in the analysis and design of the system.

Table 9.2.4.1

```
1**** Algor (c) Response spectrum analysis SSAP3H Ver 10.04-3H

INPUT FILE.............ex92
MODAL ANALYSIS FILE....ex75
-----------------------------------------------
File EX75 created by BEDIT 4.08-3H
1**** CONTROL INFORMATION
     number of node points           (NUMNP) =     15
     number of element types         (NELTYP) =    1
     number of load cases            (LL) =        1
     number of frequencies           (NF) =        2
     geometric stiffness flag        (GEOSTF) =    0
     analysis type code              (NDYN) =      -3
     solution mode                   (MODEX) =     0
     equations per block             (KEQB) =      0
     weight and c.g. flag            (IWTCG) =     0
     bandwidth minimization flag     (MINBND) =    0
     number of response spectra      (NRSC) =      0
     gravitational constant          (GRAV) =      3.8640E+02
```

DIRECTION FACTORS
X = 1.0000E+00
Y = 0.0000E+00
Z = 0.0000E+00

INDICATOR FOR SPECTRUM TYPE............... 2

EQ.0 DISPLACEMENT
EQ.1 ACCELERATION IN LENGTH/SEC.**2
EQ.2 ACCELERATION IN G'S

SPECTRA ENTERED FOR CASE.................. 1
KIND....................................... -1

1**** MODAL PARTICIPATION FACTORS

MODE	X-DIRECTION	Y-DIRECTION	Z-DIRECTION
----	-----------	-----------	-----------
1	1.2593E+01	1.7183E-11	0.0000E+00
2	-1.0032E+00	1.8832E-10	0.0000E+00

1**** MODAL ANALYSIS
mode number = 1
Displacements/Rotations(degrees) of nodes

NODE number	X- translation	Y- translation	Z- translation	X- rotation	Y- rotation	Z- rotation
6	2.1127E-01	1.6834E-10	0.0000E+00	0.0000E+00	0.0000E+00	-4.2714E-01
7	6.7577E-01	2.3502E-10	0.0000E+00	0.0000E+00	0.0000E+00	-5.6901E-01
8	1.1398E+00	1.6646E-10	0.0000E+00	0.0000E+00	0.0000E+00	-4.2622E-01
9	1.3504E+00	-2.8421E-12	0.0000E+00	0.0000E+00	0.0000E+00	0.0000E+00
10	1.3903E+00	3.8886E-11	0.0000E+00	0.0000E+00	0.0000E+00	-1.1421E-01
11	1.4779E+00	5.6052E-11	0.0000E+00	0.0000E+00	0.0000E+00	-1.5181E-01
12	1.5652E+00	4.0082E-11	0.0000E+00	0.0000E+00	0.0000E+00	-1.1350E-01
13	1.6048E+00	0.0000E+00	0.0000E+00	0.0000E+00	0.0000E+00	0.0000E+00

1**** MODAL ANALYSIS
mode number = 2
Displacements/Rotations(degrees) of nodes

NODE number	X- translation	Y- translation	Z- translation	X- rotation	Y- rotation	Z- rotation
6	-1.2920E-03	1.6222E-11	0.0000E+00	0.0000E+00	0.0000E+00	2.6073E-03
7	-4.1178E-03	2.2648E-11	0.0000E+00	0.0000E+00	0.0000E+00	3.4490E-03
8	-6.9177E-03	1.6041E-11	0.0000E+00	0.0000E+00	0.0000E+00	2.5576E-03
9	-8.1765E-03	0.0000E+00	0.0000E+00	0.0000E+00	0.0000E+00	0.0000E+00
10	-3.6657E-03	3.7474E-12	0.0000E+00	0.0000E+00	0.0000E+00	-1.2922E-02
11	6.2553E-03	5.4016E-12	0.0000E+00	0.0000E+00	0.0000E+00	-1.7219E-02
12	1.6167E-02	3.8627E-12	0.0000E+00	0.0000E+00	0.0000E+00	-1.2896E-02
13	2.0665E-02	0.0000E+00	0.0000E+00	0.0000E+00	0.0000E+00	0.0000E+00

1**** RESPONSE SPECTRUM ANALYSIS
total combined nodal deflections
Displacements/Rotations(degrees) of nodes

NODE number	X- translation	Y- translation	Z- translation	X- rotation	Y- rotation	Z- rotation
6	2.1128E-01	1.6912E-10	0.0000E+00	0.0000E+00	0.0000E+00	4.2715E-01
7	6.7578E-01	2.3611E-10	0.0000E+00	0.0000E+00	0.0000E+00	5.6902E-01
8	1.1398E+00	1.6723E-10	0.0000E+00	0.0000E+00	0.0000E+00	4.2622E-01
9	1.3505E+00	2.8553E-12	0.0000E+00	0.0000E+00	0.0000E+00	0.0000E+00
10	1.3904E+00	3.9066E-11	0.0000E+00	0.0000E+00	0.0000E+00	1.1494E-01
11	1.4779E+00	5.6312E-11	0.0000E+00	0.0000E+00	0.0000E+00	1.5278E-01
12	1.5653E+00	4.0267E-11	0.0000E+00	0.0000E+00	0.0000E+00	1.1423E-01
13	1.6049E+00	0.0000E+00	0.0000E+00	0.0000E+00	0.0000E+00	0.0000E+00

Table 9.2.4.2

1**** Algor (c) FEA Stress Processor MKNSO Ver 11.06-3H
INPUT FILE............ex92

1**** BEAM ELEMENT FORCES AND MOMENTS

ELEM. NO.	CASE (MODE)	AXIAL FORCE R1	SHEAR FORCE R2	SHEAR FORCE R3	TORSION MOMENT M1	BENDING MOMENT M2	BENDING MOMENT M3
5	1	-5.292E-04	0.000E+00	-8.952E+03	0.000E+00	6.395E+05	0.000E+00
		5.292E-04	0.000E+00	8.952E+03	0.000E+00	-3.173E+05	0.000E+00
5	2	-5.100E-05	0.000E+00	-1.003E+03	0.000E+00	7.218E+04	0.000E+00
		5.100E-05	0.000E+00	1.003E+03	0.000E+00	-3.608E+04	0.000E+00

1**** RESULTANT STRESS COMBINATIONS
No. of frequencies = 2
No. of directions = 1
Response combination method: Original SAPIV procedure
Resultant stresses will be written to mode/loadcase 3

1**** BEAM ELEMENT FORCES AND MOMENTS

AXIAL FORCE = R1
SHEAR FORCE (LOCAL 2 AXIS) = R2
SHEAR FORCE (LOCAL 3 AXIS) = R3
TORSION MOMENT = M1
BENDING MOMENT (LOCAL 2 AXIS) = M2
BENDING MOMENT (LOCAL 3 AXIS) = M3

AXIAL STRESS = P/A
BENDING STRESS (LOCAL 2 AXIS) = M2/S2
BENDING STRESS (LOCAL 3 AXIS) = M3/S3

```
EL.  N L   R1          R2          R3          M1          M2          M3
NO.  D C   P/A         M2/S2       M3/S3
------ - - ----------  ----------  ----------  ----------  --------    ---------
  5  I 3  5.317E-04 0.000E+00 9.008E+03 0.000E+00 6.436E+05 0.000E+00
          0.000E+00 6.436E+05 3.493E-05
  5  J 3  5.317E-04 0.000E+00 9.008E+03 0.000E+00 3.193E+05 0.000E+00
          0.000E+00 3.193E+05 3.493E-05
```

9.2.5 System Subjected to Design Response Spectra

This section presents the finite element solution of the frame analyzed in Example 5.15.1. The frame represents a two-story structure that is subjected to the seismic spectra defined by eqns. (4.9.8) as recommended by the UBC code for rocklike soil conditions. It should be noted that the code requires additional modifications of the spectra to account for variables related to the site and type of structural system. Such modifications have not been applied to this example. A graphical representation of the spectra is given in Fig. 4.9.3, and Table 9.2.5.1 lists the spectral values obtained from eqns. (4.9.8).

Table 9.2.5.1

No	T	S_a		No	T	S_a
1	0	1.000		10	0.60	1.625
2	0.15	2.500		11	0.65	1.500
3	0.39	2.500		12	0.75	1.300
4	0.42	2.320		13	1.50	0.650
5	0.45	2.170		14	3.00	0.325
6	0.48	2.030		15	6.00	0.125
7	0.51	1.912		16	9.00	0.108
8	0.54	1.805		17	18.00	0.054
9	0.57	1.710		18	30.00	0.032

Since the modeling aspects of the frame are discussed in Sections 7.5 and 7.5.1, we proceed to the response spectrum analysis results. Table 9.2.5.2 presents the maximum relative displacements at nodes 9 and 13. The second and third columns list the modal relative displacements, and the last column lists the maximum relative displacements determined with SRSS of the second and third columns. For example, the maximum relative displacement at node 9 is calculated from

$$4.846 = \sqrt{4.846^2 + (-0.041)^2} \tag{9.3.1}$$

Table 9.2.5.2

NODE No	MODE 1	MODE 2	RELAT. RESP.
9	4.846	-0.041	4.846
13	5.758	0.103	5.759

The "total combined nodal deflections," i.e., the maximum relative displacements, are also underlined in Table 9.2.7.1.

Notice that the maximum relative displacements are primarily the contribution of the first mode. As a general rule, due to the modal characteristics of most structures and the frequency content of most ground excitations, the maximum relative response is obtained by combining a relatively small number of lower modes.

9.2.6 Analysis with Algor

For design spectra, response spectrum analysis with Algor consists of four steps. The first two steps, that is, modal analysis and development of spectra for a harmonic ground motion with RESONATE, have been described in Section 9.2.2. Currently, Algor requires prior execution of RESONATE for a harmonic excitation before the desired design spectra are defined. It should be remembered that RESONATE creates a file with the name assigned as a response to question (b) in Section 9.2.2, i.e., "EX92. ". The third step involves modification of the "EX92. " with a text editor. In the following, the edited file is given the name "EX92M. " and is listed in Table 9.2.6.1. Editing of the "EX92." involves the following two steps:

a. The first three lines of "EX92. " are retained, while the rest of the file is deleted. The first one is a comment line that includes the modal analysis file name, i.e., "EX75. ". The second line is the "Master Control Input Line" with entries pertaining to the number of nodal points in the model, the number of element groups, the number of load cases, and the type of analysis, as specified in the Algor Processor Reference Manual. The last entry of the second line is the acceleration of gravity. The third line is the "Spectrum Control Data Line" with entries related to the "Factors for the Direction Input," the spectrum type, the modal combination procedure, and the cluster factor, as defined in the Algor Response Spectrum Release Notes.

b. The second step involves use of a text editor to input the remaining lines. The data of the remaining lines are also defined in the Algor Response Spectrum Release Notes. Specifically, the fourth line is the "Heading Data Line" with a label given by the user to best describe the type of the response spectrum. The fifth line is the "Control Data Line" that specifies the number of pairs defining the spectra and the scale factor to multiply the spectral values. The remaining lines list the pairs defining the spectra. The first entry of each line is the period and the second is the corresponding spectral acceleration S_a as a fraction of g. Notice that in our example we have defined the "Spectrum Data" with the eighteen pairs listed in Table 9.2.5.1.

The response spectrum analysis processor SSAP3H prompts to input the "EX92M. " file. Then, it requests the modal analysis "EX75. " file. As suggested in previous sections, by activating TRANS before running SSAP3H we can obtain the combined nodal displacements in the "EX92M.1" file.

Table 9.2.6.1

File EX75 created by BEDIT 4.08-3H

```
15  1 1  2  0-3     0     0    0     0  000     0    0     0     0     386.4
    1.000       .000        .000     2     1     0   -1
SA/G SPECTRUM
  18          1
  .00         1.000
  .15         2.500
  .39         2.500
  .42         2.320
  .45         2.170
  .48         2.030
  .51         1.912
  .54         1.805
  .57         1.710
  .60         1.625
  .65         1.500
 0.75         1.300
 1.50         0.650
 3.00         0.325
 6.00         0.125
 9.00         0.108
18.00         0.054
30.00         0.032
```

9.2.7 Interpretation of Results

The procedure to examine the graphical and numerical results was described in Sections 9.2.3 and 9.2.4. Figure 9.2.7.1 shows the dithered variation of the combined shear forces as well as the deformed shape for the combined deflections, i.e., the third load case. In SVIEWH, colors in the dithered plot indicate the magnitude of the shear forces as they correspond to the values in the table at the upper right corner of the figure. Next, we present selected numerical results in order to emphasize several aspects of the ASCII output files. Tables 9.2.7.1 and 9.2.7.2 list excerpts from the "EX92M.l" and the "EX92M.s" files, respectively. The "EX92M.l" provides information about the type and number of elements, the natural frequencies, the response combination methods, the modal participation factors, the spectra, and the combined response. The "EX92M.s" includes element and material information data and element forces and stresses.

Table 9.2.7.1 does not include the "response combination methods" since they are presented in the Algor Response Spectrum Analysis Release Notes, and are also discussed in Section 5.16. It lists only the total combined nodal deflections, that are the SRSS of the modal deflections. In Table 9.2.7.2 we are only listing the forces for element 5, see Fig. 7.5.1.1 for the element and node numbers. The modal forces and stresses are related to the SRSS combined forces and stresses as expressed by eqn. (5.16.1). For example, the combined shear force for node 9 of element 5 is calculated from the modal components underlined in Table 9.2.7.2, that is

$$32510 = \sqrt{(-32120)^2 + (-5019^2)} \qquad (9.3.2.1)$$

The statements made in Sections 9.2.3 and 9.1.3 about the "worst stresses," the von Mises, and Tresca stresses in identifying regions or structural members that have yielded are also valid when the excitation is expressed in the form of design response spectra.

Table 9.2.7.1

```
1**** Algor (c) Response spectrum analysis SSAP3H Ver 10.04-3H

INPUT FILE............ex92m
MODAL  ANALYSIS  FILE....ex75
-----------------------------------------------
 EX75.
```

1**** CONTROL INFORMATION

number of node points	(NUMNP) =	15
number of element types	(NELTYP) =	1
number of load cases	(LL) =	0
number of frequencies	(NF) =	2
geometric stiffness flag	(GEOSTF) =	0
analysis type code	(NDYN) =	-3
solution mode	(MODEX) =	0
equations per block	(KEQB) =	0
weight and c.g. flag	(IWTCG) =	0
bandwidth minimization flag	(MINBND) =	0
number of response spectra	(NRSC) =	0
gravitational constant	(GRAV) =	3.8640E+02

1**** CASE......................................1

DIRECTION FACTORS
X = 1.0000E+00
Y = 0.0000E+00
Z = 0.0000E+00

INDICATOR FOR SPECTRUM TYPE................ 2
EQ.0 DISPLACEMENT
EQ.1 ACCELERATION IN LENGTH/SEC.**2
EQ.2 ACCELERATION IN G'S

SPECTRA ENTERED FOR CASE...................1
KIND......................................1

1**** MODAL PARTICIPATION FACTORS

MODE	X-DIRECTION	Y-DIRECTION	Z-DIRECTION
1	1.2593E+01	1.7183E-11	0.0000E+00
2	-1.0032E+00	1.8832E-10	0.0000E+00

1**** RESPONSE SPECTRUM ANALYSIS
total combined nodal deflections
Displacements/Rotations(degrees) of nodes

NODE number	X- translation	Y- translation	Z- translation	X- rotation	Y- rotation	Z- rotation
6	7.5813E-01	6.0949E-10	0.0000E+00	0.0000E+00	0.0000E+00	1.5327E+00
7	2.4249E+00	8.5091E-10	0.0000E+00	0.0000E+00	0.0000E+00	2.0418E+00
8	4.0900E+00	6.0268E-10	0.0000E+00	0.0000E+00	0.0000E+00	1.5294E+00
9	4.8459E+00	1.0290E-11	0.0000E+00	0.0000E+00	0.0000E+00	0.0000E+00
10	4.9890E+00	1.4079E-10	0.0000E+00	0.0000E+00	0.0000E+00	4.1488E-01
11	5.3033E+00	2.0294E-10	0.0000E+00	0.0000E+00	0.0000E+00	5.5152E-01
12	5.6170E+00	1.4512E-10	0.0000E+00	0.0000E+00	0.0000E+00	4.1235E-01
13	5.7594E+00	2.5309E-12	0.0000E+00	0.0000E+00	0.0000E+00	0.0000E+00

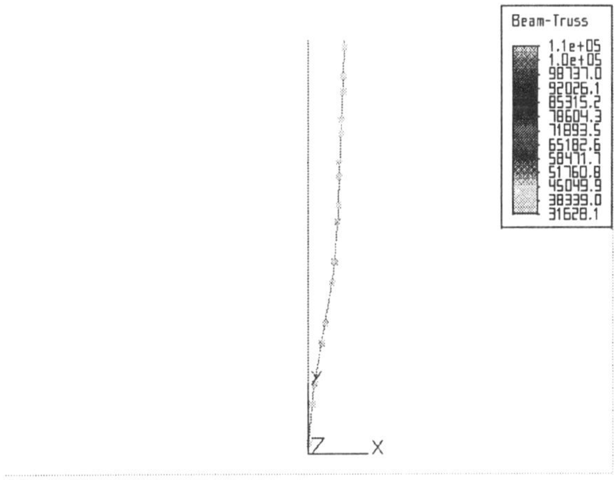

Figure 9.2.7.1 *Combined shear forces and deformations.*

Table 9.2.7.2

1**** Algor (c) FEA Stress Processor MKNSO Ver 11.06-3H
INPUT FILE............ex92m

1**** BEAM ELEMENT FORCES AND MOMENTS

ELEM. NO.	CASE (MODE)	AXIAL FORCE R1	SHEAR FORCE R2	SHEAR FORCE R3	TORSION MOMENT M1	BENDING MOMENT M2	BENDING MOMENT M3
-------	----	----------	----------	----------	----------	----------	----------
5	1	-1.899E-03	0.000E+00	-3.212E+04	0.000E+00	2.295E+06	0.000E+00
		1.899E-03	0.000E+00	3.212E+04	0.000E+00	-1.138E+06	0.000E+00
5	2	-2.553E-04	0.000E+00	-5.019E+03	0.000E+00	3.613E+05	0.000E+00
		2.553E-04	0.000E+00	5.019E+03	0.000E+00	-1.806E+05	0.000E+00

9.3 RANDOM VIBRATION ANALYSIS: NUMERICAL EXAMPLES

When a structure is subjected to dynamic loads that are expressed through known forcing functions, then its response can be determined through a deterministic analysis. Time history modal superposition, time history direct integration, frequency response, and response spectrum analysis are deterministic methods. In many instances, however, the possible dynamic loads are based on previous experience and experimental results. In these cases, the forcing functions can be better expressed through statistical parameters. Representative random phenomena include blast, earthquake, traffic, and wind loads. A discussion on the basics of random vibration is presented in Chapters 4 and 5.

The following section demonstrates the use of Algor to determine the response of the frame analyzed in Example 5.17.1.

9.3.1 Analysis with Algor

Consider the two-story frame shown in Fig. 5.15.1(a). This frame was also used in the examples on frequency response and response spectrum analysis. The foundation of the frame is subjected to the random ground acceleration described in Example 5.17.1. Assume that the damping ratio is $\xi=0.05$, and calculate the probability of the horizontal response at the top of the frame to exceed 9 in.

In random vibration with Algor, the system should always remain elastic, a requirement that is usually met in practical applications. The loading must be a ground motion acting either independently or simultaneously along any of the global X, Y and Z-axis. It should be given in the form of power spectral densities (PSD) or cross spectral densities (CSD). The output is root mean square response (RMS). As in response spectrum analysis, the RMS of the displacements is relative, that is, it refers to nodal displacements with respect to the ground excitation. A definition of the parameters used in random vibration is presented in Chapter 4.

Random vibration analysis with Algor is applicable to excitations with zero mean. Since a zero mean is assumed for the input, the mean response is also zero. Consequently, the RMS of deformation or stress is equal to the standard deviation, see eqn. (4.10.3). It should be emphasized that the output of random vibration analysis is the standard deviation of the displacements and stresses and not the actual displacements and stresses.

A three-step procedure must be followed to obtain the RMS of the response:

1. Modal analysis: The development of the finite element model and the modal analysis of the frame with an eight-element model is described in Sections 7.5 and 7.5.1. The beam model is shown in Fig. 7.5.1.1.

2. Specification of the ground motion power spectral density with the preprocessor PRESS7: The question and answer procedure prompted by PRESS7 is demonstrated for the two-story building. For the specification of the parameters inputed with PRESS7 refer to the Algor Random Vibration Release Notes.

a. Enter modal analysis file name:		**EX75**
(i.e., modal analysis file with no extension.)		
b. Enter random vibration file name (to be created):		**EX93**
(i.e, name of file with no extension.)		
c. Choose type of spectrum input desired		
Acceleration**2/Hz	(0)	
G**2/Hz	(1)	
Enter type of spectrum input:		**1**
d. Choose analysis method to be used		
Approximation method	(0)	
Numerical integration method	(2)	
Same as (2) but no cross-spectral effect	(3)	
Enter method of analysis:		**2**
e. Enter viscous damping ratio ($\xi=0.01=1$ percent):		**0.05**
f. Enter PSD or CSD index. The index		
Enter excitation PSD or CSD direction index:		**1**
g. Enter title for this spectrum input:		**White Noise**
h. Enter number of points of the spectrum data:		**2**
i. Enter spectrum amplitude scale factor:		**0.20**
j. Enter freq. (Hertz) for point 1:		**0.0**
k. Enter spectrum amplitude for point 1:		**1.0**
l. Enter freq. (Hertz) for point 2:		**10.0**
m. Enter spectrum amplitude for point 2:		**1.0**
n. Enter PSD or CSD index. The index		
Enter excitation PSD or CSD direction index:		**<CR>**

3. Execute the random vibration analysis processor PRESS7: The PRESS7 generates the "EX93. " file with information pertaining to either the PSD or the CSD of the ground excitation. The "EX93. " for the frame subjected to *white noise* excitation, see Fig. 5.17.1, is listed in Table 9.3.1.1. The description for each entry of "EX93. " is given in the Algor Random Vibration Release Notes.

The random vibration processor is the SSAP7H. First, the SSAP7H requests to input the file specified in response to question (b). Then, SSAP7H prompts for the modal analysis file, which is the response to question (a). The results generated by SSAP7H are examined in the following section.

Table 9.3.1.1

File EX75 created by BEDIT 4.08-3H

```
15    1    1    2  0-7    0    0    0    0  000    0    0    0    0   386.4
    .000       .000       .000     1     2      .1000      .0500
White Noise
 2    2.000E-01     1
   0.000E+00    1.000E+00
   1.000E+01    1.000E+00
```

9.3.2 Interpretation of Results

The SSAP7H generates as many load cases as the number of modes in the modal analysis files increased by one. For our example, three load cases can be examined in SVIEWH. The first two load cases provide the RMS of the modal responses, and the last one is their SRSS. Further details about the procedure to examine the graphical results in SVIEWH are discussed in Section 9.1.3. The combined shear forces along the X-axis are shown in Fig. 9.3.2.1.

Tables 9.3.2.1 and 9.3.2.2 present selected excerpts from the "EX93.l" and "EX93.s" files, respectively. The last part of Table 9.3.2.1 lists the modal RMS followed by the SRSS of the combined nodal displacements and rotations. For example, referring to the underlined RMS modal displacements of node 13 in Table 9.3.2.1, the "total combined nodal deflections," i.e., the RMS of the response is calculated as follows:

$$\sqrt{u_{13}^2} = 7.93 = \sqrt{7.928^2 + 0.175^2} \tag{9.3.2.1}$$

Notice that the RMS given by eqn. (5.17.1e), that is, $\sqrt{u_2^2} = 7.50$ compares well. The difference is primarily attributed to the approximation of the mean square response in step 5 of Example 5.17.1 through eqn. (5.17.1b). If we wish, by following the procedure presented in Example 5.17.1, we can use the $\sqrt{u_{13}^2} = 7.93$ to calculate the probability of the horizontal response at the top of the frame to exceed 9 in.

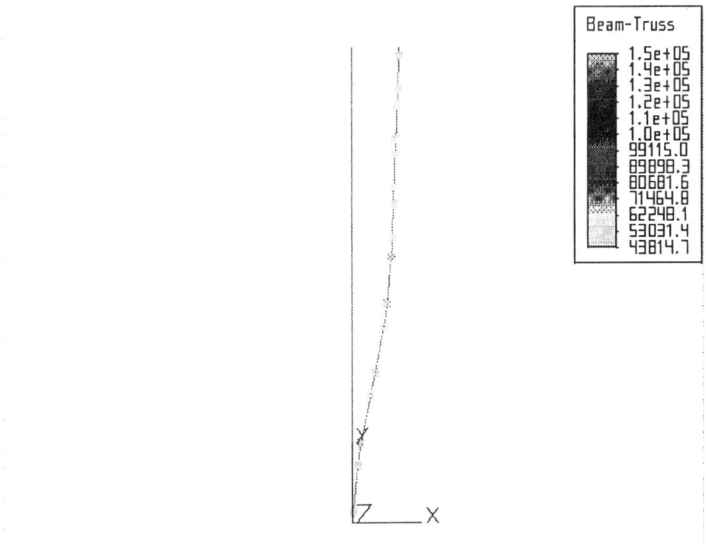

Figure 9.3.2.1 *Combined shear forces.*

Table 9.3.2.2 lists the RMS of the element forces, stresses, and their resultants. The RMS of the resultants correspond to the third load case. They are obtained as the SRSS of the previous load cases, e.g., the shear forces at node 9 of element 5, that is,

$$45030 = \sqrt{(-44220)^2 + (-8477)^2} \qquad (9.3.2.2)$$

The values in eqn. (9.3.2.2) are also underlined in Table 9.3.2.2.

Similar to displacements, the combined RMS of the stresses can be used to calculate their probability to exceed a predetermined value, e.g., the allowable or the yield stress. The remarks made in Sections 9.2.3 and 9.1.3 on the use of the von Mises and Tresca stresses for two- and three-dimensional elements as well as the "worst stresses" for frames in identifying regions that have yielded are also valid in random vibration analysis.

Table 9.3.2.1

```
1**** Algor (c) Random Vibration Processor - SSAP7 Ver. 10.04-3H
INPUT FILE.............ex93
MODAL  ANALYSIS  FILE....ex75
-----------------------------------------------

 File EX75 created by BEDIT 4.08-3H

1**** CONTROL INFORMATION

     number of node points            (NUMNP) =      15
     number of element types          (NELTYP) =     1
     number of load cases             (LL) =         1
     number of frequencies            (NF) =         2
     geometric stiffness flag         (GEOSTF) =     0
     analysis type code               (NDYN) =      -7
     solution mode                    (MODEX) =      0
     equations per block              (KEQB) =       0
     weight and c.g. flag             (IWTCG) =      0
     bandwidth minimization flag      (MINBND) =     0
     gravitational constant           (GRAV) =       3.8640E+02

1**** RANDOM VIBRATION ANALYSIS
  **** STANDARD NUMERICAL MTHD SPECIFIED FOR ANALYSIS
  **** Viscous Damping Ratio Specified =     .0500
     KIND.....................................      2
```

1**** SPECTRUM TABLE: WHITE NOISE

number of points = 2
scale factor = 2.0000E-01
index of excitation direction = 1

INPUT POINT	FREQUENCY	MAGNITUDE
-----	----------	----------
1	0.0000E+00	1.0000E+00
2	1.0000E+01	1.0000E+00

SPECTRAL MATRIX

1	7.7801E+03	0.0000E+00
2	0.0000E+00	1.7649E+00

INFORMATION PROCESSED FOR EACH NODE:
1. MODAL SQUARE ROOT DISPLACEMENT FOR EACH MODE
2. RESULTANT ROOT MEAN SQUARE (R.M.S.) DISPLACEMENTS (NOTE: RESULTANT DISP. ARE PLACED IN THE LAST MODE)

1**** MODAL ANALYSIS

mode number = 1

6	1.0437E+00	8.3162E-10	0.0000E+00	0.0000E+00	0.0000E+00	2.1101E+00
7	3.3384E+00	1.1610E-09	0.0000E+00	0.0000E+00	0.0000E+00	2.8110E+00
8	5.6306E+00	8.2233E-10	0.0000E+00	0.0000E+00	0.0000E+00	2.1055E+00
9	6.6713E+00	1.4040E-11	0.0000E+00	0.0000E+00	0.0000E+00	0.0000E+00
10	6.8684E+00	1.9210E-10	0.0000E+00	0.0000E+00	0.0000E+00	5.6419E-01
11	7.3010E+00	2.7690E-10	0.0000E+00	0.0000E+00	0.0000E+00	7.4996E-01
12	7.7323E+00	1.9801E-10	0.0000E+00	0.0000E+00	0.0000E+00	5.6069E-01
13	7.9278E+00	3.4534E-12	0.0000E+00	0.0000E+00	0.0000E+00	0.0000E+00

1**** MODAL ANALYSIS

mode number = 2

6	1.0922E-02	1.3714E-10	0.0000E+00	0.0000E+00	0.0000E+00	2.2042E-02
7	3.4810E-02	1.9146E-10	0.0000E+00	0.0000E+00	0.0000E+00	2.9156E-02
8	5.8480E-02	1.3561E-10	0.0000E+00	0.0000E+00	0.0000E+00	2.1621E-02
9	6.9122E-02	2.3138E-12	0.0000E+00	0.0000E+00	0.0000E+00	0.0000E+00
10	3.0989E-02	3.1679E-11	0.0000E+00	0.0000E+00	0.0000E+00	1.0924E-01
11	5.2881E-02	4.5664E-11	0.0000E+00	0.0000E+00	0.0000E+00	1.4556E-01
12	1.3667E-01	3.2654E-11	0.0000E+00	0.0000E+00	0.0000E+00	1.0901E-01
13	1.7470E-01	0.0000E+00	0.0000E+00	0.0000E+00	0.0000E+00	0.0000E+00

Total combined nodal deflections
Displacements/Rotations(degrees) of nodes

NODE number	X- translation	Y- translation	Z- translation	X- rotation	Y- rotation	Z- rotation
6	1.0438E+00	8.4285E-10	0.0000E+00	0.0000E+00	0.0000E+00	2.1102E+00
7	3.3385E+00	1.1767E-09	0.0000E+00	0.0000E+00	0.0000E+00	2.8111E+00
8	5.6309E+00	8.3344E-10	0.0000E+00	0.0000E+00	0.0000E+00	2.1057E+00
9	6.6717E+00	1.4230E-11	0.0000E+00	0.0000E+00	0.0000E+00	0.0000E+00
10	6.8685E+00	1.9469E-10	0.0000E+00	0.0000E+00	0.0000E+00	5.7467E-01
11	7.3012E+00	2.8064E-10	0.0000E+00	0.0000E+00	0.0000E+00	7.6396E-01
12	7.7335E+00	2.0068E-10	0.0000E+00	0.0000E+00	0.0000E+00	5.7119E-01
13	7.9298E+00	3.4998E-12	0.0000E+00	0.0000E+00	0.0000E+00	0.0000E+00

Table 9.3.2.2

1**** Algor (c) FEA Stress Processor MKNSO Ver 11.06-3H
INPUT FILE.............ex93

1**** BEAM ELEMENT FORCES AND MOMENTS

ELEM. NO.	CASE (MODE)	AXIAL FORCE R1	SHEAR FORCE R2	SHEAR FORCE R3	TORSION MOMENT M1	BENDING MOMENT M2	BENDING MOMENT M3
5	1	-2.615E-03	0.000E+00	-4.422E+04	0.000E+00	3.159E+06	0.000E+00
		2.615E-03	0.000E+00	4.422E+04	0.000E+00	-1.567E+06	0.000E+00
5	2	-4.311E-04	0.000E+00	-8.477E+03	0.000E+00	6.102E+05	0.000E+00
		4.311E-04	0.000E+00	8.477E+03	0.000E+00	-3.050E+05	0.000E+00

1**** RESULTANT STRESS COMBINATIONS
No. of frequencies = 2
Resultant stresses will be written to mode/loadcase 3

1**** BEAM ELEMENT FORCES AND MOMENTS

AXIAL FORCE = R1 TORSION MOMENT = M1
SHEAR FORCE (LOCAL 2 AXIS) = R2 BENDING MOMENT (LOCAL 2 AXIS) = M2
SHEAR FORCE (LOCAL 3 AXIS) = R3 BENDING MOMENT (LOCAL 3 AXIS) = M3

AXIAL STRESS = P/A
BENDING STRESS (LOCAL 2 AXIS) = M2/S2
BENDING STRESS (LOCAL 3 AXIS) = M3/S3

EL. NO.	N D	L C	R1 P/A	R2 M2/S2	R3 M3/S3	M1	M2	M3
5	I	3	2.650E-03	0.000E+00	4.503E+04	0.000E+00	3.218E+06	0.000E+00
			0.000E+00	3.218E+06	1.741E-04			
5	J	3	2.650E-03	0.000E+00	4.503E+04	0.000E+00	1.597E+06	0.000E+00
			0.000E+00	1.597E+06	1.741E-04			

Selected Bibliography

ENGINEERING MECHANICS

Chajes, A.: *Principles of Structural Stability,* Prentice Hall, New Jersey, 1974.

Donnell, L. H.: *Beams, Plates and Shells,* McGraw-Hill, New York, 1985.

Shames, I.H.: *Introduction to Solid Mechanics*, Prentice Hall, New Jersey, 1975.

Szilard, R.: *Theory and Analysis of Plates,* Prentice-Hall, New Jersey, 1974.

Timoshenko, S.P., Woinowsky-Krieger, S.: *Theory of Plates and Shells*, 2nd ed., McGraw-Hill, New York, 1959.

Timoshenko, S.P. and Goodier, J.N.: *Theory of Elasticity*, 3rd ed., McGraw-Hill, New York, 1970.

Tsai, W. S. and Hahn, H.T.: *Introduction to Composite Materials,* Technomic Publishing, Lancaster, PA, 1980.

Ugural A.C. and Fenster, S.K.: *Advanced Strength and Applied Elasticity*, Elsevier, New York, 1981.

Valliapan S.: *Continuum Mechanics Fundamentals*, Balkema, Rotterdam, Netherlands, 1981.

HANDBOOKS AND MATHEMATICS

Bednar, H.: *Pressure Vessel Design Handbook*, Van Nostrand Reinhold, New York, 1981.

Blevins, R. d.: *Formulas for Natural Frequency and Mode Shape*, Van Nostrand

Reinhold, New York, 1979.

Griffel, W.: *Handbook of Formulas for Stress and Strain*, Ungar Publish., New York, 1966.

Harris, C.M.: *Shock and Vibration Handbook*, 3rd edit., McGraw-Hill, New York, 1988.

Naeim, F.: *The Seismic Design Handbook,* Van Nostrand Reinhold, New York, 1989.

Steel, R. and Torrie, J.: *Principles and Procedures of Statistics,* 2nd ed., McGraw-Hill, New York, 1980.

Young, W.C.: *Roark's Formulas for Stress & Strain,* 6th ed., McGraw-Hill, New York, 1989.

STRUCTURAL DYNAMICS

Chopra, A. K.: *Dynamics of Structures, A Primer,* Earthquake engineering Research Institute, 1980.

Clough, R.N. and Penzien, J.: *Dynamics of Structures,* 2nd ed., McGraw-Hill, New York, 1993.

Craig, R.R.: *Structural Dynamics*, John Wiley, New York, 1983.

Crandall, S.H. and Mark, W.D.: *Random Vibration in Mechanical Systems*, Academic Press, New York, 1973.

Dimarogonas, A.D. and Haddad, A.: *Vibration for Engineers*, Prentice-Hall, New Jersey, 1992.

Gupta, A.K.: *Response Spectrum Method*, Blackwell Scientific Publ., Boston, 1990.

Humar, J.L.: *Dynamics of Structures*, Prentice-Hall, New Jersey, 1990.

Hudson, D.E.: *Reading and Interpreting Strong Motion Accelerograms,* Earthquake Engineering Research Institute, 1979.

Inman, D.J.: *Engineering Vibration*, Prentice-Hall, New Jersey, 1994.

Nashif, A.D., Jones, D. and Henderson, J.: *Vibration Damping,* J. Wiley, New York, 1985.

Patel, P. and Spyrakos, C.C.: *Time Domain BEM-FEM Seismic Analysis Including Basemat LIft-Off*, Engineering Structures, V. 12, pp. 195-207, 1990.

Paz, M.: *Structural Dynamics, 3rd ed.,* Van Nostrand Reinhold, New York, 1991.

Prucz, J., Kokkinos, R. and Spyrakos, C.C.: *Advanced Joining Concepts in Passive Vibration Control*, ASCE Journal of Aerospace Engng., Vol. 1, No. 4, pp. 193-205, 1988.

Richart, F.E. Jr., Hall, J.R. and Woods, R.D.: *Vibration of Soils and*

Foundations, Prentice-Hall, New Jersey, 1970.

Spyrakos, C.C.: *Strip-Foundations,* Chapter 6 in Boundary Element Techniques in Geomechanics, pp. 147-176, (G.D. Manolis & T.G. Davies Ed.), Elsevier, New York, 1993.

Spyrakos, C.C., Patel, P.N. and Kokkinos, F.T.: *Assessment of Computational Practices in Dynamic Soil-Structure Interaction*, ASCE Journal of Computing in Civil Engng., Vol. 2, No 2, pp. 143-157, 1989.

Spyrakos, C.C. and Chen, C-I.: *Power Series Expansions of the Dynamic Stiffness Matrices for Tapered Bars and Shafts*, Int. Journal for Numerical Methods in Engng., Vol. 30, pp. 259-270, 1990.

Tamma, K.K., Spyrakos, C.C. and Lambi, M.: *Thermal/Strucrural Dynamic Analysis via a Transform Method Based Finite Element Approach*, AIAA Journal of Spacecrafts and Rockets, Vol. 24, No. 3, pp. 219-226, 1987.

Thomson, W. T.: *Theory of Vibration with Applications,* 4th ed., Prentice-Hall, New Jersey, 1993.

Timoshenko, S. P., Young, D. and Weaver, W.: *Vibration Problem in Engineering*, 4th ed., J. Wiley, New York, 1974.

Wiegel, R.L.: *Earthquake Engineering,* Prentice Hall, New Jersey, 1970.

Wolf, J.P.: *Dynamic Soil-Structure Interaction,* Prentice-Hall, New Jersey, 1985.

FINITE ELEMENT ANALYSIS

Bathe, J.K.: *Finite Element Procedures in Engineering Analysis,* Prentice-Hall, New Jersey, 1982.

Cook, R. D., Malkus, D. S. and Plesha, M. E.: *Concepts and Applications of Finite Element Analysis,* 3rd. ed., J. Wiley & Sons, New York, 1989.

Desai, C. S. and Abel, J. F.: *Introduction to the Finite Element Method,* Van Nostrand Reinhold, New York, 1972.

Owen D.R.J. and Hinton E.: *Finite Elements in Plasticity,* Pineridge Press, Swansea, U.K., 1980.

Gallagher, R.: *Finite Element Analysis Fundamentals,* Prentice Hall, New Jersey, 1975.

Owen, D.R.J. and Hinton, E.: *Finite Elements in Plasticity,* Pineridge Press, Swansea, U.K., 1980.

Paulsen, W.C. et al.: *Finite Element Analysis*, Penton/IPC Education Division, Cleveland, 1983.

Yang, T.Y.: *Finite Element Structural Analysis,* Prentice Hall, New Jersey, 1986.

Zienkiewicz, O.C., and Taylor, R.L.: *The Finite Element Method, 4th ed.*, McGraw-Hill, New York, 1991.

DESIGN CODES

American Petroleum Institute: *API Standard 650*, 1988, Washington, D.C.

American Society of Mechanical Engineers: *ASME Boiler and Pressure Vessel Code*, New York, 1980.

International Conference of Building Officials: *Uniform Building Code (UBC)*, Whittier, CA, 1988.

Nuclear Regulatory Commission: *Nuclear Regulatory Guide 1.92,* February,1978.

Structural Engineering Association of California (SEAOC): *Recommended Lateral Force Requirements and Tentative Commentary*, San Francisco, CA, 1988.

Index

A

B

C

D

E

F

G

H

I

L

M

N

O

P

R

S

T

U

V

W

Y

Z